Conservation Centre

Metal Plating and Patination

Metal Plating and Patination

Cultural, Technical and Historical Developments

Edited by

Susan La Niece
and
Paul Craddock

Butterworth-Heinemann Ltd
Linacre House, Jordan Hill, Oxford OX2 8DP

 A member of the Reed Elsevier group

OXFORD LONDON BOSTON
MUNICH NEW DELHI SINGAPORE SYDNEY
TOKYO TORONTO WELLINGTON

First published 1993

British Library Cataloguing in Publication Data
Metal Plating and Patination
I. Niece, Susan La II. Craddock, Paul
739.15

ISBN 0 7506 1611 3

Library of Congress Cataloguing in Publication Data
Metal plating and patination/edited by Susan La Niece and Paul
Cradddock
p. cm.
Includes index.
ISBN 0 7506 1611 3
1. Metals – Finishing. 2. Plating. 3. Metals – Coloring. I. La
Niece, Susan. II. Craddock, Paul T.
TS653.M38 1993 93–18065
671.7–dc20 CIP

Composition by Scribe Design, Gillingham, Kent
Printed and bound in Great Britain by Martins Ltd, Berwick upon Tweed

Contents

Preface

Our whole perception of the material world, from the topography of the surface mantle of our planet to the man-made metal artefacts which are the special concern of this volume, is based upon surface appearance.

From the inception of metallurgy, man has been at pains to embellish and decorate metal surfaces. The surfaces are the first and most obvious feature of any artefact to be experienced and, until recently, superficial decoration often received far more attention than any other stage in manufacture. Even integral processes such as pattern welding or *wootz* steel damascening seem ultimately to have been decorative. In books 33 and 34 of Pliny's *Natural History*, the one major treatise on metals to survive from antiquity, the modern reader is struck by the importance attached to the *colour* of metals, especially bronzes, and by the preponderance of descriptions of surface treatments above all other metalworking processes. Both in the Old and New Worlds, the diversity of plating and surface treatments and their ubiquity from early times is very noticeable. The history of some techniques is still being extended, as exemplified by the *shakudo*-type metals which now appear to have been in use for very much longer and over a much greater part of the world, than was hitherto suspected. Indeed, electroplating and its attendant technologies are probably the only completely new major treatments of recent times. Due to greater chemical knowledge and a greater availability of materials, many of the other traditional surface treatments reached their apogee in the 19th century, paradoxically, to be largely killed off in the 20th century by the all-embracing, cheap, electrolytic techniques. Even in the *bidri* shops of remote Bidar most of the pieces on sale now are 'bronzes' of anodized aluminium.

The problem of determining the original surface appearance is complex. Surfaces by their very nature are transient, prone to wear, damage and above all corrosion, although distinguishing an attractive patina from a disfiguring corrosion can be a contentious and subjective issue in its own right. There are also the problems of determining how much of the present appearance is due to more recent restoration, alteration, or even total invention. These questions are of great importance and are thoroughly debated here. The quite deliberate but very delicate surfaces on the 19th century instruments examined by Birnie (Chapter 12) have disquieting implications for our ability to establish the original appearance of antiquities. If surfaces can be damaged, or even lost, on instruments which are relatively new and have never been buried, what hope is there of identifying original surfaces on ancient and corroded material?

The papers in this volume underline the importance of integrating the knowledge and experience of scientists, conservators and art historians. Indeed the editors have tried to encourage the contributors to reach out to disciplines other than their own. The subject may be surfaces, but the approach has been deeper here.

It was our perception some years ago that, although much useful research was being carried out, not least in the British Museum, the reports were scattered throughout the scientific, conservation and art-historical literature. There was certainly no co-ordinated approach, the study of major techniques was being compartmentalized by period, date or material, and some technologies were omitted altogether. To redress this state of affairs a meeting was held in June 1990, at the British Museum, entitled 'Surface Colouring and Plating of Metals'. The speakers were mainly invited, in order to make the coverage as balanced and complete as possible. This book is the product of that meeting although it is not just a static account of a conference proceedings. It is a mixture of review articles on major, well-known techniques such as gilding or patination, and papers on specific areas, including some never previously considered from an historic perspective, such as bower–barffing. The production and editing of the book has been very interactive to ensure continuity and an even coverage. As a direct result of the post-conference exchanges new aspects and even new materials, such as *wu tong* have come to light.

Bringing together specialists from various fields not only provides a breadth of knowledge and experience that would be impossible for one individual or even one discipline to encompass, but also highlights more subtle differences in tradition and opinion. This is exemplified by the difference of emphasis given to the discovery of electromagnetic forces and the generation of electricity, claimed for Faraday by Child (Chapter 24), and for Siemens by Raub (Chapter 23)! These differences have been allowed to remain, as have more fundamental ones such as the degree to which patination on classical bronzes was deliberate, between Born (Chapter 2) and Craddock/Giumlia-Mair (Chapter 3). Research in this area, like the surfaces themselves, is not permanent and it is valuable to bring different perspectives together, for out of this new ideas and research will be born. If that is achieved by this volume then the editors and contributors will be well pleased.

Acknowledgments

The editors would like to thank all their colleagues who have helped with this volume, especially Julie Bevan, Sheridan Bowman, Brenda Craddock, Duncan Hook, Janet Lang and Tony Milton.

We wish to express our deepest thanks to Mr and Mrs A. Haas and Mr and Mrs M.D. Schwarz of the USA for their financial assistance towards the production of this book.

S La Niece
P T Craddock

Contributors

Laurence Birnie
National Maritime Museum, London, SE10 9NF

Hermann Born
Museum für Vor- und Fruhgeschichte, Staatliche Museen zu Berlin, Preußischer Kulturbesitz, Schloß, Charlottenburg, Langhansbau, 1000 Berlin 19

Warwick Bray
Institute of Archaeology, University College London, 31–34 Gordon Square, London WC1H 0PY

Robert Child
Head of Conservation, National Museum of Wales, Cardiff, CF1 3NP

Michael Corfield
Head of Artefact Conservation and Technology, English Heritage, Fortress House, 23 Saville Row, London W1X 1AB

Paul Craddock
Department of Scientific Research, The British Museum, London WC1B 3DG

Alessandra Giumla-Mair
Via della Costa No 4, J-39012 Merano, Italy

Martha Goodway
Conservation Analytical Laboratory, Smithsonian Institution, Washington DC 20560

Victor Harris
Department of Japanese Antiquities, The British Museum, London WC1B 3DG

He Tangkun
Institute of History of Natural Science, Academia Sinica, 137 Chao Nei St, Beijing 100010, Peoples Republic of China

Richard Hughes
School of Art History and Conservation, The London Institute, Camberwell College of Arts, London SE5

Paul Jett
Arthur M. Sackler Gallery/Freer Gallery of Art, Smithsonian Institution, Washington DC 20560

Nigel Meeks
Department of Scientific Research, The British Museum, London WC1B 3DG

Ryu Murakami
Nara National Cultural Properties Research Institute, 2-9-1, Nijo-cho, Nara 630, Japan

Susan La Niece
Department of Scientific Research, The British Museum, London WC1B 3DG

Andrew Oddy
Department of Conservation, The British Museum, London WC1B 3DG

Jack Ogden
The Cambridge Centre for Precious Metal Research, PO Box 391, Cambridge CB5 8XE

Christoph Raub
Forschungsinstitut für Edalmetalle und Metalchemie, D7070 Schwäbisch Gmünd, Germany

Judy Rudoe
Department of Medieval and Later Antiquities, The British Museum, London WC1B 3DG

Susan Stronge
Indian and South East Asian Section, Victoria & Albert Museum, London SW7 2RL

Eric Turner
Metalwork Collection, Victoria & Albert Museum, London SW7 2RL

Michael Wayman
Department of Mining, Metallurgical and Petroleum Engineering, University of Alberta, Edmonton, Canada T6G 2G6

Zhu Shoukang
General Research Institute for Non-ferrous Metals, 2 Xin Jie Kou Wai St, Beijing 100088, Peoples Republic of China

Ulrich Zwicker
Universität Erlangen-Nürnberg, Martenstrasse 5, 8520 Erlangen, Germany

1

Artificial patination

Richard Hughes

Abstract

The extent to which artificial patination was practised in various cultures and periods continues to be a matter of debate. Nevertheless within the broader context of the surface treatment of colour within sculpture and architecture, a variety of different rationales for colouring can be observed. These include protection, embellishment, a desire to emulate the products of antiquity, attempts to deceive, attempts to emulate natural processes, a desire to achieve naturalistic colour in sculpture and so on. Conceptual models of colouring driven by these diverse aims have tended to shape the nature of colouring practice.

The principal techniques of patination are described and the products identified in typical cases and related to the processes of natural patination. Although the the roots of contemporary patination are relatively recent in origin, drawing heavily on the techniques and colour ranges of 18th and 19th century French sculpture and European adaptations of Japanese colouring, the broad traditions of patination persist.

Introduction

In sculpture and the decorative arts generally, the role of colour as a facet of overall aesthetic content has clearly been of great importance from earliest times, assuming a wide variety of functions. It has been used to engender representational realism, to articulate religious and secular symbolism, to support artistic conventions and more generally to unify and enhance the decorative aspects of work in most cultures. Although in one sense colour can be thought of as a secondary characteristic, in that it is often applied in a process that is distinct from creating the form of an object, in many cases it is the vital element in the visual coherence and signification of objects and therefore of prime importance.

Metals and alloys have been selected for use not only on the basis of their working properties, but also because of their natural colour. Evidence from classical sources, such as Pliny the Elder suggests that the manipulation of the composition of alloys in order to achieve particular finished colours was well known. In addition a very wide variety of methods have been employed to produce desired colours and surface finishes. These might be characterized generally, in terms of their gross effect, as constituting three principal approaches. (1) *Methods involving the removal of material*, including a wide range of mechanical methods such as polishing or matting; gross chemical action involving the dissolution of the metal or alloy such as etching; selective removal of one of the constituents of an alloy through chemical treatment as in depletion gilding. (2) *Methods involving the addition of material*, including: the application of metal leafs and foils; the deposition of metallic layers chemically or electrochemically; the application of siccative organic materials often in the form of paints or lacquers. (3) *Methods involving the chemical alteration of the metal surface*, including: oxidation, sulphidization and mineralization through the actions of chemicals and/or heat.

Patination as a process clearly belongs to the last of these broad categories, in that it generally refers to the action of chemicals in the alteration of a metal surface, and one that results in a change of colour. Nevertheless given the range of chemical agents and the diversity of processes, patination will often also involve the removal or addition of material at the surface. In many cases, chemical colouring processes include either gross or selective etching or depletion in the creation of a surface finish. Others involve the deposition of metals or metallic compounds, in addition to the chemical alteration of the surface.

Although the term patination is generally reserved for effects involving the chemical alteration of the surface resulting in a change in colour, it is used to describe both the results of deliberately applied craft processes and also those resulting from natural corrosion. Metals in general and the finished metal surface in particular represent relatively improbable states of matter, a long way distant from equilibrium conditions, which will begin to alter under the influence of the atmosphere, or the local chemical environment, as soon as they are prepared. The quality and degree of change is dependent on the particular conditions prevailing, but the resulting surface colouration of copper or bronze, for example, may include ochre, red, brown and black oxides with varying degrees of translucency, and a variety of green mineral alteration products. They may also involve intense forms of corrosion. The effects of such natural corrosion will have been observed from the earliest use of metals.

The possibility of colouring metals using heat and chemicals may well have been suggested by the role played by colour in the control of craft processes. It is often a directly observable symptom of change and has been used as an indicator of temperature and the state of the material, when no quantitative measure was available. The temperature for the annealing or soldering of copper, for example, is still judged in the workshop by observing the colour of the hot object. Most processes used in the preparation and working of metals involve heat at some stage, resulting in the conversion of the surface to oxides. Annealing operations on copper, for example, produce red, brown and black oxides, which can become tenaciously adherent. In addition, descaling and acid pickling operations will often produce deposits of variously coloured metallic compounds, if the surface is allowed to dry.

Although historically natural patination has occurred, and artificial patination has been carried out, on a wide variety of metals including silver, lead and zinc for example, it is copper and the copper alloys that have predominantly been the focus for patination. This is partly due to the central role of a range of workable and useful alloys in sculpture and decorative metalwork, but also because of the wide range of coloured products that are produced by chemical action with these metals and their relative stability.

There is little direct evidence to suggest that in early times such coloured products were commonly valued aesthetically *per se*, and indeed the creation of conceptual hierarchies of materials tended to be based on the general notion of resistance to outside influence. Materials with the greatest integrity were perceived to be those that remained unaltered, such as gold, which was reserved for use in the more important artefacts partly on account of its 'nobility' and in contrast with the more contrary 'base' metals. Nevertheless such coloured surface products and the more drastic forms of corrosion and incrustation produced in the case of buried metalwork, were valued indirectly and formed the basis for one important strand in the early development of chemical technology (Smith 1974, 1981).

In later periods there is no doubt that the patinated surface was valued and in some cases became a very highly developed aspect of craft skill. Chemical colouring was very widely used in many of the major metalworking traditions and represented in a wide range of functions and approaches, in many cases stimulated by the natural patination found on more ancient artefacts.

Patina

Etymologically the term patina has most often been linked to the latin *patena* a shallow dish, the same root as the ecclesiastical *paten*, although the reasons for the derivation remain unclear. An alternative derivation has been suggested, which links the term with the Old Italian word *patena*, a dark shiny varnish which was applied to shoes, establishing its use in connection with the notion of artificial enhancement by the application of a layer to the surface of an object (Weil 1977).

On the other hand, the earliest discovered use of the term in a printed work occurs in Filippo Baldinucci's *Vocabulario Toscano dell Arte del Disegno* published in Florence in 1681, where it is identified as a term used to describe the surface darkening of paintings (Weil 1977). The definition states that Patena is a term used by painters to refer to the dark tone (or skin) which paintings generally acquire with age, and which can improve their appearance. The definition is interesting because it includes two important and persistent elements in the concept of a patina. Firstly it points to the fact that the appearance of patina is the result of a natural process occurring over time; secondly, that such a process leads to the enhancement of the work on which it occurs.

Although the term patina has tended to be reserved to refer to a fairly narrow range of surface colouring effects on metal (and in particular the green, brown and black colourations on the copper alloys) it also has an established usage which is in some senses broader in scope, and which links it with 17th century Italian usage in relation to paintings. In that broader context the word is a complex one, seeming to connote a range of more general characteristics rather than particular surface qualities. It is, for instance, applied across a wide range of materials, generally referring to acquired characteristics resulting from gradual change and wear, particularly those which are perceived to enrich the surface quality. We refer, for example, to the patina of an old chair, whose surface has been mellowed and darkened by gradual wear and time; the patina of a silver cigarette box whose surface has been mellowed by use and handling, in the form of a lacework of very fine scratches.

When one considers the title of this paper 'artificial patination' the phrase itself links fundamental elements in the aesthetic surface history of objects. On the one hand the word artificial draws us into the context of the wide range of approaches that have been devised for the surface finishing and visual articulation of objects — into technical artifice, whilst on the other, the word patination in the context of the phrase tends, in contrast, to lead us to a sense of the results of natural process. Taken together, they additionally suggest the sense of technical artifice being used in imitation of the natural. These three elements, and the interplay between them, in many ways constitute the main strands in the history of surface colouring, which recur in a number of cultures.

The root sense of the word patina seems to be passive. A patina is the result of a natural process over time. Artificial patination is some process deliberately applied in imitation of the natural. The 'natural' dominates the concept and also seems to provide the criteria for determining a rough range of qualities associated with a good patina. The most pervasive model for patination is in one sense a rose-coloured view of the results of natural processes, which selectively ignores the more destructive traits of metallic corrosion.

The natural is of course a very slippery concept. In scientific terms the natural terminus of a process of corrosion might arrive at an electrochemical equilibrium, whatever the aesthetic result. Artistic models of the natural, on the other hand are often based on the selection of an ideal type from among the available real examples, and one which fits the prevailing aesthetic preferences of the period. In the Renaissance, for example, the dominant model was of a rich deep brown, which no doubt some bronzes did attain naturally. By the 18th century, French encyclopaedists had squarely equated the Italian term *patina* with the green colouration of copper and although some writers demonstrated a more subtle sense of the variety of qualities associated with patination (Hiorns 1892), the dominant interpretation of patina in the 19th and 20th century has similarly tended to be singular and narrow.

What then constitutes the natural processes of patination and what are the results?

Natural patination — atmospheric corrosion

Copper and copper alloys characteristically undergo a sequence of surface changes when exposed to the atmosphere, which eventually lead to the formation of a thin more or less green layer of corrosion product. The principal chemical constituents of such natural patinas were identified by Vernon and Whitby in the 1930s, correcting long-held assumptions regarding their composition (Vernon and Whitby 1929–32). More recently, a series of investigations have been carried out by Graedel and colleagues, which have provided a detailed picture of the constitution and formation mechanisms of natural patinas on copper (Graedel 1987, Franey and Davis 1987).

Clean copper surfaces exposed to the atmosphere will quickly form a thin layer of tarnish which gradually change with time to a brown colour which can vary from a dull brown to reddish brown. Continuing exposure will normally then be accompanied by a change to black, and finally the very gradual formation of the characteristic green patina. Reviews in the literature, generally of a qualitative nature, have shown that the influence of industrial and urban emissions play an important part in the extent to which the final green patina is likely to form and the time taken for complete development. The green patina may not form at all in locations which are isolated from atmospheric pollution and natural sources of airborne chemicals, and the final state of the metal surface in these cases varies from brown to black (Graedel, Nassau and Franey 1987).

The principal chemical constituents of the patina on copper and copper alloys exposed to atmospheric corrosion includes the oxides and sulphides of copper, and inorganic and organic copper salts. The initial surface product is normally cuprite Cu_2O, although in some cases chalcocite Cu_2S may form. The most common component of the green patina is nearly always the sulphate brochantite $Cu_4(SO_4)(OH)_6$ although the chloride atacamite $Cu_2C1(OH)_3$ is also commonly found. In addition the sulphate antlerite $Cu_3(SO_4)(OH)_4$ is not uncommon and traces of posnjakite $Cu_4(SO_4)(OH)_6.2H_2O)$, the nitrate gerhardite $Cu_2NO_3(OH)_3$, the carbonates malachite $Cu_2(CO_3)(OH)_2$ and azurite $Cu_33(CO_3)_2(OH)_2$ have also been reported as constituents. Other constituents of the patina complex include organic copper salts including formates, acetates and oxalates. Typically an open-air patina will also include a range of atmospheric particles including soot, alumina, iron oxide, and silica cemented into the structure.

The resulting patina will generally consist of a simple two-layered structure in which a layer of cuprite is overlaid with a green mineral alteration product (Plate 1.1). Generally speaking the constitution of the green mineral layer reflects the location of the object. In marine environments where there is a relatively high concentration of chlorides in the atmosphere the principle constituent will be atacamite together with malachite. In continental locations relatively remote from extremes of atmospheric pollution the principal constituents will include posnjakite, atacamite and brochantite. In urban and other industrially polluted environments the principal constituents include brochantite, antlerite, malachite and posjnakite, and will include significant quantities of organic and inorganic particulate matter incorporated into the patina.

The physical characteristics are that the layers may vary considerably in thickness and at the microscopic level are discontinuous rather than continuous. The green layer in particular will tend to be open textured rather than compact and will contain particulate matter gathered from the atmosphere. These physical characteristics mean that the patina layer tends to be porous and adsorbent in respect of moisture. In addition the considerable variation in thickness and composition of the patina across the surface will favour the formation of local nuclei as centres for electrochemical action.

Investigations relating to the atmospheric corrosion of copper have tended to project a picture of a relatively benign process leading to a stable and protective surface product. The extent to which this is the case, however, depends crucially on the degree of atmospheric pollution, particularly in respect of levels of sulphur dioxide, the nature and quantity of particulate matter which form the patina, and on the pattern of moisture presence over time. In addition the composition of the metal and its degree of working will also have profound effects on the outcome. Generally speaking pure copper is less subject to progressive deterioration than alloys, and wrought sheet less easily attacked than relatively unworked cast surfaces. In the case of cast bronze statuary, recent reviews have highlighted the disfiguring effects of corrosion in the form of contrasting green and black streaking and the comprehensive loss of surface that can occur in urban environments, which may be masked by these apparently superficial colour differences (Weil 1982).

The quality of the resulting metal surface is clearly very variable and dependent on location and the degree of atmospheric pollution. The results of natural open-air corrosion can vary from rich and translucent reddish or brownish patination tinged with green, to the opaque and matt green colour associated with copper roofs, but increasingly includes opaque black surfaces marred by an irregular camouflage of a granular greenish white on urban public sculpture.

Natural patination — buried and submerged metalwork

The factors which may be involved in the development of patina in the case of buried or submerged metalwork are more varied and complex, including a greater variety of chemically active agencies and far wider variation in the concentrations and acidity/alkalinity (pH) of active chemicals, and fluctuation in their availability and activity. Given that the processes involved are still the transition from relatively unstable polished metal surfaces to relatively more stable and insoluble mineral species, it is not surprising that a number of the chemical constituents remain the same as for atmospheric corrosion. Layers of cuprite are nearly always present, and the green patina products include the brochantite, atacamite and antlerite commonly found in atmospheric patinas. More common, however, are the carbonates malachite and azurite, which occur to a much lesser extent in atmospheric patinas.

However, although there is a degree of comparability with atmospheric patinas, the mechanisms and products involved in the case of buried metalwork are often distinct. Perhaps the principal difference lies in the potential for sustained electrochemical gradients and the long-term migration of ions across the surface boundary of the metal. In the case of cuprite formation for example, oxidation may penetrate along grain boundaries and seams that run deep into the core of the metal, in extreme cases resulting in the conversion of the entire core to cuprite. Similar whole body mineralization also occurs in the case of malachite formation, for example (Gettens 1970).

Considerable differences in quality are evident in the results. In some cases the original surface of the metal is preserved whilst in others it may be wholly lost. Mineralization can result in a compact stone-like surface or in an irregular and open structured accumulation of mineral efflorescence (Figure 1.1).

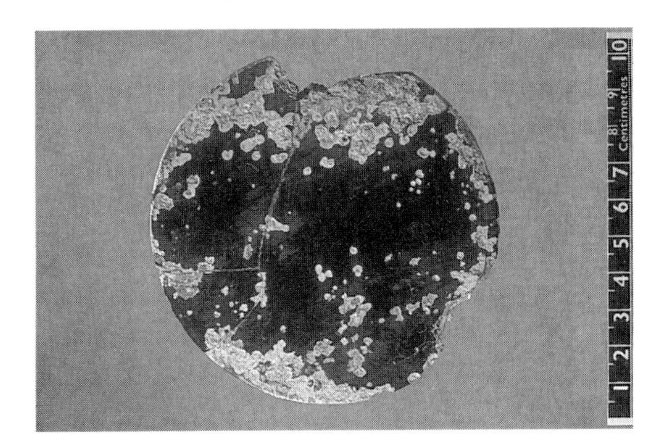

Figure 1.1. *Small mirror excavated from a Roman burial at St Albans, England. The mirror has 25% of tin and the surface has a dense compact patina, still retaining the original polish except where mineral efflorescence has erupted through.*

A particularly significant form of buried metal corrosion occurs in the case of bronzes with a relatively high tin content. The copper content of such bronzes may be progressively leached out, leaving the tin constituent of the tin-rich phase of the alloy to be converted *in situ* to cassiterite, stannic oxide SnO_2. This occurs without increase in volume, and results in the well-defined preservation of the original surface in the form of a compact and hard layer, often extending to a depth of 1 mm or more, in which the qualities of the metal surface including the lustre may be preserved. The layer tends to retain a small quantity of copper salts which tinge the naturally white or light grey tin oxide with green or bluish green colours. The resulting patinas are particularly beautiful and are well represented by their common occurrence on Chinese bronzes with a high tin content, and on Etruscan mirrors, for example (Gettens 1970).

The positive values of the 'natural' taken as models for artificial patination are highly selective, tending to focus on four principal types — the green patina found on copper roofs, the green-tinged tin oxide patina described above, the rich translucent brown patina which develops in the open air in benign environments, and the hard satin black patinas found on Chinese high tin bronze mirrors (Chase and Franklin 1979, Meeks, this volume, Chapters 5 and 6)

Historical notes on patination

The close similarity between the products of natural processes of patination and many of the products of artificial patination, means that it is extremely difficult to detect such artificial colouring on ancient artefacts. However, there is no doubt that a wide range of surface finishing techniques were used from early times, for protective or decorative purposes. Among the more notable examples discussed in the literature, include partial silvery-white coatings on Anatolian bronzes from the 3rd millennium BC and on Egyptian copperware, a range of which are reviewed by Smith (1974). Although the treatments involved in producing these finishes lie at the periphery of what we now take to be the core of chemical colouring, and there is virtually no evidence for patination in pre-classical antiquity, they are indicative of an active tradition of surface treatment at a very early date in the history of metalworking.

The weight of evidence also suggests that in the classical world of Greece and Rome, the predominant finish for bronze, for example, was an unpatinated polished surface (Craddock and Giumlia-Mair, this volume Chapter 3). Statues included in Greek and Roman wallpaintings almost invariably depict them as flesh coloured or the colour of polished bronze (Plate 3.1), and no depiction of a green statue, for example, is known. In addition a number of contemporary contracts and inscriptions referring to the cleaning and polishing of public sculpture have survived. Weil (1977) has drawn attention to the previously neglected work of both Richter and Pernice in relation to the ancient finishing of bronze. The evidence adduced by these authors includes references to extant instructions for the cleaning of bronze sculpture and detailed accounts of the way in which the design of bronze statuary was organized so that seams and joins are concealed in order to maintain an unmarred continuity in surface quality. It is argued that such attention to detail is only compatible with a clean surface, since artificial treatments would have made redundant the need for such attention to surface finish (Pernice 1910, Richter 1915). Although this is not in itself a completely convincing argument, since some coloured finishes are translucent and preserve the underlying metallic surface quality, together with any flaws, in the context of other literary evidence and the prevailing conventions of classical art, the conclusion is probably correct. The account in Plutarch's *De Pythiae Oraculis* (quoted in full in Craddock and Giumlia-Mair, this volume pp. 34–35), for example, in which visitors to the sanctuary of Apollo at Delphi discuss the blue-green patina on a bronze group, the burden of the discussion suggests that their inclination was to look for an explanation of the patina in terms of natural processes rather than craft technique. The patina is considered to be the result either of the action of the atmosphere or a natural ageing process of the bronze itself, although the part played by oil is also considered.

In the accounts by Pliny of metalworking techniques, also writing in about the 1st century AD, although the action of a variety of chemical agents on copper to produce artificial corrosion products is discussed, it is clear that these are not carried out for aesthetic purposes but rather for the production of medicinal and cosmetic treatments (Pliny XXXIV pp. 110–113). In sections where Pliny does discuss the colours that can be obtained on bronze in an aesthetic context, the most convincing interpretation is that the colour differences referred to are the natural differences that can be obtained through the manipulation of the composition of alloys (Pliny XXXIV, 8).

In addition to the literary evidence, the context of the more general artistic conventions obtaining at the time suggests that the use of artificial patination is unlikely. Although statues with polychrome detailing have survived, these take the form of contrasting metals and other materials inlaid into the main body of the bronze. Copper lips and nipples for example which, it is argued, would have relied on a polished finish for the colour contrast to work. Here again, although the argument is plausible, the evidence is not conclusive since chemical or heat treatment of the copper could have produced bright red finishes, which would have contrasted with a patinated main body.

More recent work has suggested that polychromy might well have been more extensive than had previously been thought (Born 1985 and this volume, Chapter 2), and that some particular alloys might have been valued specifically because of their potential to take on colour when patinated (Craddock 1982 and this volume Chapter 10). The example of Corinthian bronze, the traditional discovery of which Pliny relates was first produced accidentally during the sacking of Corinth, presumably as a result of the fusion of metals from artefacts melted by the fires that raged, is a case in point. The resulting alloy is supposed to have contained some silver and gold and was highly valued by collectors. Craddock has suggested that Corinthian bronze may have been representative of a family of surface treated alloys similar in composition to the Japanese *shakudo*, which is valued particularly for the rich deep purplish black colouration which it can produce on treatment with a sequence of immersions in natural acids and copper salts (Murakami, this volume Chapter 7). Although the evidence from objects is so far rather slight, the coincidence of accounts by Pliny and Pausanius with the descriptions of processes in the alchemical literature and the appearance and analysis of extant Roman objects, does at least suggest that the possibility of such a family of surface treatable alloys deserves more intensive investigation (Craddock 1982 and this volume Chapter 10). Such specialized use of alloys and colouring techniques does not appear to run counter to the idea that, generally speaking, the colours of various bronze alloys were valued in their own right and that the gross patination techniques familiar to us through their use in later centuries, were probably not generally in use.

The question of the classical finishing of bronze has been coloured by the writings of Pliny in one other respect. Although some passages in both Pliny and Pausanias do allude to certain alloys taking on colour suggesting the use of deliberate patination, such allusion is set against the background of descriptions of the natural colour of various bronze alloys and their use. Nonetheless the belief that bronzes were routinely patinated in ancient times became a prevalent view with collectors particularly in the 19th century, and the predominant belief, supported by selected examples, was clearly that the dominant colour was black, to the extent that many classical works were repatinated at that time, to restore them to their 'original' condition (Craddock 1990). In a section dealing with the treatment of bronze which includes reference to the use of vinegar and urine for cleaning bronze, and the oiling of copper and bronze utensils to protect them from corrosion, Pliny refers to the use of a protective coating of bitumen for bronze statuary. This passage has often been interpreted to mean that statues were at one time coated with an opaque black coating. The more likely interpretation has been provided by Pernice, who demonstrated the use of a thin solution of bitumen as a coating, which both enhances the natural colour of the bronze in the manner of a transparent and slightly tinted varnish and which also provides protection against corrosion (Pernice 1910, Weil 1977).

The re-emergence of bronze casting as a dominant metalworking technique for sculpture and decorative metalwork in the Renaissance particularly in the context of identifiable workshops headed by a named master, provided a stimulus for the elaboration of distinctive finishes. Although the larger scale public work such as the various projects for the production of bronze doors in the 15th century clearly created technical problems which were not fully resolved at the time and created a context in which surface treatment may well have been used to mask flaws and discontinuities in the casting (Weil 1977), smaller scale domestic work which was the subject of connoisseurship provided a forum in which more subtle surface treatments could evolve. Gilding and the contrasting natural colours of alloys were most often used to articulate such pieces visually, often in conjunction with a transparent or tinted lacquer, sometimes accompanied by the use of a range of colouring treatments. Pomponius Gauricus (1504), for example, notes that in addition to the use of the different colours of cast metal, white colouring is produced with silver leaf and yellow with gold leaf, green by applying salted vinegar, whilst black is produced by coating with a varnish of liquid pitch or by smoking the bronze with the smoke of wet straw. Vasari (Maclehose and Brown 1960) also notes the use of vinegar to produce a green colour on bronze and the use of oil or varnish to give a black colour. It is likely that chemically produced finishes would have been used in conjunction with a varnish for protection, and in many cases the use of tinted varnishes and lacquers without underlying chemical colouring was preferred. Although a variety of such lacquer finishes were developed, which tended to be characteristic of particular sculptors or workshops, such as the reddish finish favoured by Gian Bologna and his followers, the predominant conception for contemporary finishing was of a natural bronze colour which was interpreted as a rich and translucent brown. On large-scale sculpture and decorative metalwork such as Bernini's *baldachino* in St Peters, this was probably produced by the use of oil and heat.

One of the principal motives for the chemical colouring of artefacts was in imitation of the surfaces of excavated antique bronzes, which were highly valued in an atmosphere of reverence for the Classical tradition. Connoisseur-collectors were eager to acquire classical antiquities and the market for classically inspired sculpture flourished. Inevitably chemical colouring techniques were harnessed in the production of forgeries, and although this meant that such methods were subject to technical development, there is little documented evidence that this had any significant influence on the patination of contemporary sculpture.

A similar pattern of events is evident in the history of Chinese metalworking, where the evidence is that the older traditions valued the polished surface of the metal, contrasting the flat surface with linear relief often produced by modelling or cutting directly into the surface of the mould. Later variations involved the use of coloured inlay of various kinds, but nonetheless these are still thought to have relied solely on the contrast of the linear relief with the polished metal surface for artistic and decorative effect. The Chinese had a highly refined sense of history and excavated bronzes from the Chinese Bronze Age were collected and subject to scholarly and antiquarian interest from a very early date (Plate 1.2). During subsequent periods, and in particularly during the Song (AD 1137–1279) and Ming (AD 1368–1644) dynasties, imitations were often made which included attempts to imitate or replicate the patination that the classical models had acquired during centuries of burial. The more beautiful of these classical models exhibited a hard and lustrous tin oxide patina in which the surface detail of the original bronze was preserved and which were tinged with delicate blue-green colourations and compact reddish oxidization. The techniques used in imitation of these classical examples ranged from extremely crude devices such as the glueing of malachite to the surface, to refined and very involved chemical techniques. Although it is clear that many artefacts were made in respectful imitation, there is also evidence that deliberate forgery was practised as early as the Song dynasty (Kerr 1990). Several recipes for producing the mixture of emerald greens and cinnabar reds that were particularly admired have survived. Kerr cites one method recorded by the Ming connoisseur Gao Lian, which involves a complex treatment which begins by baking the bronze with an applied mixture of sal ammoniac, alum, borax and sulphuric acid, followed by placing the bronze in a pit lined with red hot charcoal which has been splashed with vinegar, and ending with the application of a variety of substances including pigment, and localized piles of salt, metal filings or cinnabar to encourage efflorescence, and sometimes further burial in acidic soil for extended periods of time (Kerr 1990, pp. 70–71. Further examples are given in Barnard 1961).

The strongest independent tradition of colouring clearly belongs to the Japanese and is associated with a particularly refined tradition of copper and copper alloy production. Although it is thought that Japanese metallurgy originated in China and Korea, the influence of Buddhism generally and of Zen Buddhism in particular at the end of the 15th century had the effect of changing the status of the minor crafts, and metalwork in particular, to a high art form. The years of peace that followed the feudal strife of the 16th century encouraged the exploration of refined metal techniques through the vehicle of the symbolically significant sword furniture. The skill which had previously been demonstrated in the production of highly refined copper was harnessed in the exploration of a very subtle and extensive range of copper alloys. These included notably *shakudo* an alloy of copper with *c.* 5% gold, which produced a deep and lustrous purple-black hue when pickled with natural acids and treated with solutions of copper salts (Murakami, this volume Chapter 7). A range of other alloys was used including *shibuichi* (copper-silver), *kuromido* (copper-arsenic), *sentoku* (a leaded tin bronze), *shintyuu* (a leaded brass) and a subtle range of composite alloys which provided a palette of colours including greys ranging from light to dark, ochres, terracottas, olive and earth browns of various hues and rich dark browns and purples after pickling. In conjunction with inlaid silver and gold these were used to great effect both in pictorially realistic styles and in abstract pattern.

The range and subtlety of finishes was enhanced by the great care taken to refine the surface of the metal by working and polishing, and the use of extremely fine punched textures (Savage and Smith 1979). Although such finishes are most often associated with sword furniture such as *tsuba* (sword guards)(Plate 1.3), they were also used on a wide variety of vessels and containers.

Whilst the Japanese during the 18th and 19th centuries were refining the use of these techniques in the development of a rich pictorial style, chemical colouring in the western tradition tended to draw on a relatively narrow range of colouring based on models taken from the late renaissance tradition, and which consisted principally of a range of brown and black colourations produced with sanguine (largely ferric oxide) and plumbago, in supposed imitation of florentine finishes, or by smoking bronzes in special ovens. A range of green patinas produced in imitation of natural antique patination was also prevalent. As the taste for small bronzes increased, so foundries became more competitive and the small bronze assumed the role of fashion vehicle, particularly in France, in an escalating race to produce ever more exotic finishes, sometimes bearing scant regard for the appropriateness of the finish. The techniques used involved a wider range of chemical treatments but also included the use of pigments and powdered metals bound in waxes and lacquers.

The revival of the lost wax process on the one hand, and the increasing use of industrial production techniques including sand casting, spinning and presswork, on the other, opened up two distinct areas for the development of colouring practice in the 19th century. *Cire perdue* came to be the pre-eminent sculptural medium, offering not only a responsive range of media for the primary processes of modelling, but also far greater precision in the reproduction of this work in metal. The development of colouring techniques in the context of sculpture reached its height in the decades surrounding the turn of the century. The use of direct modelling with coloured waxes stimulated experimentation with colour, and artists such as Degas and Rodin worked closely with founders and finishers to reproduce these tonalities in the finished bronze. In addition, the metalworking techniques and colouring methods of the Japanese tradition were brought to the attention of both the artistic and scientific communities by writers such as William Roberts Austen, who influenced the work of notable contemporary sculptors such as Alfred Gilbert (Dorment 1986).

In parallel with developments in fine art, the high volume production of commercial domestic metalwork made new demands on the metal-finishing industries. In most cases traditional finishing techniques, which were highly labour intensive and required considerable skill, were not appropriate to the newer methods of production. Colouring processes were sought which would be appropriate to batch finishing and much research was carried out on colouring copper and brass articles by immersion in baths, particularly in Germany in the late 19th century. Many of these patented processes for the production of black and a variety of brown finishes, were adaptations of colouring techniques associated with sculpture and decorative metalwork, including the processes used by Japanese craftsmen (Buchner 1907) (Plate 1.4). The commercial research promoted a greater scientific understanding of colouring which in some cases also resulted in the simplification of craft processes which had become baroque through traditional elaboration.

During the early 20th century the use of chemical colouring, both in sculpture and in industrial production, diminished in importance and many of the processes developed by artists and foundries during the previous century were lost. More recently there has been a revival of interest, particularly in the use of colour in the decorative arts, and research carried out to retrieve the colouring tradition for contemporary artist-craftsmen (Hughes and Rowe 1982).

Techniques of artificial patination

The full development of artificial patination was essentially a 19th century phenomenon, which drew on the techniques used in earlier periods and from a variety of cultures. The range of chemical treatments used in patination is very extensive and it is partly for this reason that a wide variety of colours and surface qualities can be achieved. Although it would clearly not be practicable to review the full range of such treatments, there is an underlying consistency across the chemical varieties, if one approaches patination in terms of the techniques of application. There are essentially six principal techniques: direct application, vapour colouring, immersion colouring, heat colouring, torch technique, and colouring using moistened media (particle technique).

Direct application

The direct application of chemical solutions to the surface of the metal constitutes one of the more common means of colouring. Although there are a variety of techniques used in direct application, it often takes the form of simply wiping the solution onto the surface very sparingly with a cloth moistened with chemical solutions (Plate 1.5). Alternatively the solution can be applied with a soft brush, a technique typically used in France from the mid-19th century, and thinned out by brushing with a sequence of dry brushes. In either case the result is a very thin residual film of chemical solution which is allowed to act on the metal surface, and allowed to dry completely before the procedure is repeated. Any powdery residue is rubbed away with a soft cloth, between applications. In a typical procedure the solution might be applied three times a day. Great care and skill may be required in order to produce a good coherent patina. The degree of wetness of the surface needs careful control and loose material must be burnished gently away at each stage. Over-application of the solution or attempts to speed up the process using heat or hot air drying will tend to result in the build up of superficial layers of dried-on chemicals. Although these may have a convincing appearance initially, they are generally non-adherent in the longer term, resulting in flaking and patchiness in the surface which are notoriously difficult to correct.

The length of time taken for a patina to develop is very variable, ranging from a few days to several weeks, but the pattern of colour development is similar in all cases. Initially the surface will become tarnished through the development of a thin oxide layer which is normally cuprite Cu_2O, although in some cases tenorite CuO may be present. This is followed by the gradual development of mineral alteration products which relate directly to the chemicals present in solution, usually accompanied by further development of the oxide layer. The result is normally a green or blue green patina on a ground varying in colour from red to dark brown. Products that have been identified through X-ray diffraction (XRD) analysis of typical samples have included atacamite $Cu_2(OH)_3Cl$, malachite $Cu_2(OH)_2CO_3$, antlerite $Cu_3(SO_4)(OH)_4$ and gerhardite $Cu_2(OH)_3NO_3$.

There are evident similarities between this technique and the characteristics of open-air corrosion. In both cases water is the transport medium for the chemicals and there are alternating periods of wetness and dryness, and the condition for action is generally dependent upon a thin mist-like film of chemical solution being present on the surface. In addition the more common chemical constituents are also similar, typically including sulphates, chlorides and nitrates. The principal difference is in the concentration of chemicals, which in the case of artificial patination will normally be several orders of magnitude more concentrated than for chemicals present in mist or rain. Natural corrosion in the open air may take ten years for the full development of a patina, whilst in artificial patination the time scale is likely to be closer to ten days.

Vapour colouring

One colouring technique which also has evident affinities with open-air corrosion, and which may have been derived from attempts to mimic natural conditions in the workshop, consists in exposing the metal surfaces to be coloured to the action of vapours in a closed environment. Although the technique is infrequently recorded in the literature, references do occasionally occur in text books on sculpture technique, notably in some earlier 20th century American texts (Rich 1947).

The technique has been used to produce thin layers of malachite on bronze surfaces, using carbon dioxide in a moist atmosphere, and a variety of effects using the vapour from acetic acid and ammonia, which have formed part of the stock in trade of sculptors. In the workshop the metal is generally subjected to continuous exposure to concentrated vapours. In the natural condition the action would tend to be intermittent, and the concentration relatively low. In both cases the result is a thin mist or dew of solution which condenses on the surface, and which acts more or less like a thinly applied solution. The pattern of patina development is generally comparable with other processes and natural phenomena, the chemical products tending to bear a simple relationship to the composition of the vapours present.

In practice the principal problems associated with the technique are the variations that can occur through the different orientation of surfaces within a piece, that can give rise to marked differences in colour development. In addition, the extent of condensation can vary, with run off from the surface, although these effects can usually be corrected by control of humidity and temperature. Variability of this kind is of course also evident in open-air corrosion. In the workshop these problems can usually be remedied by intermediate washing and drying, and by relieving the surface with a mild abrasive such as fine pumice, prior to further exposure.

Although the technique appears to be limited to relatively small-scale objects, requiring a closed container, the use of an inflated air bag for example, into which vapours can be introduced either continuously or intermittently, enables the technique to be applied to the larger scale.

Immersion colouring

An important technique for producing even colourations consists in immersing the whole object in a chemical solution. Cold solutions are sometimes used but more commonly they are used hot or even boiling. The more commonly used range of ingredients include oxidizing agents and copper salts, but variation is achieved through the addition of acids, alkalis and a wide range of metallic salts (Plate 1.6). Colour development will generally follow a predictable pattern. Initially the metal surface will acquire a series of lustrous colours due to interference phenomena associated with the production of very thin surface films of oxide. These quickly become more opaque as the oxide layer thickens, generally producing a more or less even layer of cuprite which can vary in colour from ochre, orange or terracotta to dark purplish brown.

If immersion is prolonged, then this oxide layer may gradually acquire stains or films of a lighter colour. These may ultimately become substantial accretions of mineral alteration products and deposits, which are most commonly green in colour, but varying from near white, through yellow green to dark bluish greens. The chemical nature of these products follows the general pattern for the corrosion products of copper alloys, typically including atacamite, gerhardite and antlerite. In some cases with particular additions of metal salts, copper–iron salt complexes and other metallic complexes may be produced.

Although some of the results of hot immersion colouring are also produced at lower temperatures, the time scales will be significantly increased. In many cases cold immersion will tend to produce soft deposits over the generally longer time scales required, as will prolonged immersion in hot solutions. Sometimes the technique may be adapted to allow the solution to be pumped and cycled over a surface, which allows it to be applied to larger-scale objects. One example of this in a conservation context was in the restoration of the statue of King Sigismundus in Warsaw, where an alkaline persulphate solution was used to generate the patina (Socha *et al.* 1980).

Nevertheless, the technique is more often used in decorative fine metalwork than in sculpture, due in part to the problems associated with scale. On the smaller scale the results can be both subtle and refined, and with a carefully prepared surface, the rich even colourations typical of Japanese sword furniture can be produced. In addition it is also possible to achieve controlled local variation of colour, in a sequence of immersions. This is achieved by selectively rubbing back areas with a mild abrasive, prior to re-immersion, and by repeating this process a number of times. Examples using sodium chlorate and copper sulphate solutions on copper, produce a dark terracotta ground which can be varied locally through to pale orange. Coupled with a similar control of green patina development, the results can exhibit a wide colour range and the qualities associated with a fine patina, including the richness of surface associated with thin layers of distinct colouring and a translucence in the final result.

Heat colouring

A wide range of colouring effects can be produced using heat alone. The use of kiln finishes on bronze such as the standard Bronze Metal Antique (BMA) and the production of fine satin black finishes has been used extensively, but is generally limited to objects of a relatively small scale. Localized heating sometimes in combination with oil burnishing or smoking has been used for small-scale industrial use (Lacombe 1910). It has also been employed for larger-scale work in architectural contexts, and was probably the finish used for Bernini's *baldachino* at San Pietro in Rome.

More specialized techniques include heating and localized rapid cooling, which was employed to great effect as one of the techniques used by the French Dinandiers of the 1920s. This group included Dunand, Mergier and Linossier, whose colourful work is well represented in the collection of the *Musee Des Arts Decoratifs* in Paris. The products of these processes are cuprite and tenorite, a wide range of colours being obtained by the thickness and natural variation in the colour of these oxides, and their balance in a given finish.

Torch technique

One of the techniques most commonly used by sculptors involves stippling or painting chemical solutions onto a metal surface which has been locally heated with a blow torch (Figure 1.2). The patina is gradually built up by alternately heating an area of the metal with an oxidizing flame, and applying the solution to the heated area. The solution is normally applied sparingly by stippling with a

Figure 1.2. *Père and Jean Limet, 'Master Patineurs', in their Paris studio, 1938, blow torch heating the bronze before brushing on patinating solution. (Photo by Malvina Hoffman, from* Sculpture Inside Out, *George Allen and Unwin, 1939, London.)*

soft cloth or brush, although individual styles vary. In principle the technique is simple, but effective results depend crucially on the skill of the patineur. The temperature of the metal and the amount of solution applied must be carefully judged, in order to avoid the production of visually convincing but loosely adherent layers, which will ultimately break down.

Although, in principle, a wide range of chemicals can be used to achieve different colouring effects, in practice a very small range has normally been used by sculptors. Copper nitrate and ferric nitrate solutions are most commonly used to produce green and reddish brown colourations respectively. Variation is achieved by using these chemicals alternately, or by modifying the results using sulphide and chloride solutions or ammonia. Copper nitrate can also be used to produce a smooth slate black layer of tenorite. A green patina is first built up and then overheated slightly, causing it to peel and reveal the black layer beneath. The thickness of this layer can vary from brownish thin and translucent to a near opaque slate black, depending on the degree to which the green patina is built up before overheating. The black colouration is commonly used as a ground upon which semi-translucent or variegated layers of green can be developed, producing a wide variety of colouring effects.

The patina products follow the same pattern as for other artificial techniques, consisting commonly of atacamite, gerhardite and antlerite on tenorite and cuprite grounds.

Particle technique (colouring with moist media)

A significant range of techniques involves wrapping or packing the metal object in some form of moisture retaining medium. The medium most often used is sawdust, but most absorbent particulate media can be used, and in some versions of the technique, cotton wool or cloths are used to wrap the object. The medium is evenly moistened with the chemical colouring solution. Particulate media tend to produce richly stippled and textured surfaces. These can be very fine, involving little loss of surface polish, but under certain conditions, particularly in the presence of chlorides, relatively coarse etching phenomena may occur involving considerable loss of surface metal.

The surface effects are the result of electro-chemical action associated with the points of contact of moistened particles or the textured weave of a cloth, with the surface. The degree to which the more extreme corrosive effects will take place depend on the scale of the particles or cloth texture and the degree to which the medium is moistened. The initial physical effects are akin to pitting corrosion at the points of contact of the medium with the metal. In the later stages the pits tend to expand laterally rather than increasing in depth, gradually extending across the whole surface. In the case of cotton wool or applied cloth, which tends to create pools of wetness and boundaries of contact with the metal surface, similar extreme corrosion occurs at the boundaries. The effects can occur with a wide range of chemicals, but are most aggressive where significant concentrations of chlorides are present.

Although the closest analogy in natural patination for this technique might appear to be the corrosion of buried metalwork, it is likely that the mechanisms involved are more akin to certain types of open-air corrosion. In the case of buried metalwork the time scale and the typical pattern of corrosion involves molecule by molecule replacement of the substrate metal with mineral products, and migration of metal ions through the original surface layer. In time this can lead to the mineralization of the whole body of the metal, although surface layers may remain. In the case of particle techniques this does not occur. Processes appear to be either self-limiting in some way, or alternatively result in corrosion products which are leached out during the process, so that the net result is loss of metal from the surface in the form of gross etching.

The nature of the corrosion and the removal of soluble mineral products perhaps bears closer resemblance to some forms of open-air corrosion. In some cases accretion of particulate matter on the surface of metal in the open air creates local nuclei or centres of action which cause extensive pitting corrosion. In conjunction with streaming or run off from surfaces which creates similar effects at the edges of streaming, this can leave islands of coloured metal in a general field of substantial corrosion, where the surface of the metal may be significantly lowered (Weil 1982)

The chemical products obtained with this technique generally follow the same pattern as those previously discussed, including cuprite as the principal oxide, and a range of mineral alteration products including atacamite, antlerite and gerhardite. In some cases the principal mineral alteration production has been connellite $Cu_{19}Cl_4SO_4(OH)_{32}2H_2O$.

Approaches to patination — colouring traditions

The nature of colouring practice in different cultures and contexts has been shaped in part by local conceptions of what constitutes a 'natural' finish, and by attention to selected ideal models of antiquity. The other elements in the balance of forces that constitute the conceptual framework for a given tradition of patination, include the *uses of colouring*, and the *pragmatics of technique*. A relatively consistent picture of the variety of uses for chemical colouring emerges from a consideration of the history of colouring. The more notable of the broad categories of use include protection, modifying surface properties, mimicry and deception, creating visual consistency within a surface and creating visual difference within a surface.

Protection

Colouring techniques have to some extent been used to protect the underlying metal from further corrosion, although their capacity to do so is by no means comparable with other finishing methods such as tinning or electroplating. One example of a range of colouring treatments which offer some protection from corrosion are the blueing and browning processes used to colour firearms (Angier 1936). Similarly the artificial black patination of cast iron by the bower–barffing process (Goodway, this volume Chapter 13). The natural patina produced by the atmospheric corrosion of copper also offers some protection to the underlying copper, and a range of artificial treatments have been developed to induce such patination on copper sheet (Hemming 1977).

Modifying surface properties

The use of colouring techniques to alter the surface properties of an object are clearly more common. The colouring of firearms is said to have been developed partly to create a surface that would be less subject to adventitious reflection. Similarly, in the 19th century matt black finishes were developed for use on brass optical instruments to minimize internal reflection (Birnie, this volume Chapter 12). Treatments have also been used widely to modify surfaces where the choice of materials for the purposes of production has meant that the resulting surface lacks the appropriate visual finish, for example the bright dipping of brass presswork. To some extent the gilding of statues in antiquity and the use of coloured varnishes to change the tonality of sculpture in the Renaissance can be viewed simply in terms of the desire to modify visual properties, although the reasons were probably more complex and often linked to mimicry or imitation.

Imitation of other surfaces/materials

Surface treatments can clearly be used to change the visual properties of an object, in imitation of other materials. Examples include the use of depletion gilding to produce a gold or silver surface from an alloy which contains a relatively low percentage of precious metal, as in the surface-enriched metalwork of prehispanic America (Bray, this volume Chapter 16), or the relatively recent coloured chromate deposits produced commercially on brass to give the appearance of gilding. Small brass articles such as domestic fittings and picture hooks are often still finished by dipping to produce a bronzed finish, which is protected with lacquer. Treatments have also been used to create a context in which objects would be read in a different way, for example, in the ubiquitous use of antique effects. Although these have often been used without intention to deceive, they rely on reference to genuine antique examples or conceptions of the antique, to provide a modified context in which objects may be read.

Creating surface continuity

One of the most important uses of colouring treatments is in the creation of visual unity or continuity in a surface. Metalworking techniques, in all but the simplest pieces, tend to create local differences in surfaces for a number of reasons — casting may create differences in surface composition through differential crystal structure, through localized porosity and through sweating, etc. In addition the scale or complexity of a given piece may mean that it is composed of a number of parts, which are subsequently joined together by, for example welding, casting on, soldering, mechanical joining. These processes will tend to create local visual differences which subsequent finishing techniques can be designed to minimize.

Creating differences in surfaces

Perhaps the most important use of chemical colouring techniques lies in the creation of differences in a surface which can be harnessed to a variety of artistic intentions. Colour can clearly be used to support representational realism by matching colour in the real world. The most developed examples of the use of patination to this end are found in the subtle pictorial styles in the later periods of Japanese *Tsuba* production, although the use of selective patination to create distinctions between flesh and clothing in bronze sculpture are also evident in the western tradition in the late 19th century, for example in the work of Alfred Gilbert. Parallel with the use of colour to engender pictorial realism, is its use to articulate symbolic codings, as in the highlighting of halos and auroras with gilding, and in religious conventions such as the rendering of the clothing of the Madonna in blue.

Although it is tempting to regard the evocation of realistic or symbolic colour as the dominant factor in the colouring of sculpture, the use of chemical colouring treatments are more often associated with formal conventions. By this is meant the ways in which colouring can be used to create emphasis or to play off against the qualities of a sculpture which are connected with its form. The simplest examples are the use of colour or tonal difference to create distinctions between figure and ground, to emphasize or de-emphasize the modelling of form, and to create surface contrasts. The colouring and relieving of areas to create highlights is probably the most pervasive colouring technique in figurative sculpture.

It is clear from the history of sculpture that the interplay between the factors of representational realism, symbolic coding, and formal convention is highly complex. The development of artistic conventions consists in the definition of the framework within which visual problems are to be solved.

One of the roots of sculptural tradition consisted in working within the framework of a given material. In the early classical tradition elements that either defeated the attempts of sculptors at representation, or were thought to be inappropriate to depict in the given material, were depicted outside it. So glass eyes were inserted into bronzes, red lips were added and trappings of real materials were used. In later sculptural tradition it was felt that this was to avoid the deeper questions for sculpture — the point was not to create a facsimile using any means going, but rather to try to capture qualities within a material that were not inherent in it. The stance taken constituted the framework for sculpture — it defined the nature of sculpture at a given time. There is of course no natural or correct framework, and in the 19th century, classical traditions came to appear too restrictive and resulted on the one hand in the emergence of mixed material sculpture, and on the other in forms of modelling that retained the quality of sculptural material, rather than yielding entirely to the quality of what was depicted. Parallel considerations are evident in patination. The Renaissance tradition, for example, tended not to use naturalistic colour, but rather to opt for a narrow colour range based on a conception of the colour that was perceived to be 'natural' to bronze, and to use variation in that colour for general modelling effects and local changes in tonality. Clearly this was not the case in the 19th century or in the pictorial tradition of later Japanese metalwork. The 18th and 19th century French traditions, used the techniques in order to articulate formal properties, but had a much broader conception of what constituted an appropriate colour range for bronze.

The remaining factor in the development of colouring traditions is associated with the pragmatics of technique — the constraints on the selection of techniques.

Although the range of techniques and chemical solutions potentially applicable in the chemical colouring of metals is very extensive (Hughes and Rowe 1982), in practice these will be considerably reduced in the light of both pragmatic and conceptual criteria. The single most significant constraint is the scale of the object to be coloured. Many techniques, in particular immersion colouring, pose such great problems of handling and involve the use of large quantities of costly chemicals, that they are effectively precluded from use on all but domestic-scale items. Although in some cases techniques may be modified for use on the large scale, for example, the use of solutions cycled over the surface of a large bronze as an alternative to immersion, generally speaking the results will not be identical.

A second constraint which is also of great significance is the compatibility of the colouring techniques with the overall techniques used in the production of a piece. The simplest example is in the development of techniques applicable to volume industrial production, where most other considerations give way to the need for simple colouring operations which will yield repeatable and consistent results. In the context of fine art the considerations may be more complex. Firstly the physical quality of the colouring process may be an important factor for the colourist. Not only are individual sculptors or finishers likely to show a general preference for techniques which are susceptible to ongoing modification and which allow for a more direct contact with the metal, but they are also likely to adopt techniques which reflect their more general approach to sculpture production. The speed with which some indication of surface colour and quality can be achieved, for example, may well influence the choice of technique. The techniques which have been favoured by sculptors in the late 19th and 20th century, such as the use of the torch technique have tended to reflect the characteristics of the modelling process; direct contact with the material and an immediate sense of the visual result.

The interplay between the natural/artificial balance, the formal/ representational balance, and the compatibility of colouring processes with the more general production of metalwork has tended to crystallize out into a few types of colouring tradition. There are essentially four types of colouring tradition, which are primarily characterized by the interaction between the artistic conventions and practice of metalwork, and the practice of colouring.

Complex metallurgy — simple colouring technique

The development of refined approaches to the development of alloys, allows for the possibility of a naturally broad colour range using simple and consistent colouring techniques. The classical sculptural tradition as described by Pliny, for example, used a variety of bronzes for their natural colour, which were protected and enhanced by the use of simple organic coatings. The Japanese tradition is a more highly developed example, which illustrates the advantages of this approach. The use of a wide range of alloys which are coloured with a relatively standard colouring process which is selected for the compatibility of its coloured products with the natural alloys. The approach to metalwork and the scale allows for the consistent, even and high quality colouring produced by hot immersion.

Simple metallurgy — complex colouring

This is essentially the approach taken in the 19th century western tradition, where cast bronze of a relatively standard composition is used to produce sculpture, and where the development of surface and colour variation depends on the elaboration of patination techniques. The preferred colouring techniques such as direct application and torch technique colouring are compatible with a trend away from classicism and towards a more gestural approach to modelling and finishing.

Eclecticism — a mixture of materials and colouring techniques

This approach is represented by elements of late 19th century British sculpture including, for example, some of the work produced by Alfred Gilbert, and also by French Art Deco metalworkers and sculptors, and the decorative metalwork of the Dinandiers. Essentially these are pluralistic responses in reaction to a classical tradition, which all interestingly hover on the edges of kitsch.

Simple metallurgy — simple colouring

The predominant approach to the use of patination in decorative and architectural metalwork rather than sculpture. It does not really appear in the form of strong localized traditions, but is rather the background low-level use of colour, including the use of 'oxidized' finishes in silver, the bronzing of architectural metalwork and the colouring of domestic brass and copperware including bronzed finishes on small-scale architectural fittings, which occurs throughout the period that patination has been practised.

References

Angier, R.H. (1936) *Firearm Blueing and Browning*, Stackpole, Pennsylvania

Barnard, N. (1961) *Bronze Casting and Bronze Alloys in Ancient China*, Monumenta Serica Monograph XIV, Canberra

Born, H. (1985) Polychromie auf prähistorischen und antiken Kleinbronzen. In *Archaeologische Bronzen, Antike Kunst Moderne Technik* (ed. H. Born) Staadtliche Museen Preussischer Kulturbesitz, Berlin

Buchner, G. (1907) *Die Metallfärbung*, Verlag von M. Krayn, Berlin

Chase, W.T. and Franklin, U.M. (1979) Early Chinese black mirrors and pattern etched weapons. *Ars Orientalis*, **11**, 216–258

Craddock, P.T. (1982) Gold in antique copper alloys. *Gold Bulletin*, **15**(2)

Craddock, P.T. (1990) The art and craft of faking, copying, embellishing and transforming. In *Fake? The Art of Deception* (ed. M Jones) British Museum Publications, London

Dorment, R. (1986) *Alfred Gilbert - Sculptor and Goldsmith*, Royal Academy, London

Franey, J.P. and Davis, M.E. (1987) Metallographic studies of the copper patina formed in the atmosphere. *Corrosion Science*, **27**(7), 659–688

Gettens, R.J. (1970) Patina noble and vile. In *Art and Technology – A Symposium on Classical*

Bronzes (ed. S. Doeringer) MIT Press, Cambridge, Mass, pp. 57–68

Graedel, T.E. (1987) Copper patinas formed in the atmosphere – a qualitative assessment of mechanisms. *Corrosion Science*, **27**(7), 721–740

Graedel, T.E., Nassau, K., and Franey, J.P. (1987) Copper patinas formed in the atmosphere. *Corrosion Science*, **27**(7), 640

Hemming, D.C. (1977) The production of artificial patination on copper. In *Corrosion and Metal Artefacts*, National Bureau of Standards Special Publication 479, Washington DC, pp. 93–102

Hiorns, A.J. (1892) *Metal Colouring and Bronzing*, Macmillan, London

Hughes, R.S. and Rowe, M. (1982) *The Colouring, Bronzing and Patination of Metals*, Crafts Council, London

Kerr, R. (1990) *Later Chinese Bronzes*, Victoria and Albert Museum, London

Lacombe, M.S. (1910) *Nouveau Manuel Complet du Bronzage des Métaux et du Platre*, Debonliez et Malpeyre, Nouvelle Édition, Revue, Classée et Augmentée. Encyclopédie Roret, Paris

Maclehose and Brown (eds) (1960) *Vasari on Technique*, New York

Pernice, E. (1910) Bronze Patina und Bronzetechnik in Altertum. *Zeitschrift für Bildende Kunst*, **21**, 219–224

Pliny (1968) *Natural History XXXIV*, Loeb Classical Library, London

Pomponius Gauricus (1504) *De Sculptura*, A. Chastel and R. Klein (eds), Geneva

Rich, J.C. (1947) *The Materials and Methods of Sculpture*, Oxford University Press, New York

Richter, G. (1915) *The Metropolitan Museum of Art, Greek, Etruscan and Roman Bronzes*, Metropolitan Museum of Art, New York

Savage, E.I. and Smith, C.S. (1979) The techniques of the Japanese *tsuba*-maker. *Ars Orientalis*, **11**, 291–328

Smith, C.S. (1974) Historical notes on the colouring of metals. In *Recent Advances in Science and Technology of Materials* (ed. A. Bislay) Plenum Press, New York, pp. 157–167

Smith, C.S. (1981) Some constructive corrodings. In *A Search for Structure*, collected papers by C.S. Smith, MIT Press, Cambridge, Mass, pp. 332–344

Socha, J., Lesiak, M., Safarzynski, S. and Lesiak, K. (1980) Oxide coating in the conservation of metal monuments: the Column of King Sigismundus III in Warsaw. *Studies in Conservation*, **25**, 14–18

Vernon, W.H.J. and Whitby, L. (1929–32) The open air corrosion of copper, a chemical surface patina. *J. Inst. Metals*, **42**, 181–195 (part 1); *J. Inst. Metals*, **44**, 389–396 (part 2); *J. Inst. Metals*, **49**, 153–161 (part 3)

Weil, P.D. (1977) A review of the history and practice of patination. In *Corrosion and Metal Artefacts* (eds B.F. Brown, H.C. Burnett, W.T. Chase, M. Goodway, J. Kruger and M. Pourbaix) National Bureau of Standards Special Publication 479, Washington DC, pp. 77–92

Weil, P.D., Gaspar, P., Gulbranson, L., Lindberg, R. and in memoriam, Zimmerman, D. (1982) The corrosive deterioration of outdoor bronze sculpture. In *Preprints of the Contributions to the Washington Congress: Science and Technology in the Service of Conservation*, IIC. London, pp. 130–134

2

Multi-coloured antique bronze statues

Hermann Born

Abstract

The controversy over possible artificial patination and painting of antique bronze sculpture extends back into the 19th century. It has been revived in the last few years, partly as a result of the interdisciplinary investigations and restoration of several Greek and Roman statues and the attempt at reinterpretation of ancient writings. It has been categorically denied that painted decoration or artificially applied patina exists on large bronzes. To add to the difficulties of interpretation, many of the large bronzes are sea or river finds with homogeneous black copper sulphide layers on the surface.

The theoretical debate can only be sustained in the future through the new study of classical literary sources and contemporary vase paintings and frescos, and, most important of all, chemical and mineralogical investigations of the hundreds of unrestored fragments of Greek, Hellenistic and Roman bronze statues to be found in our museums.

The controversy about the existence of artificial patination and painting of Greek and Roman bronze statues goes back to the 19th century (Reutersward 1960). A handful of archaeologists and classical philologists proposed new interpretations of passages in classical texts which referred to the surface appearance of bronze statuary claiming that they were not bronze coloured as had hitherto been presumed, but rather that they had a patinated, naturalistic surface colour. This served to expand the canon of colouration then known from monumental statuary which had included amber or bone inlayed eyes, copper nipples and lips as well as silver teeth etc. moving towards a concept of total polychromy; a phenomenon which has been definitely established for marble statuary and is demonstrated by a unique Greek vase painting on a 4th century crater from the Metropolitan Museum (Inv. Nr. 50. 11. 4) in New York. This shows an artist or craftsman painting the lion's hide on a Herakles statue (Born 1990).

The ensuing controversy was, however, purely literary in character as a mere handful of the early monumental bronzes, or fragments thereof, were known before 1900. Indeed many of the now world famous sculptures were discovered much later (Bol 1985, Gazda and Hanfmannn 1970, Finn and Houser 1983). The publication in 1927 of K. Kluge and K. Lehmann Hartleben's comprehensive standard book on the classical bronze statue seemed to put a preliminary stop to the discussion on monumental antique bronzes including the topic of polychromy, with a resounding no to the thesis of large-scale patination and colouration. Not even P. Reutersward's remarkable monograph from 1960, which raised the possibility of colouration anew, could shake the entrenched view that the appearance of bronze statuary was bronze metallic monochrome.

Thus we still must consider three pertinent subjects which, then as now, need closer and differentiated analysis. These are:

1. The classical sources which refer to bronze casting and the appearance of antique bronzes.
2. The illustrations of classical bronzes on contemporary vase paintings and frescos. And finally and most importantly:
3. The technological analysis and study of the surface of the classical sculptures themselves.

The discussion of these three areas has hitherto been almost exclusively the domain of classical philologists and archaeologists. This has led to a one-sided picture of the appearance of classical bronze sculpture and has so dominated past and present archaeological literature that it may well dominate scholarly opinion for decades to come. It seems an unsatisfactory solution to depend on mainly late antique authors for our present day knowledge about the appearance of classical bronze statues. If we take the sources literally, we find nothing in them that would not support both sides of the argument after interpretation. Thus literary sources from the classical world cannot serve alone as primary sources for our knowledge of antique reality. Indeed without supporting archaeological evidence, almost all of these literary passages remain shadowy and ambiguous, regaining usefulness when confronted with physical evidence which allows us to understand ancient technology within its historical development.

There is no surviving 'technical literature' for craftsmen from the classical period. A portion of the texts referring to contemporary craftsmanship are contained in the context of 'exempla'. These are passages which were intended to illustrate abstract thoughts with examples from daily life. Many ancient authors borrowed these examples from previous sources or described technical processes of which they had little or no detailed knowledge.

Figure 2.1. *The Youth from Marathon, 4th century B.C. , The National Museum of Athens. (Photo from D. Finn and C. Houser, 1983.)*

Thus these works are only a qualified help in understanding ancient manufacturing techniques (Zimmer 1985).

A consideration of a particularly contentious classical description helps to emphasize this situation. That is, the interpretation of the so-called *EKPHRASEIS* (a Greek word which is translated as 'descriptions') which was written by the Greek author Kallistratos in the third or fourth century AD and is seen to be a prosaic description of 14 bronze and marble statues (Altekamp 1988). The quality of this source seems to be a matter of controversy, and Kallistratos' writings have yet to be investigated in the light of recent progress in the understanding of antique bronzes. Yet at first glance these quotations seem to be significant or at least worth mentioning. An interesting example is the mention of a bronze Dionysis ascribed to Praxitheles: 'Even though (it was

Figure 2.2. *The Artemision Zeus, 5th century* BC, *The National Museum of Athens. (Photo from D. Finn and C. Houser, 1983.)*

made) of ore, he was red in colour. The hide of a doe covered him, but it was not a real one, as Dionysis uses to cover himself but made of bronze transformed to imitate the hide.' Up to this day archaeological scholarship rejects notions of painting and coloured patination on bronze statuary still. The classical sources which allow such interpretations are treated with utmost reserve.

This is different in the case of marble. Kallistratos describes an Indian or Ethiopian whose hair, 'had tips with a sheen like Tyrian purple'. In this case Kallistratos is treated as a reliable witness since there is a large relief from the Kolonos Square in Athens (National Museum of Athens, Inv. -Nr. 4464) with a splendid horse which is being held by a lad with crinkled hair. This youth has blue skin and cinnibar red hair. In this case scholarship accepts the quotation as a parallel and footnotes contain references to a possible antique tradition of displaying red hair in this context. This subject of compiling evidence for the total polychromy of bronze statuary is so complex, that, as we have seen, a simple theoretical analysis would reach the proportions of a monograph if done properly. Thus we can only deal with a few interesting pieces which call previously held notions of the appearance of classical bronze statuary into

Figure 2.3. *A detail of the so-called 'Foundry Cup', Red Figured Attic Cup, c. 480 BC, Antikenmuseum (F 2294), Berlin. (Photo from G. Zimmer, Antike Werkstattbilder, Berlin, 1982.)*

Figure 2.4. *Hephaestus Making Armour for Achilles: probably showing the polishing or cleaning of the shield with pumice. Attic Red Figure Nolan Amphora, The Dutuit Painter, 5th century BC. (Photo courtesy of Boston Museum of Fine Arts, Inv. -Nr. 13. 188.)*

question. As early as 1929, C. Zenghelis called attention to the high sulphur content of the hair of the bronze Youth from Marathon (Figure 2.1) which dates to the 4th century BC. He argued that an artificial blacking was intended and similar claims were then made for the Cape Artemision bronzes (Figure 2.2). It is indeed commonplace that when the Greeks illustrated statues they gave parts of the bodies naturalistic colour, for instance in the medium of vase painting. This can be successfully demonstrated in the case of the so-called Foundry Cup in Berlin's *Antikenmuseum* (Inv. -Nr. F 2294), a Greek drinking cup which dates around 480 BC and has often been used to illustrate the segmented construction of bronze statuary (Figure 2.3). The subject of polychrome surfaces is not only addressed by the naturalistic black colouring of the hair of both bronze figures but also a detail which has so far escaped serious attention: the vase painting shows that after the separate segments of the monumental sculpture had been connected, it stood in a scaffolding and was treated with a scraper (a so-called 'strigilis') and oil. This is the final treatment which served to smooth out unevenesses — the sharply faceted, parallel tracks of these scrapers are repeatedly found on statues and statuettes. This is certain proof that the metallic sheen of the bronzes was not kept intact through continual cleaning and polishing as interpretations of ancient sources suggest (Figure 2.4). Continual cleaning would certainly have destroyed these marks and rubbed out details such as the hairs, veins and fingers (Figure 2.5).

Figure 2.5. *The veins of the right hand of a statuette of Antiochus IV, Greco-Roman 50* BC–AD *50. Nelson-Atkins Museum of Art, Nelson Fund 46–37. (Photo from The Gods Delight, the Cleveland Museum of Art, 1988.)*

This shows that for all practical purposes the onset of natural corrosion happened shortly after they were set up and changed the character of the sculptures. It seems clear that unbridled natural patination (corrosion) was not desired and it is exactly this which the appropriate sources confirm (Pernice 1910). This throws references in ancient texts which deal with pitch or bitumen coatings into a new light (Plinius *Naturalis Historiae* 34,15/35,182. Pausanias, *Description of Greece* 15,4). The interpretation of Greek statuary is moving towards the acceptance of the fact that it was not the intention of the classical craftsmen to maintain or even increase the golden metallic glint of the metallic surface. Indeed they wished to imitate suntanned skin with a thin bitumen varnish. This is an organic coating that would simultaneously protect the surface from corrosion and could also be easily washed off with turpentine, for example, if it was soiled, a process that would neither polish or scratch surface details. Indeed it is obvious that suntanned skin and black hair play their role in the chromatic realism of classical sculpture. Of course monumental sculptures with a metallic surface also existed, a point which need not be debated. We must also always differentiate between the treatment of bronze household utensils (which were certainly kept polished), weapons and statuary.

If we consider the 'golden statue' proposition we should keep a few facts in mind (Born 1988). For instance that:

1. The early Greeks were certainly in the position to gild bronzes, a method which was put into practice from the Hellenistic period.
2. Fragmentary finds of early monumental sculpture covered with gold sheathing have been discovered.
3. However, gold, which was considered the most noble of metals, was not employed to decorate statues of common soldiers or sportsmen in sanctuaries or public places.

Gold as well as ivory had its place in sacred artistry, it was the 'stuff of the gods' in ancient Greece. Famous sculptors shaped the so-called chryselephantine sculptures, that is monumental gold and ivory statues such as Phidias' image of the enthroned Zeus in his temple in Olympia (Pausanias, 11,2). The Greeks mastery of the casting process in the 6th century (Mattusch 1988) enabled them to solve the problems which were inherent to wooden and stone sculpture. With the use of bronze, sculptors had finally found a material which enabled them to achieve a perfect naturalism in the form and surface texture of their product. There are many indications, which have been found on those few early Greek statues which have been comprehensively studied and restored to date, that point to artificial patination.

The Riace bronzes (Mello, Parrini and Formigli 1984) are a case in point (Figure 2.6). The nearly unnoticed fact that a fine copper sulphide layer was found on many areas of the carefully restored statue A lying directly on

Figure 2.6. *The statue of a warrior. The black patina lies directly on the metal surface. Statue A, Riace, 5th century* BC. *(Photo from E. Mello, P. Parrini, and E. Formigli, 1984.)*

the metal surface (Formigli 1985) and that delicate surface details, such as the veins on the hands, showed no signs of polishing, indicates one of two possibilities:

1. The statue was coloured naturalistically with a sulphur compound through the use of an unknown ancient technique. The change in colour to black could have resulted during its watery deposition.
2. The statue was treated with an oil or asphalt glazing and landed in the sea without previous oxidizing corrosion, whereby the constant reductive conditions under water would have lead to the formation of an even copper sulphide layer on the surface.

In either case the sculpture received 'body treatment' without the use of aggressive polishing agents such as that employed on a pumice polished Greek shield (see Figure 2.4). The resulting scouring traces can always be easily identified as such.

Similar observations have been made on other restored bronzes which, however, mainly date to the Hellenistic and Roman Imperial periods. Just as Alexandrine negro statuettes were naturalistically coloured black, other bronze statues and sculptures were also naturalistically tinted. This can be shown to be the case particularly for Roman and Late Antique statuary — thus an illumination from the pre-Medieval Vatican Virgil Manuscript (Cod. Vat. lat. 3225 'Vaticanus', around AD 400) may illustrate artificial black patination on statues (Kömstedt 1929).

An interesting example of complete polychromy is shown by the Dionysis statue (Figure 2.7) in Rome's Thermal Museum, which was completely restored and investigated a few years ago (di Mino 1988, Carruba, Formigli and Micheli 1988). This beautifully preserved bronze which was found in the Tiber has a fine brown-black exterior. The hair of the statue is dark black and contrasts vividly with the reserved patination of the skin. This contrast is heightened by the copper inlayed ribbon and grapes in his hair. The antique patination at least of the hair in this case seems to be unequivocal.

The two monumental bronzes from the Quirinal in Rome are worth mentioning here, i.e. the so-called Ruler and the Boxer which were shown in a special exhibition in Bonn in 1989. The fact that they were recovered from an earth burial context makes them particularly interesting for future research on surface structures. The results of the restoration and analysis of these bronzes which has recently been completed, have yet to produce evidence for naturalistic patination. Yet, particularly in the case of the Boxer (Figure 2.8), this must certainly have been the case, as the antique artist even went to the extreme of showing the blood of the vanquished fighter in gruesome detail through the use of copper inlays (Himmelmann 1989).

Finally, the first incontrovertible evidence for complete patination comes from the so-

Figure 2.7. *The statue of Dionysos, Thermal Museum, Rome. (Photo from Wissenschaften in Berlin, 1987.)*

Figure 2.8. *The head of the so-called Boxer from the Quirinal, probably 1st century* BC *– 1st century* AD, *Thermal Museum, Rome. (Photo from Himmelmann, 1989.)*

Figure 2.9. *The Youth from Eleusis (Saburoff-Youth), 1st century* BC, *Antikensammlung Berlin. (Photo from G. Zimmer, Antike Werkstattbilder, Berlin, 1982.)*

called Ephebes from the 1st century BC. These statues of youths served as 'dumb waiters' in Roman villas and carried lamps or trays in their hands. Well-known representatives of this group of functional bronzes are found in Germany in the Pergamonmuseum, Berlin, originally from Xanten (Hiller, in press), another one in the Antikensammlung Berlin, from the sea near Salamis in Greece (Heilmeyer 1985), one in Spain, from Antequera and one in Italy, from Naples.

The Saburoff Youth (Figure 2.9), also known as the Youth from Salamis, in the Berlin Antikenmuseum is the best understood member of this group. It was found in the sea near Salamis and was freed from incrustations in an amateurish way about 100 years ago, shortly after its discovery. The original surface has survived in a few places and shows that a glossy black patina, copper sulphide in this case, lay immediately upon the metal surface. This black patina is over 1 mm thick and is totally homogeneous in section. Particularly in the case of the right foot (Figure 2.10) we can see that this layer quite clearly covers antique repairs, for instance the soldered seam in the middle of the foot (Born 1988). We know of an *aes nigrum* from late antique Roman sources and from unambiguous findings from contemporary statues (Philostratos *c.* 170–249 AD). This 'black ore' or 'black bronze' was well known to the Egyptians since the 18th dynasty on small-scale bronzes (Cooney 1966).

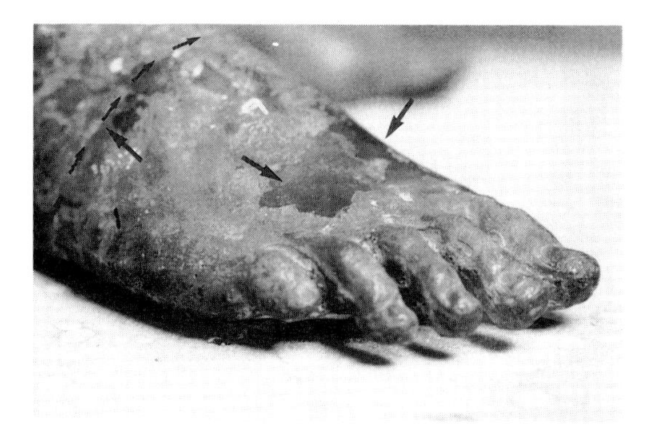

Figure 2.10. *Detail of Plate 2.9. The right foot shows a thick layer (c. 1 mm) of black patina (copper sulphide – covellin) also covering a soldered joint (arrows). (Photo from H. Born, Berlin.)*

The evidence for black surface colour on statues was not taken seriously because most monumental bronzes and their fragments were found in water. This lead automatically to the assumption that the copper sulphide coating resulted from corrosion in an anaerobic environment through the transformation of organic material to hydrogen sulphide.

A fresco fragment from Pompeii which now lies in Naples offers clear evidence for complete patination of a Roman bronze youth (Figure 2.11). This central decorative fragment from the Triclinium of the house Pompeii V2/4, shows a symposion (drinking party) in progress. This badly damaged fresco shows the right-hand side of a room with an Ephebe standing on a round pedestal holding a tray, and has an exact parallel in the youth from Xanten/Germany (Figure 2.12) and in the youth from Antequera/Spain (Figure 2.13). The depiction of the darkened bronze is quite definite and easy to see. It is not yet possible to say if the coloured pigment chosen was dark green or black. The published interpretation of this fresco is that an older and thus patinated antique bronze statue is being used as interior decoration in this scene (Moormann 1988). Functional bronzes such as the lamp- or tray-bearing youths were designed to be set up in

Figure 2.11. *A fresco fragment from the Triclinium of the house Pompeii V2/4 showing a symposion (drinking party) in progress with a dark bronze statue of a tray bearer on his base on the right-hand side. National Museum of Naples. (Photo from Thomas Fröhlich, Freiburg.)*

Figure 2.12. *Detail of the Youth from Xanten, 1st century* BC, *Pergamonmuseum, Berlin. (Photo from H. Born, Berlin.)*

Figure 2.13. *The Youth from Antequera, 1st century* BC, *Antequera, Spain. (Photo from Wissenschaften in Berlin, 1987.)*

living rooms from the beginning and could thus not possibly corrode. It is unthinkable that their owners would have permitted costly gold coloured, i.e. metallic statues to be covered with a soot and fat stained household patina and would surely have given them preferential cleaning. The Salamis youth which we considered earlier demonstrates a further reason for intended patination. This involves the almost unbelievable number and shapes of repairs carried out on miscasts on the surface of this and other thinly cast Roman mass produced tin–lead bronzes. The hundreds of repairs would have been visible within a very short time through the accumulation of dirt and oxidation boundaries on their rims which would have disfigured the product. A black patination would have served to cover this up even in cases where larger scars and cracks had to be sealed with the use of tinted wax (in fact this is a simple method of refining surfaces). Thus these light- or tray-bearers must be seen as a popular 'stage effect' in the interior of Roman villas. In the glow of candles or oil lamps their black surface would be illuminated and show the startling colour contrast between it and the inlayed nipples, eyes and lips.

Acknowledgements

I would like to express my thanks to Louis Nebelsick, Berlin, for his translation and who, as an archaeologist, contributed to many discussions on details of this manuscript. The author also wishes to thank Thomas Fröhlich from the Archaeological Institute of the University of Freiburg, for his reference to the Roman fresco in Pompeii. I also wish to thank Dr Steven Bianchi from the Egyptian Department of The Brooklyn Museum, New York, for stimulating discussions on Egyptian black bronzes.

References

Altekamp, S. (1988) Zu den Statuenbeschreibungen des Kallistratos. *Boreas*, **11**, 77–154

Bol, P. C. (1985) *Antike Bronzekunst*, C.H. Beck, München

Born, H. (1988a) The technology of ancient bronzes: direct and indirect methods of constructing bronze statues. The J. Paul Getty Museum (unpublished lecture with new results on the research of ancient statues)

Born, H. (1988b) Zum derzeitigen Stand der Restaurierung antiker Bronzen und zur Frage nach zeitgenössischen polychromen Oberflächen. In *Griechische und römische Statuetten und Großbronzen* (eds K. Gschwantler and A. Bernhard-Walcher), Kunsthistorisches Museum, Wien, pp. 175–180

Born, H. (1990) Patinated and painted bronzes: exotic technique or ancient tradition? *Small bronze sculpture from the Ancient World.* The J. Paul Getty Museum, Malibu, pp. 179–196

Carruba, A. M., Formigli, E. and Micheli, M. (1988) Indagini tecniche sul Dioniso dal Tevere. In *Griechische und römische Statuetten und Großbronzen* (eds K. Gschwantler and A. Bernhard-Walcher), Kunsthistorisches Museum, Wien, pp. 167–171

Cooney, J. D. (1966) On the meaning of 'black bronze'. *Zeitschrift für Ägyptische Studien*, **93**, 43–47

Di Mino, M. R. (1988) Dioniso dal Tevere. In *Griechische und römische Statuetten und Großbronzen* (eds K. Gschwantler and A. Bernhard-Walcher), Kunsthistorisches Museum, Wien, pp. 165-167.

Finn, D. and Houser, C. (1983) *Greek Monumental Bronze Sculpture*, The Vendome Press, New York

Formigli, E. (1985) Die Restaurierung einer griechischen Großbronze aus dem Meer von Riace/Italien. In *Archäologische Bronzen. Antike Kunst - Moderne Technik* (ed. H. Born) D. Reimers Verlag, Berlin, pp. 168–175

Gazda, E. K. and Hanfmann, G. M. A. (1970) Ancient bronzes: decline, survival revival. In *Art and Technology* (eds. S. Doeringer, D. G. Mitten and A. Steinberg) M.I.T., Cambridge, pp. 245–270

Heilmeyer, W.-D. (1985) Neue Untersuchungen am Jüngling von Salamis im Antikenmuseum Berlin. In *Archäologische Bronzen. Antike Kunst - Moderne Technik* (ed. H. Born) D. Reimers Verlag, Berlin, pp. 132–138

Hiller, H. Zum Xantener Bronzeknaben. 10. Internationale Tagung über antike Bronzen, Freiburg (in press)

Himmelmann, N. (1989) *Herrscher und Athlet*, A. Mondadori, Mailand/Bonn

Kluge, K. and Lehmann-Hartleben, K. (1927) *Die antiken Großbronzen, I-III*, Walter de Gruyter, Berlin

Kömstedt, R. (1929) *Vormittelalterliche Malerei*, Filser Verlag, Augsburg, Figure 2. 34

Mattusch, C. C. (1988) *Greek Bronze Statuary*, Cornell University Press, Ithaca

Mello, E., Parrini, P. and Formigli, E. (1984) Alterazioni superficiale dei bronzi di Riace: Le aree con patina nera della statua A. In *Due bronzi da Riaca I.* Ministero per i Beni Culturali e Ambientali Florence, pp. 147–156

Moormann, E. M. (1988) *La Pittura Parietale Romana come Fonte di Conoscenza per la scultura Antica.* Universitätsdruck, 72, 166

Pausanias, *Description of Greece*, 15, 4

Pernice, E. (1910) Untersuchungen zur antiken Toreutik V: Natürliche und künstliche Patina im Altertum. *Österreichische Jahreshefte*, **13**, 102–107

Philostratos, Flavius, c. AD 170–249, Greek writer from Lemnos, worked in Athens and Rome. He mentioned 'aes nigrum' in a collection of painting descriptions

Plinius, C. *Naturalis Historiae*, 34, 15/35, 182

Reutersward, P. (1960) *Studien zur Polychromie der Plastik (Griechenland und Rom)*, Scandinavian University Books, Stockholm. With a bibliography of the subject up to 1954

Zenghelis, C. (1929) Contribution a l'etude des bronzes antiques. *Mouseion*, **3**, 113

Zimmer, G. (1985) Schriftquellen zum antiken Bronzeguß. In *Archäologische Bronzen. Antike Kunst - Moderne Technik* (ed. H. Born) D. Reimers Verlag, Berlin, pp. 38–49

3

Beauty is skin deep: evidence for the original appearance of classical statuary

Paul Craddock and Alessandra Giumlia-Mair

Abstract

In this chapter the evidences for the original appearance of classical bronze statuary is considered. Although it seems that the metal was usually kept polished and substantially free of corrosion, the literary sources do suggest that treatment with oil gave a distinctive patina.

Introduction

An alternative title for this chapter could have been 'Beauty is in the mind of the beholder'. Our perception of an object is largely based on its surface appearance and in the case of an antiquity the surface is likely to have undergone substantial changes from the original. The situation is more complicated with ancient bronzes because the patination they have acquired since antiquity is itself valued and admired in its own right. Thus for example a recent advert for a patinating process could claim that 'The "finishing patina" for bronze sculptures are the keys that breathe history, nobility and life into a newly cast creation' (Anon 1992) – added history is now part of the manufacturing process! This, coupled with familiarity, has encouraged the tendency to assume that the bronzes would have been covered in their present patina or one similar in antiquity. The problems are yet further compounded by the attentions of later collectors and restorers who frequently gave real ancient pieces quite spurious surfaces either in the belief that this restored the original appearance, or simply that it was an improvement. But how did they really look? This question has been tackled over the last century, notably by de Villenoisy (1896), Pernice (1910 a, b), Richter (1915), Hill (1969), Weil (1977), and Born (1985, 1990, and this volume, Chapter 2). This chapter is a consideration of the various strands of evidence for the original surface appearance of classical statuary and other works of art made of ordinary bronze, setting aside the special alloys such as Corinthian bronze which are considered elsewhere in this volume.

The problem can be approached from three principal sources of evidence, the surviving metalwork itself, ancient illustrations of statuary, and the relevant comments to be found in the contemporary literature. These are considered below.

The surviving metalwork

Here we should be dealing with reality but the present appearance of ancient metalwork can often owe more to recent antiquarian taste than even the appearance when first unearthed, let alone the original appearance in antiquity. The general opinion amongst collectors up to the early 19th century was that classical bronzes originally had a black patina, apparently based on some remarks in the *Natural History* of Pliny, discussed in more detail below. Thus the hundreds of bronzes from the collection formed by Richard Payne-Knight in the late 18th and early 19th century, now in the British Museum are easily recognized by the distinctive black patina that he applied to them (Craddock 1990 p.258). Similarly the great collection of bronzes found in 1764 at Chalon-sur-Saone were claimed by the collector Caylus, who acquired them immediately after their discovery, to have been found coated in the black lacquer that still covers them (Kent Hill 1969 p.65). However, no other bronzes have been excavated in this state and it seems intrinsically unlikely that a lacquer could survive prolonged burial under normal soil conditions. Even the opinion that ancient bronzes were originally patinated was not universal: for example Mantegna, the epitome of the scholarly accurate Renaissance-painter, in his *Triumph of Ceasar*, now in Hampton Court Palace, London, depicted the bronzes very clearly as polished and bronze coloured, not patinated.

Figure 3.1. *Lifesize head of a young man from the Yemen, on loan to the British Museum from the Royal Collections (WAA L 26). The hair is now heavily patinated but the flesh is bronze coloured. Close examination reveals that selective stripping of the face has given the head a polychromy it never had in antiquity.*

The bronze head of a young man found in the Yemen and now on loan to the British Museum from the Royal collections (WAA L.26, Hinks 1936-7) (Figure 3.1) provides a good example of the attentions of restorers to 'improve' an antiquity. The head is now polychrome, with flesh of bright golden bronze contrasted against the hair of dark, heavily patinated metal. However, the head is one casting and examination shows that the bright shiny bronze of the flesh was once also heavily patinated but has been selectively treated thereby creating the present appearance.

Bronzes have also been deliberately repatinated in order to provide a protective layer against corrosion when standing in polluted atmospheres. The best example known to us concerns a statue of the baroque period, but it seems very likely that classical statuary will have experienced similar vicissitudes. The bronze statue of Sigismundus III of Poland, dating from 1644 was originally gilded, the gilding was lost in 1887 when the statue was given a vert-antique patina with acetic acid. The polluted atmosphere in the centre of Warsaw caused extensive sulphidic corrosion in this century which was countered in 1977 by removing the sulphides and inducing a patina of cuprous oxide with an aqueous solution of ammonium persulphate, sodium hydroxide and mercuric chloride (Socha *et al.* 1980). The result was an attractive golden patina colour, but which bore no resemblance to the original, and the possible presence of traces of mercury retained from the patinating solution could be interpreted as evidence for mercury gilding in some putative scientific investigation that was ignorant of the statue's history. In this case the conclusion would actually be historically correct, but for the wrong reason.

However, polychromy certainly was quite common in both Greek and Roman statuary bronze, although normally achieved by the use of contrasting metals rather than different patinas. Thus the lips, eyelashes and nipples were frequently rendered in copper to contrast with the bronze of the main body, as exemplified by three Hellenistic bronze heads now in the British Museum. The head of a Berber (Walters Cat 268) has lips cast from copper and the head of Sophocles (Walters Cat 847) has lips cast from copper containing about two per cent of tin (a classical 'compromise' alloy encountered elsewhere that had the colour of copper but was much easier to cast than copper itself (Craddock 1985 p.62)), whilst the Chatsworth Apollo (GR 1958 4-18.1) has lips inlaid with sheet copper (Craddock 1977 p.113). The details at least on these statues must have been kept polished and metallic or else no colour contrast would have been visible at all, as is now the case. Further, just to have kept the details polished and of the correct colour but to have allowed the rest of the bronze to patinate would have produced what to us would appear a most strange and unpleasant effect. This can be appreciated to some degree from the present condition of the head of Augustus found at Meroe in the Sudan (GR 1911. 9-1.1) where the eyes, inlaid with glass and shell, retain the realistic colours, set in a face that is now green. Similarly the well-known pair of bronze statues recovered from the sea off Riace in Calabria have lips, nipples and eyelashes of copper, together with teeth of silver and ivory eyeballs, which would have looked strange, if, as claimed, the statue was originally patinated black (Formigli 1985a p.171, and 1985b). Many of the bronzes that now have a black suphidic patina come from waterlogged environments and it is possible that the patinas have developed naturally with sulphur-fixing micro-organisms, but the original surface appearance of some major works such as the Riace bronzes is still far from certain.

There are a few pieces of quite complex polychromy where coloured metallic and patinated inlays were used in combination. The best surviving example is on the extraordinary Roman bronze drapery found at Volubilis in Morocco discussed in more detail elsewhere in this volume (p. 115), where silver, brass and copper inlays were complemented by black, orange and brown patinations all set against the bronze of the body metal (Boube-Piccot 1972 p.54–62).

Figure 3.2. *Detail of heavy scoring on the gilded surface of the Horses of San Marco (from Marchesini and Badan 1979). Corrosion and overgilding make it difficult to decide when and for what reason the scoring was done.*

The practice of gilding statues became prevalent in later Roman times. This would indicate that the fashion then was for the metal to be in a bright untarnished state and it has been suggested that the gilding was in part functional, making the bronze much more corrosion resistant, and consequently easier to maintain especially if the statue stood outside, (Kent Hill 1969 p. 64). One might have thought that the original appearance of a gilded surface should not be the subject of debate, but this is not necessarily so. For example, the famous Horses of San Marco, in Venice were cast from copper with about one per cent of tin, the usual alloy for castings that were to be mercury-gilt (Oddy *et al.* 1990), indicating that gilding was intended from the start. The surfaces have been very heavily scored with rather random, quite widely spaced deep lines (Figure 3.2). These lines are now filled with corrosion products that have leached through from the underlying bronze (Marchesini and Badan 1979), and are thus of some age. They could have been cut to key the surface ready for regilding with gold leaf, during one of the many restorations of the bronzes but equally they could be original, with the function of matting the surface to reduce the glare from the gilding (rather in the manner of the carefully ruled lines on some of the optical instruments described elsewhere in this volume p. 149, although the lines on the San Marco Horses are much more coarse). Unfortunately it is not possible to be sure at what stage the lines were scored as the immediate surface beneath the gilding is corroded and the gilding itself has been repaired, once again illustrating some of the problems encountered when dealing with surface phenomena.

Contemporary illustration

Classical mosaics and wall paintings have left us with a wealth of genre scenes; landscapes and streets appear in full colour formed of inorganic pigments that should have undergone little if any colour change since the day they were made.

Here we see antiquity, not as it was but as its artists and their patrons wanted it portrayed, which is not necessarily quite the same thing. Where bronze statuary is portrayed in an outdoor setting it is always bronze coloured, never green or black (Plate 3.1). Similarly illustrations of bronze vessels show them invariably appearing as golden or brown, apparently not patinated even when forming part of an architectural scheme (Riz 1990). The implications of this are significant, the argument so far in this paper has been that there is little evidence for surface treatment, from which it could be inferred that little attention was paid to the surfaces of the bronzes, but on the contrary the bronzes must have been regularly cleaned in order to maintain the bronze appearance and keep corrosion in check. The ancient contemporary sources document evidence for the regular cleaning of bronzes as well as descriptions of the appearance of both architectural and sculptural bronzework, and to these we must now turn.

Literary evidence

Such ancient descriptions as exist which specify the appearance of outdoor architectural copper and bronzework suggest that it was maintained in a metallic condition. Thus, for example, in Egypt the tops of obelisks were often copper covered, as were the temple doors, and these are sometimes referred to as reflecting 'rays like the solar disc' (Anon 1927 p. 33).

The doors of major buildings were frequently clad with copper or bronze throughout antiquity, and in some cases even the walls were covered, and in every case they are described as being shiny, golden or ruddy, there is never any suggestion that they could have been patinated. Thus in the *Odyssey* (Murray 1927), Homer described the Palace of Alcinous as possessing brazen walls surrounding pilasters of silver and doors of gold such that there was a gleam as of the sun or moon over the high-roofed house (VII 84–93). Similarly in the *Critias* (Bury 1929), where Plato described the capital of Atlantis, he states that the wall around the Acropolis was plated with *oreichalkos*, or brass 'which had a fiery resplendence' (114) (Caley 1964 p.20). Admittedly both of these descriptions are of imaginary places, but they were grounded in the experience and expectations of their times. This tradition continued into the early Medieval period, thus for example in 1076 Pantaleone d'Amalfi gave instructions that the bronze doors which he had donated to the church of S. Michele at Monte S. Angelo were to be cleaned annually so that they would always appear shiny and bright (Weil 1977).

Copper and its alloys do not retain a 'fiery resplendence' for long without constant attention, as evidenced by Bol's experiments (1985 p.155–161) where very frequent polishing is required to keep his reproductions bright and shiny, admittedly, in a temperate climate. There is some direct evidence, this time from the ancient world, for the cleaning of statuary. In the accounts surviving from some temples, expenditure on cleaning the bronzes is itemized; thus in AD 215 the temple of Jupiter Capitolinus at Arsinoe in Egypt paid out 28 drachms for oil and another 8 drachms for a workman to apply it to the bronzes. Note it was a workman who made the application, not a priest, there is no suggestion of ritual annointment (Pernice 1910a p.107, Johnson 1936 p. 662-3). Similarly an inscription from Erythrae on Chios dating to the 4th century BC records that the *agoranomoi* were paid to keep the image of Philitos bright and shiny, and clean from patination (Dittenberger 1960 p.496, 284, 15 and 22).

One of the very few references to patinated bronzes occurs in Plutarch's *Moralia* as part of a detailed description of the deliberate patination of Corinthian bronze and the causes of corrosion on bronzes generally (Richter 1915, Zimmer 1985) This extremely important discussion seems little known amongst conservators and scientists alike and so it is given here in full. A party of visitors at Delphi are admiring the famous statues of the sea captains and admirals erected by Lysander in 403 BC to commemorate the Spartan victory over the Athenians at Aegospotami; and their unusual colour attracts attention and sparks off the following discussion [the Greek texts used here are those of the Loeb (Babbitt 1936) and Belles Lettres editions (Flacelière 1972) but a more literal, if less elegant translation is given here by the authors]:

> but he admired the patina of the bronze because it did not resemble either rust nor corrosion but was shining because of a dyeing (βαφῇ, *baphe*) of *kyanos* (κύανος).
> So he made a joke about the sea captains, for he had begun the visit with them, who because of their colour really stood there like creatures of the deep sea.
> 'Wasn't there' he said 'some alloy and a (colouring) treatment used on bronze by the ancients, like the so-called tempering of swords, which being abandoned, bronze had a respite from employment in war? And actually they say that Corinthian ware acquired the beauty of colour not through art, but through accident, fire destroyed a house that had some gold and silver and a lot of stored copper (or bronze), these (metals) being melted and fused together, the large quantity of copper gave the name because it was the major quantity.'
> And Theon beginning to speak said 'but we have heard a more complicated story, that a bronze worker at Corinth, having found a casket full of gold, afraid of being detected, divided it a little at a time, quietly adding it to the bronze which acquired a singular alloy, he gained a lot because (its) colour and beauty were appreciated.'
> But both this and that story are a myth: however, it seems there was some mixing and treatment (seasoning), because in some way now they also alloy gold with silver and produce some extraordinary, and to me excessive and wondrous sickly paleness, and an ugly shading of the colours.
> 'What do you think' said Diogenianos 'could be the cause of the colour of the bronze here (in Delphi)?' and Theon said: 'When of the things that are considered and really are the first and most natural, fire, earth and air and water, none other comes close to the bronze nor is in contact except only the air, it is evident that it is affected by this and that because of this it has acquired this difference it has, because it is always on it and all around it. But indeed "I knew this before Theognis was born" as the comic poet said. And do you desire to learn what nature the air has and by what power of the touch it coloured the bronze?' As Diogenianus assented, he said 'I also want to, my boy, let us then try and discover together, and first of all, if you agree, by what cause olive oil, most of all liquids, soils (bronze) with patina (*ios*): for certainly oil of itself does not put the patina on the bronze, being pure and uncontaminated when applied'. 'In no way', said the young man 'it seems to me that something else causes this, for the *ios* becomes visible, when it comes in contact with it (the oil), which is thin and pure and transparent, (however) it does not appear in the other liquids'.

And Theon said 'Very good, my boy, and nicely put, but consider, if you will, the reason given by Aristotle.' 'I certainly will' he said.
'Now he says that the thinness of all other liquids, being irregular and not compact, penetrates invisibly and disperses the patina, but it is retained by the density of the oil and remains accumulated. And so if we ourselves are really able to formulate some similar hypothesis, it will certainly not be too difficult to find a charm or remedy for this problem. Then since we urged him on, he continued and said that the air in Delphi, being thick and dense and having vigour because of the repulsion and resistance it encounters from the mountains, it is also thin and sharp, as in some way the facts about the digestion of the food bears witness. Therefore penetrating and cutting the bronze because of its thinness, it scrapes a great quantity of earthy patina from it, but holds it and tightens it again, the density not allowing the diffusion, the deposit blooms and takes sheen and splendour on the surface. As we agreed, the foreigner said that just one hypothesis was sufficient for the question. 'The thinness' he said 'will seem to be in opposition to the said density of the air, but it is not necessary to introduce this. Indeed the bronze of itself, becoming old, exudes and releases the patina which the density of the air holds and makes solid and renders visible because of the abundance.'
Taking this up Theon said: 'O foreigner, indeed, what prevents something from being both insubstantial and dense, like the silken and linen cloths of which Homer said "Streams of the liquid oil flow off from the close-woven linen", showing the precision and fineness of the cloth, because the oil does not stay there but runs off and slides away, the thinness and the density (of the fabric) not allowing (the oil) to run through and then the thinness of the air can be used (as an argument) not only for the loosening of the patina, but also it seems to make the colour itself more pleasant and bright by mixing light and sheen with the blue.'

A great deal of information is contained here (that relating to Corinthian bronze is discussed separately elsewhere in the volume (p. 110)). First the statues are free of ordinary corrosion and dirt but have a blue colour, which significantly is regarded as highly unusual, and the suggestion is that the visitors were used to something very different, not even the blue-green of ordinary malachite/azurite corrosion.

Secondly, the peculiar properties of the air at Delphi, high up in the mountains is considered as a likely cause of the unusual patina.

Finally treatment with oil was clearly fairly usual, and equally clearly was routinely associated with some chemical change on the surface of the bronze, in addition to the binding properties of the oil which gave cohesion to the surface.

Apart from the special treatments for alloys such as Corinthian bronze and the rare examples of polychromy found on bronzes such as the Volubilis drapery, the only regular treatment of bronzes in general does seem to have been with oil.

The two most detailed references to its use are to be found in Plutarch's *Moralia*, discussed above, and in Pliny's *Natural History* (Rackham 1968) specifically at Bk. 34.15, 95, 99 and Bk. 35.182.

At 34.15 Pliny states that in early days people used to stain their statues with bitumen, which makes it remarkable that afterwards they became fond of covering them with gold. This quotation is in a section on the history of bronze statuary at Rome. The reference to bitumen is a little surprising, however, Pernice (quoted in Richter 1915 p.xxx) painted a thin coat of pitch diluted in turpentine onto bronze and discovered that 'the wash increased rather than diminished the brightness of the bronze, and at the same time protected the surface from atmospheric effects'. Born (personal communication) has carried out similar experiments applying bitumen to bronze and found they developed a fine golden lustre. Thus Weil would suggest that Pliny was expressing surprise that people should go to the trouble of gilding bronzes when the same effect had previously been obtained much more cheaply with bitumen, a very different interpretation from the usual and more obvious one contrasting black bitumen with gold. We should perhaps be a little cautious in accepting what Pliny *believed* people did in the past, rather than what he *knew* to be current practice where he was usually much more reliable.

At 34.95 as part of the description of various types of bronze, Pliny states that Capuan bronze has *plumbum argentarium* added which gives it the agreeable colour imparted to ordinary bronze by oil and salt. Adding lead would make the alloy greyer, tone down the golden colour, and make it appear darker. The treatment with oil and salt is difficult to reconcile with this, salt would promote the formation of copper chlorides which are green, nothing remotely like the colour brought about by the addition of lead to bronze. However, the Latin *sal* for salt is very close to *sol* for sun, and some editions (Detlefsen 1873) of the *Natural History* do indeed read oil and *sun*. This makes much more sense and moreover concurs well with the comments in the *Moralia* on oil and light.

At 34.99 Pliny notes that bronze gets covered more quickly in corrosion or tarnish (*robigo*) when kept polished than when neglected, unless they are well treated with oil; and it is said that the best way to preserve them is by application of a coating of *liquida pix* which is translated by Rackham as liquid vegetable pitch. It could be this or any other heavy oily material, light pine pitch or mineral oil. Pliny is using the term *robigo* which implies a coarse, disfiguring rust or corrosion, and this could be prevented with oil.

In 35.182, part of a section on the uses of bitumen, Pliny states that bitumen is used to coat bronze vessels. He then repeats the information given at 34.16 that it used to be applied as a coating to bronzes, and additionally that it was still used to coat ironwork, the heads of nails etc.

The first statement is translated by Rackham as rendering the bronze vessels fireproof which must be wrong; in fact the text is rather loose and is capable of several interpretations. That it makes the radiance or glow of the bronze vessels more durable seems a more sensible translation than fireproofing in this context! In the second statement Pliny is once again speaking of what he believed *used* to take place, and by implication was not practised in his day; although bitumen was currently applied to ironwork, presumably to increase resistance to corrosion rather than for decorative purposes.

There are a few other references to the use of pitch in the ancient literature. Thus Pausanias (1,14,4; Jones 1918) mentions the bronze shields in the *Stoa Poikile* which were 'smeared with pitch in order that time and corrosion (*ios*) do not damage them.' Pernice (1910a p. 105) quotes an account from a temple at Delos which specified that pitch was to be smeared on the statue of the god.

There is only one other reference in the classical literature known to us concerning the application of bitumen to statuary bronze. Lucian, a writer of the 2nd century AD, in his *Zeus Rants* (II 33–34 Harmon 1925 pp.139–141) has the characters describe how the famous *Hermes Agoraios* bronze statue in Athens (Fraser 1898 p.131) was forever covered in pitch (*pitta*) by sculptors wishing to take casts. Presumably the pitch must have acted as a releasing agent. (We are grateful to Ian Jenkins of the British Museum for bringing this reference to our attention and discussing its significance.)

Discussion

From these various sources there does come the distinct impression that in Roman times at least, oil was regularly applied to bronze statues, and moreover this treatment was doing more than just cleaning and protecting the metal from corrosion. There is a definite implication that a distinctive colour or patina was created. Common experience tells us that metal coated with oil darkens, and this was utilized by sculptors in more recent times to darken their bronzes. Thus Vasari (Melita 1991, given here in English translation by Maclehouse and Brown 1907) wrote in the 16th century that:

> Bronze which is red when it is worked assumes through time a colour that draws towards black. Some turn it black with oil, others with vinegar make it green, and others with varnish give it the colour black, so that everyone makes it come as he likes best...

By far the most serious study of the effect of oil on bronzes was carried out by the Society for the Encouragement of the Arts, Berlin in 1864 (reported in Hiorns 1892 pp.169–173), where regular washing and treatment with oil produced a green patina, although it is not certain from Hiorn's report whether the bronzes were given a chemical patina before the experiment. They were concerned with the problem of the patina on bronze statuary exposed to modern polluted city atmospheres. As part of their research they placed a series of bronze busts in various parts of Berlin where pollution was especially bad and bronzes ordinarily just went black. They also noted that where bronzes were accessible and could be handled they developed a more beautiful green patina, and conjectured that this might be due to the grease from human contact. Accordingly, to test this, they set up four test bronzes; one was washed daily and well rubbed with bone oil once a month, one was washed daily and rubbed with oil twice a year, the third was washed daily, but no oil was applied, and the last was left unwashed and untreated. The bust that had been oiled monthly developed a fine dark green patina, the bust that had been treated twice yearly had a less beautiful appearance, the bronze that had been merely washed had none of the beautiful features and the last bronze predictably, just got dirty.

The committee concluded that a bronze exposed to air and rubbed with oil monthly would become covered with a beautiful patina (NB this treatment is different from their experimental piece which was also washed daily). The bi-annual treatment was tried because it was clearly neither wise nor practical to rub down the bronzes too often. Unfortunately it is not clear from Hiorn's description whether the bronzes used in the experiments were in a perfectly clean metallic state or whether they had been given the usual chemical patination from which the other patinas developed.

The committee also commented on the function of the oil:

> The manner in which the oil acts in inducing the formation of the patina is difficult to determine. The experiments, however, have demonstrated that it is necessary to avoid excess of oil, and that it should be rubbed off as far as convenient. If the oil is in excess it attracts dust, and the bronze takes a poor appearance. It cannot be supposed that the minute quantity of oil that remains forms a chemical combination with the layer of oxide of the bronze, since bone oil was shown by experiment to be as advantageous as olive oil. It is probable that the minute layer of oil has no other effect than that of preventing the deposition and adherence of moisture, which so easily fixes the the dust, adsorbs gases and vapors, and in which vegetation is so frequently developed. In every case in these experiments, whatever the mode of action, it was abundantly proved that greasy matter contributed to the formation of the patina.

It is fascinating to compare the conclusions of the Berlin committee's experiments with the musings of the visitors to Delphi in Plutarch's *Moralia*, and to speculate if the committee knew of this passage when drawing up their programme of experimentation. There were, of course, important differences between the Berlin experiments and conditions in the Mediterranean. The former were carried out in a polluted temperate climate, and wetted daily which would favour the formation of the familar green-blue copper hydroxycarbonates, malachite and azurite, whereas in the much drier Mediterranean climate the patination would be more likely to stop at the copper oxide stage, red cuprite. Only in unusual climates such as that at Delphi, high in the mountains, would high humidity promote the formation of the hydroxycarbonates on bronzes standing in the open, even though treated with oil.

Conclusion

Weighing all the evidence it would seem most probable that the artistic bronze work of classical antiquity was kept clean and largely free of deep patination caused by chemical attack on the surface. Regular treatment with oil, or just possibly with a light pitch would build up a dark hue, resulting from the formation of a thin oxidization layer of dark red cuprite, further darkened by the oil and minute particles of dirt rubbed in and collecting in scratches and small holes on the surface.

Perhaps the Renaissance collectors were not so far out after all.

References

Anon (1927) *Copper in Architecture*, Copper and Brass Extended uses Council, Birmingham

Anon (1992) The art of 'distressing' bronze *Sciental* **4** Trade magazine for Merck. London. p. 18

Babbitt, F.C. (trans.) (1936) *Plutarch's Moralia* (Loeb edn.) Heinemann, London

Bol, P.C. (1985) *Antike Bronzetechnik*, C.H. Beck, München

Born, H. (1985) Polychromie auf prähistorischen und antiken Kleinbronzen. In *Archäologische Bronzen*, Staatliche Museen Preussischer Kulturbesitz, Berlin, pp. 71–84

Born, H. (1990) Patinated and painted bronzes. In *Small Bronze Sculpture from the Ancient World*, Paul Getty Museum, Malibu, pp. 179–196

Boube-Piccot, C. (1972) *Les Bronzes Antiques du Maroc*, Études et Travaux d'Archéologie Marocaine, Rabat

Bury, R.G. (trans.) (1929) In *Plato: The Critias* (Loeb edn.) Heinemann, London

Caley, E.R. (1964) *Orichalcum and Related Ancient Alloys*, American Numismatic Society Monograph 151, New York

Craddock, P.T. (1977) The composition of the copper alloys used by the Greek, Etruscan and Roman civilisations II. *Journal of Archaeological Science*, **4**, 103–123

Craddock, P.T. (1985) Three thousand years of copper alloys. In *Application of Science in Examination of Works of Art* (eds P.A. England and L. van Zelst) Museum of Fine Arts, Boston, pp. 59–67

Craddock, P.T. (1990) The art and craft of faking. In *Fake* (ed. M. Jones) BMP, London, pp. 247–274

de Francis, A. (1963) *Il Museo Nazionale di Napoli*, Di Mauro, Naples

de Villenoisy, F. (1896) *Revue Archéologique*, **28**(3), 67 and 191

Detlefsen, D. (1866–1873) *Plinii Naturalis Historiae Libri XXXVII*, Weidmann, Berlin

Dittenberger, G. (1960) *Sylloge Inscriptionum Graecarum IV* (ed. G. Olms) Verlagsbuchhandlung, Hildesheim

Flacelière, R. (trans.) (1974) *Plutarque Oeuvres Morales*, Tome 6 (Dialogues Pythiques), Les Belles Lettres, Paris

Formigli, E. (1985a) Die Restaurierung der griechischen Grossbronzen aus dem Meer von Riace. In Born (1985) pp. 168–176

Formigli, E. (1985b) Alterazioni Superficiali dei Bronzi di Riace: Le Aree con Patina nera della Statua A. In *Due Bronzi da Riace*, Istituto Poligrafico e Zecca dello Stato, Roma, pp. 147–156

Fraser, J.G. (1898) *Commentary on Pausanias*, Macmillan, London

Harmon, A.M. (trans.) (1925) (II) and (1936) (V) *The works of Lucian* (Loeb edn.), Heinemann, London

Hiorns, A.H. (1892) *Metal-colouring and Bronzing*, Macmillan, London

Johnson, A.C. (1936) *Roman Egypt II*, John Hopkins Press, Baltimore

Jones, H.L. (trans.) (1918) In *Pausanias: Description of Greece* (Loeb edn.) Heinemann, London

Kent Hill, D. (1969) Bronze Working. In *The Muses at Work* (ed. C. Roebuck) MIT, Cambridge, Mass

Maclehouse, L.S. and Brown, G.B. (eds) (1907) *Vasari: On Technique*, Dent, London

Marchesini, L. and Badan, B. (1979) Corrosion phenomena on the Horses of San Marco. In *The Horses of San Marco* (ed. G. Perocco) Thames and Hudson, London, pp. 200–210

Murray, A.T. (trans.) (1927) *Homer: The Odyssey* (Loeb edn.) Heinemann, London

Oddy, W.A., Cowell, M.R., Craddock, P.T. and Hook, D.R. (1990) The gilding of bronze sculpture in the Classical World. In *Small Bronze Sculpture from the Ancient World*, J. Paul Getty Museum, Malibu, California, pp. 103–124

Pernice, E. (1910a) Untersuchungen zur antiken Toreutik, V. Natürliche und künstliche Patina im Altertum. *Jahreshefte des Österreichischen Archäologischen Institutes in Wien*, **XIII**, 102–7

Pernice, E. (1910b) Bronze Patina und Bronzetechnik in Altertum. *Zeitschrift für Bildende Kunst*, **XXI**, 219–224

Rackham, H. (trans.) (1968) *Pliny: The Natural History* (Loeb edn.), Heinemann, London

Richter, G.M.A. (1915) *The Metropolitan Museum of Art, Greek Etruscan and Roman Bronzes*, Metropolitan Museum, New York

Riz, A.E. (1990) *Bronzegefässe in der Römisch-Pompejanischen Wandmalerei.* Deutsches Archäologisches Institut Rom, Philipp von Zabern, Mainz

Socha, J., Lesiak, M., Safarzynski, S. and Lesiak, K. (1980) Oxide coating in the conservation of metal monuments. *Studies in Conservation*, **25**,. 14–18

Vasari, G. (1991) *Le Vite dei piũ Celebri Pittori, Scultori e Architetti*, Melita, La Spezia

Weil, P.D. (1977) A review of the history and practice of patination. In *Corrosion and Metal Artifacts* (ed. B. Floyd Brown, M. Goodway, H.C. Burnett, J. Kruger, W.T. Case and M. Pourbaix) National Bureau of Standards Special Publication 479, Washington, pp.77–92

Zimmer, G. (1985) Antike Quellen. In Born 1985, pp. 38–51

4

Aesthetic and technical considerations regarding the colour and texture of ancient goldwork

Jack Ogden

Abstract

The present surface of an ancient gold object, typically matt and often purer than the underlying metal, will not always reflect the originally intended appearance. The two variables are colour and texture. Colour could be controlled by choice of alloy (extreme varieties such as red copper/gold alloys are less common after the Bronze Age) or by surface treatments including etching to enrich the surface. Texture ranged from a bright shiny surface – naturally resulting from the soldering and other processes used or deliberately produced by abrasive polishing or burnishing – to a matt surface caused by chemical etching or, in theory, coarser abrasion. The processes used were often limited by the nature of the object, for example a granulated Etruscan ornament could not be burnished. This chapter considers the evidence, which although limited so far, tends to suggest that ancient gold jewellery was more often intended to be bright and shiny than its present appearance might suggest.

Gold and silver are completely miscible and form a continuous series of solid solutions. In nature, gold is found with a variable silver content ranging from under 1% to 50% or more. The colour of the alloy becomes greenish and then grey as the proportion of silver increases. The pale alloy with over about 20 to 25% silver in it is generally referred to as electrum. Silver can also be intentionally added to gold for economic or aesthetic reasons and opinion is divided as to whether the term electrum should be reserved for the man-made alloys and that the native silver-rich gold alloys should be called white gold (Healy 1980). Since white gold is a common term for a totally different type of gold alloy in the modern jewellery business, and since it is not always possible to distinguish native electrum from the man-made alloys, the present author prefers to keep the term electrum for all pale gold-silver alloys with over 25% silver. The alloy can be called native electrum or man-made electrum if there is sufficient evidence to do so. The means of distinguishing native from man-made electrum are generally based on other trace elements in the alloy. For example, it has been suggested that the presence of traces of lead is expected in refined silver which could be alloyed with the gold, but not in native electrum (Healy 1980). However, this is not an infallible guide: 0.2% lead was reported in a sample of native electrum from West Africa (Palache, Berman and Frondel 1944 p. 91). Small traces of lead have been found in several categories of Prehistoric European goldwork and in some Bronze Age Anatolian goldwork. The small lead content of some early Lydian and other Anatolian electrum coinage could also reflect a small lead presence in some local gold deposits.

The other main alloying metal with gold is copper. With increasing copper content, the alloy becomes redder and with judicious balancing of silver and copper content, an acceptable 'gold' colour can be retained with even quite high silver contents. Copper frequently occurs in native gold, but very rarely more than about 2.5%. Thus gold objects with more than this amount of copper are usually considered to be deliberately alloyed. Again, this is not an absolute rule, there are records of native golds from Borneo and the Urals with 5% or more copper present (Palache, Berman and Frondel 1944 p.91).

The hardness of the alloy increases with both silver and copper additions and the melting point drops. The latter feature would have been noted at an early period when goldsmiths used natural, and then artificial, gold alloys for soldering.

In the earliest periods the jewellers would have used the native gold or electrums, but then the possibilities of alloying were discovered, probably by some time in the third millennium BC. The opposite process, the removal of the silver or gold from the alloy by refining, allowed further flexibility of alloy composition. There is literary evidence that refining was understood in the second half of the second millennium BC. For example, Ur III texts seem to refer to the weight loss when gold was melted and an often quoted letter from Burraburias to Akhenaton in Egypt in the 14th century BC describes how 20 minae of gold were reduced to only 5 minae when they were taken from the fire (Knudtzon 1915, documents 7 and 10; Forbes 1971). So far there is little evidence available that refining of any sort became common until around the mid-first millennium BC, possibly this was connected to the introduction of precious metal coinage which required stringent quality control. Once refined gold became the norm for state refined ingots or for coinage, lower purity alloys were produced by debasing the refined gold.

We can therefore distinguish four basic stages in man's utilization of gold:

1. The use of native gold and electrum as found.
2. The intentional addition of silver and/or copper to the native alloy.
3. The refining of the natural alloys to increase their intrinsic value or to standardize them.
4. The intentional addition of silver and/or copper to high purity refined gold for economic, including fraudulent, reasons.

Changing composition results in changes in working properties, such as melting temperature and malleability, but the most obvious difference as far as the jewellery owner would be concerned is that of colour. In some instances the copper content is so high that the alteration of colour must have been a major, if not the only, motivation for its production. Prime examples are the red, high-copper gold alloys used for gold signet rings in the 18th dynasty in Egypt. One example analysed by the present writer several years ago contained about 40% each of gold and silver and 20% copper (Ogden 1977). Another example analysed by Lucas contained as much as 75% copper (Lucas and Harris 1962 p. 229).

We can also mention here the so-called rosy or pink gold of the Egyptian New Kingdom. This, however, was a surface effect, even though it might have been produced by deliberately alloying iron or an iron mineral with gold (Lucas and Harris 1962 pp. 234–5, Frantz and Schorsch 1990). This rosy gold is abundantly present in the goldwork from Tutankhamun's tomb where its careful positioning in combination with normal yellow gold is ample proof of its intentional nature. The earliest published example is from the tomb of Queen Tiye, the wife of Amenhotep III, the latest dateable example is from the time of Rameses XI. 'Pink' gold seems to be limited to Royal burials.

Other instances when it is possible to determine that alloys were deliberately chosen for aesthetic reasons include those objects with a variety of alloys providing colour contrasts. In most cases these are inlaid objects, usually copper alloys inlaid with gold or silver alloys.

A fine example is a dagger blade from Mycenae of about 1500 BC, now in the National Museum in Athens (no. 395) most recently published by Xenaki-Sakellariou and Chatziliou (1989). The blade is inlaid with lions cut from sheet gold, there are also silver (or quite possibly electrum) inlays. The manes of the lions are of noticeably redder colour than their bodies, a colour contrast that must be the result of an intentional aesthetic choice. Other spectacular inlaid copper-alloy objects are of the Egyptian Third Intermediate period, *c*. 1000–700 BC, where we also find inlay combinations of yellow and red gold, as well as electrum (or silver) and copper.

Since such inlaid objects can have two or three different colours ranging from pale gold to red gold, and even copper, against a copper-alloy background, it is as evident to us as it must have been to the ancient craftsman, that the background had to be treated in some way to provide contrast. Such basic considerations have received far too little attention. For example, museum catalogues and other publications frequently refer to bronzes inlaid with gold or to iron objects inlaid with silver, without any reflection as to how the objects were originally intended to appear. Clearly silver will not show up against polished iron, or gold against many copper alloys. Presumably the silver was blackened or the iron blued or rusted. In the case of gold inlays in copper alloys, the background would have to be fairly pure copper – and thus reddish – or artificially darkened.

The use of the darkening of a copper alloy background to show off inlays has been discussed over the years, particularly in relation to Mycenaean dagger blades, such as that referred to above. These inlaid daggers were found in the Mycenae shaft graves (Karo 1930) and in Mycenaean burials at other sites (Laffineur 1974, Xenaki-Sakellariou and Chatziliou 1989). These daggers, of which fewer than 20 have survived, are dated to around 1500 BC. Most of the surviving examples are in the National Museum, Athens, but others include one in the Danish National Museum, Copenhagen (Vermeule 1964 pl. 13) and at least two in private hands, including the example discussed below and one other illustrated by Vermeule (1964).

From the time that Schliemann first discovered the inlaid Mycenaean daggers in the late 19th century, there has been debate as to the nature of the blackened central strips into which the precious metal alloys are inlaid, but minimal analysis work to resolve the question. Over the years, the material has been variously identified as bronze, oxidized silver, blackened copper, a vague 'non-ferrous metal alloy', a silver-iron alloy, and some type of niello (Laffineur 1974). The term 'painting in metal' has also been suggested which, in the present writer's view, has nothing to recommend it.

The blades are usually considered together with the electrum vessels which have gold and a black inlay decoration and which have also been found at Mycenae, Pylos, Enkomi, Dendra, and other mainland sites (Karo 1930, Blegen and Rawson 1966, Xenaki-Sakellariou and Chatziliou 1989). Recently the black inlays on two of these vessels have been analysed and are reported to be some type of metallic sulphide mixture, in other words true niello (Xenaki-Sakellariou and Chatziliou 1989 p. 34). This early use of true niello might be matched on an unusual triple gold vessel from Vulchitrun, now in Bulgaria, dated to the late second millennium BC. According to some reports, the triangular and rectangular recesses on the handles are filled with a black substance identified as a type of niello (Venedikov and Gerassimov, 1975).

Research is in progress on examples of Mycenaean inlay work but so far true niello has not been found on the daggers, despite even the most recent statements to the contrary (Athens 1990 cat. no. 2). There appears to be more than one assembly method for the daggers. The lion blade mentioned above is made in what is perhaps the most developed technique. The basic blade is made from a copper alloy which was then inlaid along each side with separate copper-alloy strips that were artificially blackened. These strips, with a flat or rounded profile, perhaps depending on place or origin, were then in turn inlaid with the representational scene in precious metals.

A detailed study of an inlaid Mycenaean dagger blade has recently been carried out and reveals the inlaying technique. This is a dagger from a private collection, almost certainly originally from mainland Greece and dating to around 1500 BC (IAR 1990, Habsburg and Feldman 1990). Detailed technical and analytical study, plus examination by art historians specializing in Classical Bronze Age metalwork, tend to dispel the doubts about authenticity voiced when this unprovenanced object first appeared on the market. More recent technical studies of other provenanced blades support this. This dagger, missing its handle, is 20.6 cm long with a maximum width of 4.6 cm (Plate 4.1). The blade itself is a simple tin bronze with about 87% copper and 13% tin. Each side of this blade is inlaid with flat strips of a copper alloy. A microscopic sample was removed and proved to be composed of 92.3% Cu, 5% Sn, 0.52% As, 1.72% Au and 0.53% Ag. Surface analyses by energy dispersive X-ray fluorescence spectrography (EDX) gave various compositions due to the surface corrosion effects but confirmed the presence of gold in proportions up to about 5%. The absence, or near absence, of sulphur means that the black surface is not any type of niello. X-ray diffraction analysis (XRD) of the black layer identified it as essentially cuprite. The precious metal inlays in this Mycenaean dagger depict fallen warriors and are shaped from sheet metal with remarkable precision (Plate 4.2). The gold inlays are composed of 75.9% Au, 15.5% Ag, 8.6% Cu (the surface had 8% more Au) while the 'silver' parts on the inlaid design are actually of silver rich electrum with 23.4% Au, 65.9% Ag and 10.7% Cu (the surface had 2% more gold). The high copper content of both the gold and the electrum inlays indicate that they were artificial alloys. (Since the main analyses were taken from samples of the interior metal, contamination from the underlying copper alloy is improbable.) The use of electrum rather than silver for the inlays is of some interest since it is known that the Mycenaeans were obtaining silver from the Greek mines at Laurion (Stos-Gale and Gale 1982).

The presence of the small proportion of gold in the copper alloy is of considerable significance and interest since it brings to mind Japanese *shakudo. Shakudo* is a traditional Japanese copper alloy that typically also contains a small addition of gold. The analyses of seven Japanese shakudo alloys with between 0.5% and 4.2% are given by Gowland (1977) who says that the Japanese recognized 15 grades of *shakudo*, the finest with about 4% gold. More recently Untracht (1982) has described *shakudo* alloys with up to 25% gold. Untracht notes that the presence of the gold not only allows the production of the black patina but results in a fine grain size in the metal which suits inlay and other detailed work. After careful cleaning and polishing, the object is subjected to the so-called *nikomi-chakushoku* process, a form of pickling by immersion in fruit juice or chemical mixtures, and the fine purplish-black surface colour develops. As Gowland says 'Its deep rich tones of black, and the splendid polish which it is capable of receiving, render it alike a perfect ground for inlaid designs of gold, silver, and copper, and for being similarly inlaid in them.' X-ray diffraction and auger electron spectroscopy studies of the black layer on Japanese *shakudo* has identified it as mainly cuprite which includes some gold (Murakami, Niiyama and Kitada 1988 and this volume, Chapter 7).

The early history of *shakudo* is not known for certain, but it was in use by medieval times in Japan. There is also evidence for use of a similar process in China and Tibet (Craddock 1981, Chuan and Kuang 1987). There are several Roman literary references to 'Corinthian bronze' which was apparently a bronze alloy which could have a liverish or black colour and which contained a small proportion of gold (Craddock 1982a, b). According to one ancient source, the colour was imparted by means of immersion in rhubarb juice. Paul Craddock at the British Museum Research Laboratory suspected that this 'Corinthian bronze' might well prove to be akin to the Japanese *shakudo*. This is supported by the study of a Roman plaque with gold inlays and an 'unusual black patina', which was found to be an alloy of copper with a small amount of gold. The black surface proved to be similar, if not identical, to *shakudo* (Craddock 1982a, b). Since Craddock's analysis of the Roman plaque, other possible classical instances have come to light, for example, a Roman period bracelet recently examined by Bayley (1992) (see also this volume, Chapter 9).

The presence of what appears to be a Mycenaean instance of a *shakudo*-like technique dates the process back to the mid-second millennium BC. Current research on a provenanced Mycenaean blade in Athens has revealed a similar composition and structure and further results and full publication are awaited with interest (E. Photos-Jones, personal communication). It might not be coincidence that the culture that appears to have made the most decorative use of contrasting red gold/silver/copper alloys – as in the lions' manes above – also experimented with small amounts of gold in copper alloys and produced a decorative and useful effect.

Of course, such a decorative technique is unlikely to have been limited to mainland Greece. A probable Egyptian example recently studied by the writer is a gold-inlaid copper alloy uraeus (royal cobra) attachment from an Egyptian statuette, of the Third Intermediate Period (*c.* 1000 – 850 BC). The copper alloy, which has a smooth, shiny black patina, was composed of 96% Cu, 3% Sn and 1% Au. The gold inlays were 89% Au, 8% Ag and 3% Cu. There is also a bronze figure of Tutankhamun in the University Museum Philadelphia (FUM. acc E14295), described by Fishman and Fleming (1980) which has an 'odd dark patina'. This statuette is composed of 88.7% copper, 4.6% tin and 4.7% gold. The publishers suggest that the gold was an accidental impurity caused by recycling scrap metal, but this now needs some reconsideration. A number of other objects from Tutankhamun's tomb have been described as 'black copper' and work in progress by Paul Craddock and others shows that some other magnificent inlaid Egyptian bronzes are of *shakudo*-type alloys. More work is needed on the chronology of these bronzes since many museums automatically tend to place most Egyptian bronzes into the 'Late Period'. It is the present writer's belief that many of the finer bronzes, particularly the inlaid ones, might be better placed into the Third Intermediate Period. Western Asiatic examples of *shakudo* type alloys have not yet been recorded but some early first millennium BC examples could be expected: inlaid copper alloy weights might be a fruitful category of object to examine.

From what was said above regarding the red-gold rings and the various inlaid precious metals, we can see that the production of deliberate copper-rich gold alloys for aesthetic reasons can be established from about 1500 to 700 BC even though the alloying of copper with gold dates back to the third millennium BC, in Egypt if not elsewhere. Over 10% copper was found in a sheet gold bead of Predynastic date examined as part of the present study (it also had about 15% silver) and Eluère has recently stated that alloys of gold and copper were being made by the beginning of the Egyptian dynastic period (Eluère 1989). Copper was also added to other alloys for less obvious reasons. The presence of the copper in the electrum inlays of the Mycenaean dagger is not unexpected. Early Bronze Age electrums quite frequently contain an intentional addition of copper for example, an electrum spear-head from Ur contained 30.3% gold, 59.7% silver and 10.35% copper (Woolley 1934). The addition of copper to gold alloys was a characteristic of the Late Bronze Age in Europe (Hartmann 1970). We can note that as a rule of thumb few genuine ancient Old World objects from the Bronze Age through to late Roman times have both Cu > 2% and Cu > Ag, this is not true of many fakes. Again there are exceptions, including gold inlays and overlays from copper alloy objects (study of long-term diffusion would be of interest here) and the occasional gold ornaments. One example is such a Phoenician earring in the Louvre with 2.0% Ag and 5.7% Cu (Quillard 1987 p. 74). Possibly relatively higher copper contents will prove to be more typical of some particular classes of ornament (see for example la Niece in Kidd 1987). Some early silver also contains copper. For example an early New Kingdom Egyptian silver statuette in the Metropolitan Museum of Art, New York contains copper (Hayes 1959) and the average copper content of New Kingdom Egyptian silver objects published by Lucas is about 5% with a maximum of 9% (Lucas and Harris 1962).

The addition of copper to gold alloys continued right through until Roman and later times, but after about the time of the early first millennium BC, deliberate red-gold alloys no longer appear and the copper was probably always added for economic or technical reasons, to provide a cheaper and harder alloy but with enough copper to counteract the paling effect of silver. (We can note that red copper/gold alloys became favoured again in the Medieval European and Islamic worlds.)

Electrum, natural and man-made, was used extensively for coinage from the 7th century

BC until the 4th century BC. However, electrum was less often used for jewellery, exceptions include some Etruscan jewellery and some Greek and East Greek signet rings. There appears to be minimal electrum jewellery of the Hellenistic, Roman or Early Byzantine periods. One exception is a pair of earrings of Syrian 2nd century AD type which contained 72.02% gold, 26.17% silver and 1.81% copper (EDX analysis of an abraded area). These earrings have a distinctly pale colour and, under the microscope, exhibit intergranular corrosion cracking.

Perhaps we can suggest that in early times there was a genuine love of interesting colours and contrasts and that a red copper-rich gold alloy could be admired on its own merits. Then, with the general introduction of refining and the circulation of high purity gold coinage from the mid-first millennium BC onwards, people became aware of what pure gold looked like and then, rather unimaginatively, spurned the coloured variants, however attractive they might be.

Of course the desire for high purity gold was tempered by technical and economic considerations and gold was carefully debased, and probably sometimes subjected to surface treatments to 'improve' the outward appearance.

Over the last few years the writer has carried out several hundred analyses of ancient and fake goldwork, mostly of Hellenistic and Roman types. Most of these analyses were carried out by energy dispersive X-ray fluorescence spectrography. The Link analytical process with the ZAF computer program provides a quantitative idea of the main components, but is not suitable for accurate trace element determination. Repetitive analyses on the same object plus comparative analyses by atomic absorption and wavelength dispersive XRF, confirm the general accuracy of the results. The analyses are either carried out on the object or on microscopic samples removed from the object and mounted for examination. The latter process is often less disfiguring than the former which requires local abrasion to produce an area free from surface enrichment and contamination. The mounted samples can be kept for later re-examination.

Although the number of analyses carried out so far are too few to produce precise statistical information, they can be taken in conjunction with other published analyses to indicate certain chronological and geographical patterns.

The variations possible can be seen from an examination of ancient classical goldwork. Etruscan jewellery analysed in the present survey is typically between about 70 and 85% gold. This compares well with the large number of analyses by Cesareo and von Hase (1973) which showed purities between about 62% and 85% gold. In many cases early Etruscan jewellery contains enough silver to be considered electrum – one ring examined had about 30% silver. The analyses carried out as part of the present study found no examples of Etruscan gold with copper contents higher than is likely with a natural alloy. However, the analyses by Cesareo and von Hase do reveal up to 4% or more copper in some Etruscan goldwork, perhaps deliberate additions. Possibly this copper content became more frequent as the Etruscan period progressed (Eluère 1989). Gold coinage was not in circulation in Etruria, but it is possible some imported electrum and gold coins were used as raw material for jewellery. Other jewellery of around the mid-first millennium BC, from the eastern Mediterranean, also contains enough silver to be classed as electrum and some does contain noticeable copper additions – up to 9% has been found in analyses so far. Again, coinage might have been used as raw material in some instances.

When we turn to Hellenistic jewellery there seems to be a more definite pattern. Of the analyses so far carried out by the writer, goldwork that can be stylistically assigned to the early Hellenistic period contains between about 90% and 99% gold and has a copper content under about 2.7%. On the other hand, goldwork that can be dated to the later Hellenistic period on stylistic grounds often contains between 80 and 91% gold (but occasionally is far purer) and has copper contents varying from under 2% up to 6% or so. Here, as noted above and as with almost every category of ancient Old World gold, the silver content exceeds that of the copper.

Far more analyses need to be carried out before we have a firm picture of Hellenistic Greek gold purities, but these recent analyses do largely match those carried out by Young some years ago (Hoffmann and Davidson, 1966) and the occasional other analyses scattered in the literature. For instance, a Late Hellenistic earring from Egypt, in the Brooklyn Museum, contains 82.14% gold, 17.86% silver and a trace of copper (Williams 1924). It is too early to postulate a deliberate compositional change from early to later Hellenistic gold jewellery, but the analysis results must provide at least evidence for a trend towards the greater debasement of gold in the later Hellenistic period. A fuller study of the composition of Hellenistic coinage is vital since this was almost certainly a major source of gold for jewellers and their patrons.

In the Roman period, from the 1st to early 3rd century AD, gold coinage remained at a remarkably high level of purity – often 99% or higher. On the other hand, Roman gold jewellery shows a remarkable variation in fineness which ranges from gold over 99% or more pure down to the earrings with only 72% gold, mentioned above. Most Roman jewellery seems to be composed of between about 80 and 97% gold although the present surface purity can be greater due to natural or artificial effects. As with the Hellenistic period, the trend is for gold to be more debased as the Roman period progressed. Generally speaking Roman goldwork of the 1st to 3rd centuries AD has low copper contents. On the basis of analyses so far the main exception seems to be goldwork from Egypt, where up to 10% copper has been noted, although 2%–5% is more typical. Even so the silver content still typically exceeds the copper content.

Some debasement with copper, up to 8% or so, is also found in Early Byzantine goldwork. It is noteworthy that Byzantine gold solidi coins tend to be slightly less pure than the Roman aurei. This is probably deliberate and the Byzantine mint still could produce pure gold for special functions – such as a medallion of Constantine which was over 99% pure (Delbourgo and Lahanier 1980). This medallion is mounted in an openwork setting which is of only 92.8% gold – a common purity level for much Early Byzantine gold jewellery (Oddy and La Niece 1986).

Considering the range of gold alloy compositions in Hellenistic and Roman jewellery (and in other categories of ancient goldwork), do we assume that some variations in colour were accepted or that treatments were carried out to produce a uniform surface of higher purity? The evidence is not at all clear cut.

With ancient gold objects it is not unusual to find differences between surface and interior gold purities. In the baser Roman jewellery examined, for example, the surfaces are quite often up to 10% or so purer in gold than the interiors. However, it is difficult to determine whether such compositional gradients are due to natural or man-made causes. Long burial in the ground can result in extensive surface enrichment effects by the leaching out of silver and base metals (Ogden 1983, 1985). On the other hand, surface 'pickling', usually in an acid solution, is used by goldsmiths all over the world today to enhance the appearance of their wares. Actual recipes have survived from Roman Egypt that describe how to improve gold surfaces by means of a variety of chemical etchants. The Leiden papyrus no. 10, dating to the 3rd century AD or earlier, provides numerous goldsmiths' recipes which include two for solutions in which the goldwork had to be heated to improve its colour (Caley 1926). One recipe was a mixture of *misy* (almost certainly a sulphide or sulphate, probably iron pyrites) plus salt and vinegar. The other recipe was of *misy*, alum and salt. Both would produce quite strong acids that would etch out the silver and base metals.

We might expect that artificial surface 'improvement' was commonly used on ancient goldwork, but it is difficult to distinguish natural from artificial enrichment by simple visual examination. In theory, ancient gold jewellery that had a certain amount of wear in antiquity should be able to supply some clues. If the surface had been artificially treated at the time of manufacture we would expect to see a difference today between the colours and textures of worn and non-worn surfaces such as chain links or edges. If the enrichment was due to burial effects there should be no difference between worn and non-worn surfaces. Despite examining thousands of ancient gold objects microscopically, the present writer has seen no certain visible indications of intentional surface enrichment.

Analysis in conjunction with studies of assembly technique provide conflicting evidence. A pair of simple tapered hoop earrings of about 2nd century AD date from Egypt is an example of the difficulty of determining deliberate surface treatment (Figure 4.1). The hoops are of sheet gold. The exterior of the earrings were 97% gold, 3% silver and under 0.2% copper. The inside – totally enclosed at the time of manufacture but revealed by a hole caused by wear or damage – is only 85% gold, 10% silver and 5% copper. However, although this hole might have existed at the time of burial, the original sandy filling and the limited possibility for interior ground water circulation would be expected to limit interior compositional change in burial. Thus there is no way to determine whether the richer outer surface was due to burial or intentional surface treatment.

Figure 4.1. *A pair of Romano-Egyptian gold earrings of thin sheet gold.* c. *2nd century* AD.

Another piece of Romano-Egyptian goldwork had a similar variation in composition: exterior 96% gold, 3% silver and 1% copper, interior 83% gold, 14% silver and 2% copper, but this is easier to explain as natural enrichment due to burial. The object is an open-topped box-like necklet component that originally held a piece of glass on a plaster bedding (one of the sections in Figure 4.2). The object could not have been subjected to any surface treatment based on heat and/or caustic substances after the glass and plaster were in place. However, if it was treated before setting with the glass, the interior would have been just as enriched as the outside. In this case natural surface enrichment appears to be the probable process.

Figure 4.2. *A group of Romano-Egyptian gold necklet elements.* c. *3rd century* AD.

In some cases a good idea of original intended surfaces can be obtained from goldwork with a protective coating such as calcite or copper-corrosion products resulting from burial phenomena (Ogden 1983). These coatings protect the gold from surface enrichment effects and can also preserve the underlying surface in near-pristine condition. On the basis of observation of protected areas of ancient gold objects, it would seem that much ancient jewellery had a shiny appearance.

Such a shine can, in theory, derive from the nature of the metalworking processes, including soldering techniques, or from subsequent polishing or burnishing. In practice there are technical and stylistic constraints. For example, intricately granulated Etruscan jewellery or Hellenistic work with its characteristic filigree could not be burnished or polished to a high shine. However, as observation of partly encrusted objects shows, this was quite frequently shiny (Figure 4.3) and, in the case of most Etruscan and some Hellenistic work, of a distinct pale lemony colour. We must assume that either the soldering techniques and conditions employed left a shiny surface or that the objects were washed in an acid solution that resulted in a shiny rather than dull 'etched' surface. In either case it would appear that a shiny lemony surface was tolerated, if not intentionally produced. Clearly this contrasts with the rich, matt surfaces of most of the gold objects now in museums. On the other hand, the fine shine that would best flatter the gently curving, undecorated surfaces so typical of much early

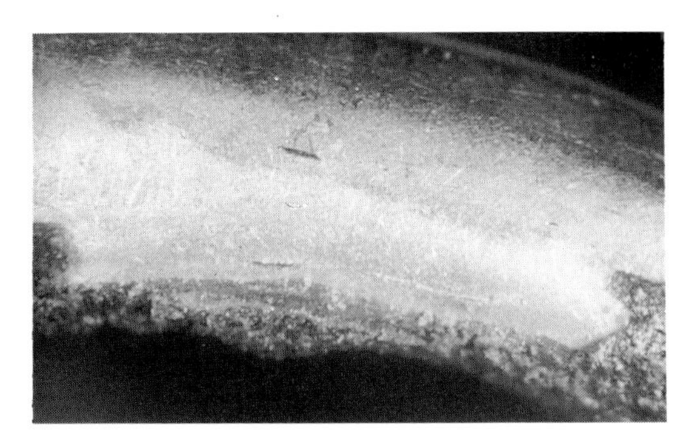

Figure 4.3. *Detail of a Greek gold ring with copper alloy core. Where corrosion has been chipped from the gold surface, the gold retains a high polish.*

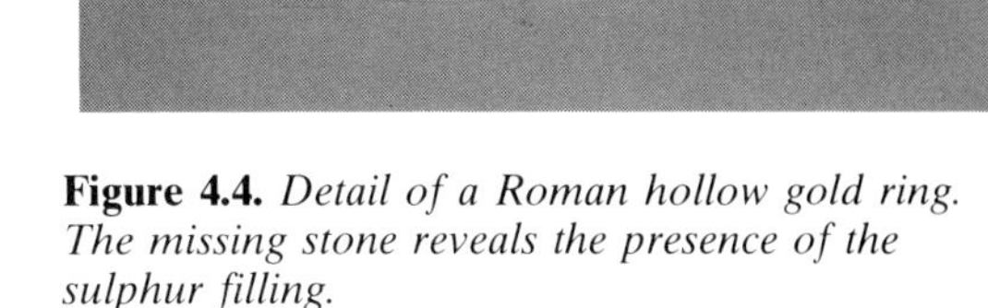

Figure 4.4. *Detail of a Roman hollow gold ring. The missing stone reveals the presence of the sulphur filling.*

Roman jewellery could have been produced by burnishing or polishing, perhaps after surface enrichment.

Burnishing was probably far commoner than abrasive polishing for ancient goldwork, indeed burnishing was probably often an integral part of the shaping process. There is little surviving visual evidence for abrasive polishing of ancient gold but, again, this could have been largely disguised by subsequent burial effects. We can also note that even quite minimal surface polishing will remove an enriched surface, while careful burnishing will actually improve it. Ancient sheet gold components often still retain surface impressions of the leather used to protect the gold during the beating of the sheet or indentations caused by the anvil surface. The presence of such marks must show that in these cases there was minimal if any subsequent polishing or burnishing.

The deliberate mechanical alteration of the surface texture of gold in antiquity is hard to demonstrate apart from the fine surface stippling with a small chasing tool or punch sometimes used to provide a contrasting background on some Hellenistic repoussé goldwork. The very fine 'powder' granulation on some Etruscan work also has the effect of producing a rough, even though originally not necessarily matt, surface. There is no evidence for deliberate abrasive 'matting' of ancient gold – equivalent to the use of the glass brush or sand blasting by some modern goldsmiths. However, we can note that forgers use a variety of methods to impart a rich matt surface to their wares including various abrasive treatments.

The order of work is important. Any heating operations such as soldering or enamelling, must precede artificial surface enrichment since heat will rapidly cause the diffusion of silver from the interior of the alloy to the surface. The rapidity with which this paling takes place has astonished some recent restorers who have attempted to solder or anneal ancient gold objects.

In some cases we might deduce the likely original surface. For example, Roman gold rings are often hollow and, like much Roman jewellery, made of sheet gold filled with sulphur. In many cases these have stones held in place by burnishing the edge of the setting over the stone. The sulphur is poured into the hollow ring through the opening that will eventually hold the stone (Figure 4.4). Surface enrichment using a chemical etchant in combination with heat, if employed, must have preceded the filling of the ring with sulphur since heat might melt the sulphur and dislodge the stone. In such a case the eventual burnishing of the edge of the setting to hold the stone in place would have imparted a shine to the gold bordering the stone. Since surface variation is unlikely to have been favoured, we must assume that either the whole ring was then burnished or that chemical surface enrichment was not used, in either case the ring would have been delivered to its first owner with a shiny surface.

Sulphur fillings can be added quite early in the manufacturing process. If molten sulphur is poured into a rudimentary hollow gold object and then the whole object suddenly cooled by plunging into water, the sulphur changes into a rubbery state that can act as a support for repoussé and chasing work. This rubbery sulphur eventually hardens to provide a firm filling material. Incidentally, we might expect some reaction between the sulphur and the silver present in the gold alloy, but so far I have seen no sign of discolouration of the sulphur even when it has been in contact with the gold alloy for nearly two millennia. It must be assumed that sulphur fillings were reserved for the higher purity gold objects. The lowest quality gold filled with sulphur found in the present study contained about 90% gold.

Ancient gold jewellery survives in such good condition that we often forget to enquire as to its originally intended surface appearance. The limited evidence so far tends to suggest that much ancient goldwork was intended to have a shiny surface and that deliberate surface enrichment might have been less common than might be thought. Clearly there is much work to be done on the ancient aesthetic and technical approaches to gold alloys, their surface colours, and finishes. Experiments need to be carried out to determine how choice of gold alloy composition and manufacturing techniques would be influenced by the desire for particular surface texture or colour. The possibilities are limited by various factors. Composition, decoration and intended surface appearance might be more interrelated than hitherto assumed.

References

Athens (1990) *Troy, Mycenae, Tiryns, Orchomenos – Heinrich Schliemann: the 100th Anniversary of his Death* (ed. K. Demakopoulou), Athens

Bayley, J. (1992) Non-ferrous metallurgy in England. PhD. Thesis, University of London

Blegen, C.W. and Rawson, M. (1966) *The Palace of Nestor at Pylos in Western Messenia*, Princeton University Press, Princeton

Caley, E. R. (1926) The Leyden papyrus X: an English translation with brief notes. *J. Chem. Ed.*, **3**(10), 1149–1166

Cesareo, R. and von Hase, F-W. (1973) Non-destructive radioisotope XFR analysis of early Etruscan gold objects. *Kerntechnik*, **12**, 565–569

Chuan,Y. and Kuang, W. (1987) A study of the black corrosion-resistant surface layer of ancient Chinese bronze mirrors and its formation. *Corrosion Australasia*, **12**(5), 5–7, 11

Craddock, P. T. (1981) The copper alloys of Tibet and their background. In *Aspects of Tibetan Metallurgy* (ed. W. A. Oddy and W. Zwalf) British Museum Occasional Paper no. 15, pp. 1–31

Craddock, P. T. (1982a) Corinthian Bronze: Rome's purple sheen gold. *MASCA Journal*, **2**(2), 40–41

Craddock, P. T. (1982b) Gold in antique copper alloys., *Gold Bulletin*, **15**(2), 69–72

de Cuyper, F., Demortier, G., Dumoulin C. J. and Pycke J. (1987) La croix byzantine du Trésor de la Cathédrale de Tournai. *Tornacum*, **1** (= *Aurifex*, **7**), Louvain-la-Neuve

Delbourgo, S. and Lahanier, C. (1980) L'analyse scientifique des bijoux antiques. *Dossiers d'archéologie*, **40**, 6–21

Eluère, C. (1989) *Secrets of Ancient Gold*, Trio, Guin-Düdingen (Switzerland)

Fishman, B. and Fleming, S. J. (1980) A bronze figure of Tutankhamun: technical studies. *Archaeometry*, **22**(1), 81–86

Forbes, R. J. (1971) *Studies in Ancient Technology* VIII, 2nd ed. Brill, Leiden

Frantz, J.H and Schorsch, D. (1990) Egyptian red gold. *Archaeomaterials*, **4**, 133–152

Gowland, W. (1977) Metals and metal working in Old Japan. In *Japanese Crafts: Materials and their Applications* (ed. B. Hickman) Fine Books Oriental, London

Habsburg, Feldman (1990) *Auction sale of Gold: Important ancient and ethnic jewellery and works of art in precious metal*, Geneva, 14 May

Hartmann, A. (1970) *Prähistorische goldfunde aus Europa*, Mann, Berlin

Hoffmann, H. and Davidson, P.F. (1966) *Greek gold: Jewelry from the age of Alexander*, Museum of Fine Arts, Boston

Hayes, W.C. (1959) *The Scepter of Egypt: 2, The Hyksos Period and the New Kingdom*, Metropolitan Museum of Art, New York

Healy, J.F. (1980) Greek white gold electrum coin series. *Metallurgy in Numismatics*, **1**, 194–215 (Royal Numismatic Society, London)

IAR (1990) Unpublished Independent Art Research Ltd report no. 89086 (a summary of this report is given in Habsburg, Feldman 1990)

Karo, G. (1930) *Die Schachtgräber von Mykenai*, Bruckmann, Munich

Kidd, D. (1987) 'Some Observations on the Domagnano Treasure', *Anzeiger des germanischen Nationalmuseums*

Knudtzon, J.A. (1915) *Die el-Amarna-Tafeln*, 2 vols, Hinrichs, Leipzig

Laffineur, R. (1974) L'incrustation à l'époque mycénienne. *Antiquité Classique*, **43**, 5–37

Lucas, A. and Harris, J.R. (1962) *Ancient Egyptian Materials and Industries*, Arnold, London

Murakami, R., Niiyama, S. and Kitada, M. (1988) Characterisation of the black surface layer on a copper alloy coloured by traditional Japanese surface treatment. In *The Conservation of Far Eastern Art, Preprints of the Contributions to the Kyoto Congress*, 19–23 September 1988, London, pp. 133–136

Oddy, W. A, and La Niece, S. (1986) Byzantine gold coins and jewellery. *Gold Bulletin*, **19**(1), 19–27

Ogden, J. (1977) Platinum group metal inclusions in ancient gold artifacts. *Journal of the Historical Metallurgy Society*, **11**(2), 53–72

Ogden, J. (1983) Surface changes in buried gold items. In *Preprints 2nd International Symposium Historische technologie der edelmetalle*, Meersburg, 25–28 April, pp. 24–26

Ogden, J. (1985) Potentials and problems in the scientific study of ancient gold artifacts. *Application of science in examination of works of art* (eds A.E. England and L. van Zelst) Proceedings of the Seminar 7-9 September 1983, Museum of Fine Arts, Boston

Palache, C., Berman, H. and Frondel, C. (1944) *The System of Mineralogy*, 7th edition, vol. 1, Wiley, New York

Quillard, B. (1987) *Bijoux Carthaginois* **II** (= *Aurifex* **3**), Louvain-la-Neuve

Stos-Gale, Z.A. and Gale, N.H. (1982) The sources of Mycenaean silver and lead. *Journal of Field Archaeology*, **9**(4), 467–485

Untracht, O. (1982) *Jewelry: Concepts and Technology*, Hale, London

Venedikov, I. and Gerassimov, T. (1975) *Thracian Art Treasures*, Caxton, London

Vermeule, E. (1964) *Greece in the Bronze Age*, University of Chicago Press, Chicago and London

Williams, C.R. (1924) *Catalogue of Egyptian antiquities numbers 1-160: Gold and silver jewelry and related objects*, New York Historical Society

Woolley, P.L. (1934) *Ur excavations: II, the royal cemetery*, British Museum, London

Xenaki-Sakellariou, A. and Chatziliou, C. (1989) *Peinture en Metal à l'époque mycénienne*, Ekdotike Athenon, Athens

5

Studies of ancient Chinese mirrors and other bronze artefacts

Zhu Shoukang and He Tangkun

Abstract

This chapter investigates the shiny black patina on ancient Chinese bronze mirrors and other wares. Seventy-two mirrors and a few score of weapons, utensils and ornaments have been examined by optical microscopy, scanning electron microscopy, X-ray diffractometry, etc. The chemical composition of the surface layer or patina is found to be distinctly different from that of the bulk alloy. The very top film is transparent and compact, and is highly corrosion-resistant. The chief component of this film is tin oxide.

It is documented in the 2nd century BC that mirrors and some other wares were rubbed with a powder called *Xuan Xi* to make the surface bright and white, and this was the cause of formation of the original tin-rich layer. This study examines the evidence for *Xuan Xi*.

The results of examination of the black patina, now on the objects, show that it was formed naturally over a long period by oxidation of the original tin-rich mirror surface.

Introduction

Bronze wares, especially bronze mirrors, with black surfaces can often be found in archaeological excavations. They are glossy and shiny as if coated with lacquer, so they were named *Hei-Qi-Gu* (black lacquer antique), a name which appeared in the Song dynasty (AD 960–1279) but people had noticed the beautiful ancient black mirrors much earlier. (The Chinese dynasties are chronologically listed in Table 5.1.) The coating of *Hei-Qi-Gu* is a special, strongly corrosion-resistant substance.

To study the composition of *Hei-Qi-Gu* and explore the mechanism of its formation, we

Table 5.1 Chronological table of Chinese dynasties

Xia Dynasty	21st–16th century BC
Shang Dynasty	16th–11th century BC
Zhou Dynasty	
Western Zhou	11th century–771 BC
Spring and Autumn Period	770–476 BC
Warring States	475–221 BC
Qin Dynasty	221–207 BC
Han Dynasty	
Western Han Dynasty	206 BC–AD 25
Eastern Han Dynasty	AD 25–220
Three Kingdoms	AD 220–265
Jin Dynasty	AD 265–420
Northern and Southern Dynasties	AD 420–589
Sui Dynasty	AD 581–618
Tang Dynasty	AD 618–907
Five Dynasties	AD 907–960
Song Dynasty	AD 960–1279
Yuan Dynasty	AD 1271–1368
Ming Dynasty	AD 1368–1644
Qing Dynasty	AD 1644–1911

examined and analysed more than 130 bronze mirrors dating from late Western Zhou dynasty to mid-Qing dynasty (9th century BC to 18th century AD). The surface layers of 72 of these mirrors have been studied in detail using various analytical techniques (microscopy, SEM analysis, etc.). In addition, we studied a few score other ancient bronze wares with shiny black (or green, grey, etc.) surface layers. These bronze mirrors and wares were unearthed in various parts of China, both in the south and the north. Progress has been made in probing the mechanism of formation of *Hei-Qi-Gu* (black lacquer antique) through our research which is continuing.

Figure 5.1. *The earliest bronze mirror* (c. *2000* BC) *unearthed in Guinan, Qinghai.*

Chinese bronze mirrors

Brief history

The history of mirror manufacturing in China is long, though the exact beginning is still obscure. Ancient texts, such as *A Collection of Plates of Ancient Mirrors* edited in the Song dynasty, unfortunately give little explanation about procedures for manufacturing mirrors. There appears to be no good reference work on Chinese mirrors generally. Some ancient books talked about the method for manufacturing mirrors, but the descriptions are very simple and not very specific. However, according to the historical records, the outward appearance of mirrors were all bright and white. The earliest mirror known to date was unearthed in Guinan county, Qinghai province in 1977 dated to 2000 BC. The diameter of the mirror is 90 mm. On the back there is a seven-pointed star in relief for decoration (Figure 5.1).

From the Warring Period onwards, bronze mirrors were widely used in south China. The state of Chu used to be the centre of mirror manufacturing. During the Han dynasty mirrors became popular throughout the country. The designs on the back were varied and interesting. Large mirrors appeared; a gigantic rectangular mirror (115.5 cm long, 57.7 cm wide and 1.2 cm thick, weighing 56.5 kg) with a dragon design made in the early Western Han (*c.* 2nd century BC) was excavated in 1980 (Museum of Zibo City, 1985).

After the Song dynasty, the composition and the design of mirrors changed dramatically and very few that have been excavated are found uncorroded, they are of various colours, some are green, others are brown, greyish-yellow, yellowish-brown, etc.

Bronze mirrors began to be gradually replaced by glass mirrors in mid-Qing dynasty (*c.* 18th century). The bronze mirror industry, which had lasted several thousand years, came to an end.

Technology of mirror production

The procedures for manufacturing bronze mirrors were as follows:

1. *Purification of copper*

The copper used for casting mirrors had to be repeatedly fire-refined to remove the impurities before it was melted with tin and lead to form bronze. After refining, the purity of copper was very high. Take iron, the main impurity, as an example; the copper smelted from ore contained about 5% iron, while in bronze mirrors, iron is found to be only about 0.1% and some Han dynasty mirrors contain

Figure 5.2. *Clay mould for mirror casting (the Warring States).*

Figure 5.3. *Clay mould for mirror casting (the Western Han dynasty).*

only 0.04–0.05% of iron. The terms such as *Sanlian, Jiulian* and *Bailian* (three-, nine- and a hundred-refining) were used in the literature of the Western Han dynasty to indicate the quality of bronze. *Bailian* mirrors were regarded as the best quality mirrors.

2. *Casting*

Unfired clay moulds were mostly used for mirror casting (Figures 5.2 and 5.3), although occasionally stone moulds were used in the Han dynasty. All the excavated moulds were of two parts, i.e. front and back pieces. On the back-moulds there were always very exquisite designs. The clay moulds were used dry and were preheated before the bronze alloy was poured in. Occasionally, stone moulds were used in the Han dynasty. Lost-wax casting was the most likely method used for mirror production from the Tang dynasty.

3. *Polishing*

Polishing is the last procedure for making a mirror. According to *Huai Nan Zi* (a book written by Marquis Liu An of Huai Nan, 177–122 BC), the compound used for polishing is *Xuan Xi* (black tin). The book says, 'After being rubbed with *Xuan Xi*, the mirror is so bright that it can reflect even a single hair or an eyebrow.' [See later for details of *Xuan Xi*.]

Composition of mirrors

In the early stages, the tin content of mirrors was low and variable. Between the Warring Period and the Tang dynasty, the composition of mirrors was mostly in the following range:

	wt%	wt%
Cu	66–78	mainly 69–72
Sn	18–26	22–25
Pb	1–9	5–6

This alloy appears silver-white in colour when polished.

From the Song dynasty the composition changed greatly, with tin content declining (some mirrors only contained 2% Sn), and lead content increasing (some contained as high as 26% Pb). Brass mirrors appeared (with about 20% zinc content, a few with 33%), the earliest brass mirror excavated so far being dated to between the Tang and Song dynasties.

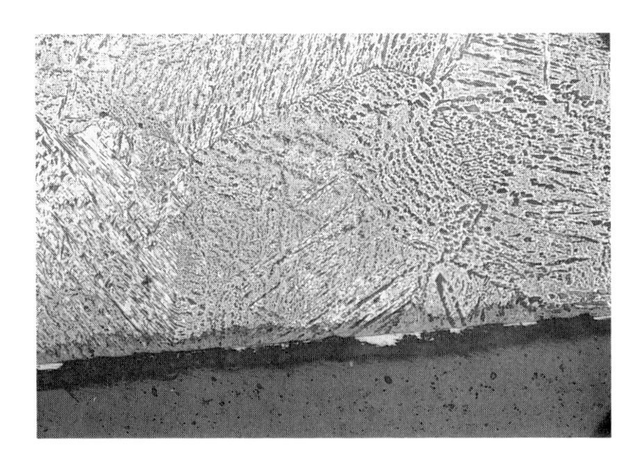

Figure 5.4. Hei-Qi-Gu *mirror no. 1, cross-section detail of the surface, × 320, unetched, optical micrograph.*

Figure 5.6. Lu-Qi-Gu *mirror no. 1, cross-section detail of the surface, × 320, unetched optical micrograph.*

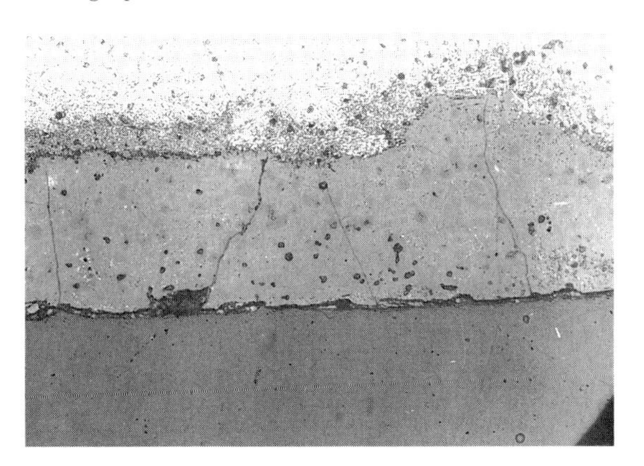

Figure 5.5. Hei-Qi-Gu *mirror no. 2, cross-section detail of the surface, × 320, unetched optical micrograph.*

Figure 5.7. Lu-Qi-Gu *mirror no. 2, cross-section detail of the surface, × 320, unetched optical micrograph.*

The surfaces of these mirrors appear reddish-yellow or brownish-yellow, and it is believed that brass mirrors were also treated with *Xuan Xi* to make them bright and white.

The surface layers of bronze mirrors

Among the bronze mirrors excavated from the tombs dated between the Warring Period and the Tang Dynasty, there were some which exhibit shiny jet black, emerald green, dark grey or silver-white surfaces. In ancient China (at least beginning in the Song dynasty), these mirrors were named *Hei-Qi-Gu* (black lacquer antique), *Lu-Qi-Gu* (green lacquer antique), *Qian-Bei* (lead grey) and *Shui-Yin-Qin* (quick-silver white) respectively. Among them, *Hei-Qi-Gu* is of most interest. Geographically, *Hei-Qi-Gu* and *Qian-Bai* appear mostly in south China, especially in Hunan, Hubei and Anhui provinces, while *Shui-Yin-Qin* mainly appear in north China.

In order to study the composition, structure, property and mechanism of formation of the surface layer, we examined in detail a total of 72 ancient mirrors.

Microscopic studies

Microscopic examinations of the cross-sections of *Hei-Qi-Gu*, and *Lu-Qi-Gu* mirrors (Figures 5.4–5.7) showed that the surface layer can

Table 5.2 Chemical composition of the bronze mirrors

Mirror No.	*Dynasty*	*Place of excavation*	*Composition of Bulk Alloy, wt%* Cu	Sn	Pb
1	Warring Period	Jiangling, Hubei	78.83	21.17	–
2	Warring Period	Changsha, Hunan	77.29	21.34	1.36
3	Warring Period	Changsha, Hunan	73.48	18.78	7.60
4	Warring Period	Changsha, Hunan	78.11	20.62	1.26
5	Warring Period	Zouxian, Shandong	83.97	15.97	0.06
6	Early Western Han	Echeng, Hubei	69.50	25.90	3.00
7	Western Han	Changsha, Hunan	70.70	25.50	5.00
8	Eastern Han	Echeng, Hubei	70.60	23.80	5.70
9	Three Kingdoms	Echeng, Hubei	71.30	24.80	6.00
10	Six Dynasties	Echeng, Hubei	67.60	28.20	5.40
11	Early Eastern Han	Mengcheng, Anhui	77.20	21.68	1.12
12	Early Eastern Han	Mengcheng, Anhui	70.40	24.80	5.90
13	Eastern Han	Echeng, Hubei	75.77	20.23	4.00
14	Three Kingdoms	Echeng, Hubei	70.43	23.95	5.38
		Average	73.94	22.62	3.70

generally be divided into three regions; Figure 5.5 is a typical example. The very top region is a colourless transparent film which is compact and no metal structure can be seen within. This region is about 10–20 microns thick. The middle region is a black penetrating zone, 50–150 microns thick. In this region, there are many defects and cracks, without any clear metal structure. The third region is a black interim zone, 20–80 microns thick, and a metal structure can be seen. This region penetrates the uncorroded bulk metal in an irregular fashion.

The total thickness of the surface layer is 100-250 microns. Usually, the coating is thicker on the face of the mirror than on the back, but the situation is complicated, in that the surface layer is of uneven thickness at different points (see Figures 5.4–5.7). This structure is typical of the *Hei-Qi-Gu*, *Lu-Qi-Gu* and *Qian-Bei* surface structure. Some mirrors have only two regions, a transparent film and a penetrating zone (Figure 5.6), or a transparent film and a very thin interim zone (Figure 5.7). Occasionally, the surface film is so compact it prevents further oxidation of the alloy. In this case only one black layer can be seen, although in fact there exists the transparent film and a very thin interim zone (Figure 5.4)

By comparison, *Shui-Yin-Qin* has a much thinner film (several score to several hundred nanometres). Usually the layer can be divided into two regions, the top region is a film (mainly SnO_2), and the other is an oxidized zone, silver-grey in colour with a rather high tin content.

Scanning electron microscopy (SEM), and energy dispersive X-ray (EDX) analysis

We chose 14 mirrors as examples to illustrate the results of our examination. No.1 to 10 are typical *Hei-Qi-Gu*, No. 11 to 13 exhibit glossy and shiny dark grey surfaces with emerald green or bright yellow spots, and No.14 is *Qian-Bei*, (see Table 5.2).

First, the mirror fragments were placed into the SEM to examine and analyse the top surfaces. The SEM operating voltage was 25 kV and the depth of electron penetration was 3–4 microns. The composition (relative content) of the transparent surface layer or film of the mirrors are shown in Table 5.3.

We can see from Table 5.3 that the chemical composition (normalized) of the top film is distinctly different from that of the bulk alloy. Compare the average composition of Tables 5.2 and 5.3; copper, which comprises 73.94% in the bulk alloy, decreases to 12.93% in the surface film; tin, on the other hand, increases from 22.62% in the bulk to 68.68% in the surface. There are substantial amounts of Fe, Si, Al and P (and unanalysed oxygen) in the

Table 5.3 Chemical composition of the top oxide surface film of the bronze mirrors (results normalized to 100%). (Typical, non-normalized total is about 70%)

Mirror No.	*Location of Surface Analysis*	*Composition wt%*						
		Cu	*Sn*	*Pb*	*Fe*	*Si*	*Al*	*P*
1	Face of mirror	19.50	69.85	4.03	5.37	1.20	–	–
2	Face of mirror	9.55	64.29	20.52	1.30	4.34	–	–
2	Back of mirror	5.48	73.87	10.99	4.17	2.94	2.55	–
3	Face of mirror	16.59	67.87	8.01	1.28	2.25	2.29	1.71
4	Face of mirror	8.61	70.58	6.06	1.68	5.87	7.20	–
5	Face of mirror	21.60	65.76	1.01	3.62	8.02	–	–
6	Back of mirror	5.68	75.58	2.71	8.13	5.46	2.44	–
7	Face of mirror	18.01	65.66	1.80	5.88	3.35	4.40	0.90
7	Back of mirror	10.19	69.60	3.65	7.12	4.25	4.59	0.60
8	Face of mirror	41.39	49.76	4.92	0.97	2.96	–	–
8	Back of mirror	7.37	70.36	7.65	4.34	4.91	3.59	1.86
8	Back of mirror	6.03	72.06	6.59	4.45	4.57	3.57	2.75
9	Face of mirror	18.15	61.38	13.94	0.76	5.46	–	0.31
9	Back of mirror	8.84	72.54	6.81	6.12	4.76	0.93	–
10	Face of mirror	8.36	63.35	12.05	0.42	11.40	1.84	1.17
11	Face of mirror (on a grey background)	20.53	65.81	7.95	1.93	3.78	–	–
11	Face of mirror (on a green background)	15.25	73.01	7.76	0.69	3.29	–	–
11	Back of mirror (on a grey background)	16.10	68.96	11.22	–	3.71	–	-
12	Face of mirror (on a green background)	7.93	75.98	9.94	2.24	3.53	0.38	–
		2.53	79.23	6.42	1.99	3.83	–	–
12	Face of mirror (on a grey background)	21.01	64.12	7.42	2.16	3.51	1.72	–
		6.89	79.08	8.27	2.00	3.75	–	–
13	Face of mirror (on a grey background)	11.55	67.89	8.96	2.07	4.94	2.48	–
13	Face of mirror (on a green background)	7.67	75.97	7.19	1.79	3.71	2.12	–
		5.48	74.09	9.15	1.72	4.91	4.16	0.49
13	Back of mirror (on a grey background)	11.13	53.29	8.66	3.56	13.58	9.65	0.13
13	Back of mirror (on a yellow background)	9.57	65.62	9.23	2.43	8.03	5.08	0.04
14	Face of mirror	21.08	67.64	4.29	–	3.75	3.24	–
	Average	12.93	68.68	7.76				

surface films, which can hardly be detected in the bulk metal. We can also see in Table 5.3 that the chemical composition of the surface changes greatly at various parts of the same mirror.

Next, the cross-sections of the mirrors were examined to determine the composition of the penetrating zone and interim zone (Figures 5.4 and 5.5). Mirrors No.8 and 13 are taken as examples and the results are shown in Table 5.4. The chemical composition changes gradually with the increase of depth from the surface and finally approaches that of the bulk alloy. At the same time, the content of iron and silicon decrease.

Auger electron spectroscopy (AES) studies

A few of the mirror surfaces were analysed by AES. The results show that the surface layer of a typical mirror contains copper, tin, a certain amount of oxygen, small amounts of lead, iron, silicon, aluminium, phosphorus,

Table 5.4 Chemical composition along the cross-section of the mirrors (results normalized to 100% for relative concentration)

Mirror No.	*Location of analysis point, distance from the surface in microns*	*Cu*	*Sn*	*Composition, wt%* *Pb*	*Fe*	*Si*	*Al*	*P*
8	10	45.96	44.93	4.55	1.17	2.85	–	0.54
	20	49.23	42.46	4.64	0.55	2.75	–	0.38
	30	52.42	40.66	4.18	0.36	2.38	–	–
	40	55.45	38.77	3.34	0.36	2.08	–	–
	60	62.40	34.47	1.55	0.34	1.24	–	–
	80	62.96	34.15	1.18	0.27	1.44	–	–
	100	64.51	33.16	1.08	–	1.25	–	–
	150	64.38	31.86	2.54	–	1.24	–	–
13	48	26.41	61.20	7.29	–	5.10	–	–
	72	27.83	61.09	5.46	–	5.62	–	–
	96	35.33	55.69	4.72	–	4.26	–	–
	120	47.87	45.28	3.44	–	3.41	–	–
	200	77.93	19.51	2.07	–	0.49	–	–

carbon and traces of mercury, silver, antimony, arsenic, zinc, sulphur and chlorine.

X-ray diffraction studies

X-ray diffraction measurements were made using a diffractometer directly on the surface of two mirrors, the sample size was about 3 cm^2. It is difficult to identify the crystal structure as the composition of the surface layer is very complicated. The main constituent is thought to be SnO_2. Other compounds that are present may be $CuCO_3$, SiO_2, PbO_2, and Fe_5Sn_3 (according to the American Society for Testing and Materials, ASTM).

Etching experiments

Etching experiments were carried out to test the corrosion resistance of the various types of mirror. Concentrated hydrochloric acid was dropped on the mirror surfaces. *Hei-Qi-Gu* was not attacked but *Lu- Qi-Gu, Qian-Bei* and *Shui-Yin-Qin* all exhibited white spots on their surfaces. Only nitric acid tarnished *Hei-Qi-Gu* and made a slight greyish spot. The experiments show that *Hei-Qi-Gu* mirrors are unusually corrosion resistant.

The surface layers of other bronze wares

Some other ancient bronzes (of lower tin content than the mirrors) unearthed in China have shiny black surfaces, just like the *Hei-Qi Gu* mirrors. They include sacrificial vessels, weapons, utensils and ornaments (Han Rubin *et al.* 1983, Ma Zhaozeng and Han Rubin 1987). Some artefacts of the Shang Dynasty have beautiful shiny black surfaces. A typical example is the well-known *Si Yang Zun* (wine vessel with a four-ram design, 58.3 cm high, weighing 34.5 kg) excavated at Ningxiang, Hunan province, in 1938. But this vessel has not been examined.

We studied a few score bronze wares dating back to the period from early Western Zhou to Western Han dynasty. The surface layer of the bronzes are shiny black, green, yellow or grey. Microscopic studies show that these bronzes have the same surface coating as the *Hei-Qi Gu* mirror (Figures 5.8 and 5.9). The thickness of each region varies greatly, but mostly in the following range: transparent film from 5 to 20 microns; penetrating zone from 50 to 300 microns and interim zone from 20 to 200 microns. The X-ray diffraction pattern of the transparent film shows SnO_2 as the main compound, the pattern is almost identical to

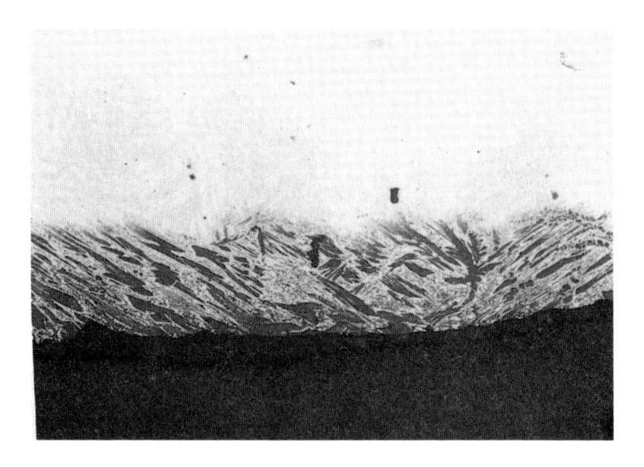

Figure 5.8. *Bronze scraper (see group 2, No.3), cross-section detail of the surface, × 320, unetched optical micrograph.*

Figure 5.9. *Bronze bracelet (see group 3, No.3), cross-section detail of the surface, × 320, unetched.*

that of the mirrors. There were no remarkable differences between XRD patterns from the various coloured bronze wares.

We carried out bulk alloy and surface film SEM analytical studies on three main groups of bronzes, the results of which are shown in Tables 5.5–5.8:

Group 1. In the 1980s, archaeologists excavated more than 200 tombs of the early Western Zhou dynasty at Liulihe, Beijing and a large number of bronzes were unearthed. Among them were some which had glossy surfaces with different colours. Results of the examination of some of these bronzes are shown in Table 5.5.

Group 2. A large number of bronzes of the late Warring Period (*c.* 3rd century BC) were unearthed at Emei, Sichuan province in 1963. A few of them exhibit shiny black or green on their surfaces. The examination results are shown in Table 5.6.

Group 3. Some bronzes of the late Warring Period to Western Han dynasty which were unearthed at Dianchi, Yunnan province in 1972 also exhibit shiny surfaces. The examination results are shown in Table 5.7.

To determine the composition of the centre of the penetrating zone, SEM examination was carried out on the cross-section of the samples. The results are shown in Table 5.8.

From Tables 5.5 to 5.8 (normalized analytical results) we can see similar results to those we get from the mirrors: the chemical composition of the surface coating is distinctly different from that of the bulk alloy. The coating contains relatively high Sn (and Pb) but much less Cu, and also contains some Fe, Si, Al and P.

We think that these bronze wares were also polished with *Xuan Xi*. The mechanism of patina formation is the same as for the mirrors. The lower tin content of the bronze wares does not affect the results.

A preliminary exploration of the mechanism of formation of *Hei-Qi Gu* (black lacquer antique)

The polishing powder *Xuan Xi*

In ancient China, the surface of the bronze mirror was rubbed with a powder called *Xuan Xi* to make it *bright and white*. The function of *Xuan Xi* was not only to polish but also to plate a thin silvery coloured layer on the mirror surface. We can often observe polishing marks under the shiny surface layer.

What is *Xuan Xi*? We consulted a lot of ancient Chinese literature and managed to find several passages with accounts of its ingredients. For example, *Duo Neng Bi Shi* by Liu Ji of the Ming dynasty recorded the composition of *Xuan Xi* (weight ratio):

Tin 1, Mercury 1, Alum 6, Deerhorn ash 1

Table 5.5
Composition of bulk alloy and surface film of group 1 bronzes

No.	Type	Surface colour	Cu	Composition of bulk alloy, wt% Sn	Pb
1	Rectangular decoration	Black	81.09	13.13	5.78
2	Bulb	Greenish-grey	87.55	11.77	0.68
3	Semi-spherical decoration, No. 1	Green	81.96	8.39	9.65
4	Semi-spherical decoration, No. 2	Yellow	87.10	7.17	5.73
5	Chariot fitting	Grey	84.05	15.95	–
6	Mouth of pot	Black	86.99	11.00	2.00
7	Ornament	Green	88.97	11.03	–
8	Shield	Black	84.38	15.27	0.36
9	Dagger-axe, No. 1	Greenish-grey	89.31	10.69	–
10	Dagger-axe, No. 2	Black	84.93	14.86	0.21

Composition of surface film, wt% (normalized to 100% for relative concentrations)

No.	Surface colour	Cu	Sn	Pb	Fe	Si	Al	P
1	Black	34.14	39.25	22.59	0.52	1.66	–	1.84
2	Green-grey	23.83	61.24	10.88	–	3.98	–	0.07
3	Green	16.75	37.96	37.69	–	7.60	–	–
4	Yellow	3.35	66.19	6.12	15.65	7.66	–	–
5	Grey	47.85	44.81	0.34	0.22	6.11	–	–
6	Black	54.43	34.56	8.15	–	2.86	–	–
7	Green	41.83	53.34	0.79	–	3.16	–	0.88
8	Black	40.12	47.47	–	0.35	12.06	–	–
9	Green-grey	16.98	56.04	8.74	1.17	16.37	–	–
10	Black	56.86	31.96	0.76	0.96	9.46	–	–

Xuan Xi is a greyish-white powder whose colour changes gradually to black when it is rubbed onto the bronze surface, hence the name *Xuan Xi* which translates to black tin, although this is not the direct cause of the black mirror surfaces. The interpretation of the literature makes it clear that tin amalgam is the main constituent of *Xuan Xi*. When a mirror is rubbed with *Xuan Xi* it is then heated to eliminate the mercury, and a bright silvery coloured film of tin is left on the mirror surface. However, we are not sure that the mirrors were necessarily heated to eliminate mercury as rubbing with *Xuan Xi* makes the mirrors bright and white, so it is not necessary to drive off the mercury (see Meeks pp. 78–81). We discovered recently that when a mirror rubbed with tin amalgam was laid aside for 3 years, nearly all the mercury on the surface layer vaporized and could not be detected by ordinary methods. The mirror surface tarnishes after it has been exposed in air for a long time and should be rubbed with *Xuan Xi* again.

Recent experiments by Chen *et al.* (1987) have proved that after a bronze mirror is treated with tin amalgam, a thin, silvery surface layer is formed and merges smoothly into the bulk alloy. No mechanical interface lies between. The chemical composition of this layer is high-tin and low-copper. The micro-hardness of this metallic surface layer is higher

Table 5.6
Composition of bulk alloy and surface film of group 2 bronzes

No.	*Ware*	*Surface colour*	*Cu*	*Composition of bulk alloy, wt%* *Sn*	*Pb*
1	Knife	Black	78.31	11.98	9.71
2	Scraper, No. 1	Black	77.47	8.25	14.24
3	Scraper, No. 2	Black	81.34	18.66	–
4	Chisel	Green	84.00	10.39	5.61
5	Small knife	Black	78.76	10.92	10.32
6	Pot	Greenish-grey	70.50	12.30	17.00

Composition of surface film, wt% (normalized to 100% for relative concentration)

No.	*Surface Colour*	*Cu*	*Sn*	*Pb*	*Fe*	*Si*	*Al*	*P*
1	Black	23.84	43.33	17.69	3.77	6.51	2.01	2.81
2	Black	12.35	27.51	39.09	9.82	4.82	1.54	4.86
3	Black	25.56	44.85	1.84	7.15	11.65	8.25	0.70
4	Green	13.93	32.83	29.61	10.30	5.80	2.83	4.73
5	Black	7.47	28.18	42.69	1.96	4.70	12.03	–
6	Green-grey	24.47	44.58	21.60	2.46	2.21	1.16	3.52

Table 5.7
Composition of bulk alloy and surface film of group 3 bronzes

No.	*Ware*	*Surface colour*	*Cu*	*Composition of bulk alloy, wt%* *Sn*	*Pb*	*Fe*
1	Armour	Milky white	92.95	7.05	–	–
2	Bracelet, No. 1	Black	79.07	20.93	–	–
3	Bracelet, No. 2	Yellow	82.77	14.48	2.49	0.15
4	Lance	Greenish-grey	90.10	9.90	–	–
5	Sword	Black	87.80	12.20	–	–
6	Ornament on sheath	Yellow	92.37	6.51	–	1.12

Composition of surface film, wt% (normalized to 100% for relative concentrations)

No.	*Surface colour*	*Cu*	*Sn*	*Pb*	*Fe*	*Si*	*Al*	*P*
1	Milky white	62.04	34.32	0.89	1.00	0.73	–	0.98
2	Black	35.23	60.99	0.72	0.41	1.35	–	1.29
3	Yellow	12.61	64.49	17.70	3.31	1.24	–	0.65
4	Green-grey	23.96	63.83	–	1.03	4.59	5.92	0.67
5	Black	47.87	48.07	–	0.48	1.47	1.45	0.66
6	Yellow	20.72	57.02	4.03	14.65	1.23	1.44	0.91

Table 5.8 Composition of penetrating zone of group 3 bronzes

Composition, wt% (normalized to 100% for relative concentrations)							
No.	*Cu*	*Sn*	*Pb*	*Fe*	*Si*	*Al*	*P*
1	48.61	46.29	1.55	0.45	1.09	–	2.01
2	47.14	51.15	–	–	1.71	–	–
3	41.18	49.51	6.05	0.99	1.17	–	1.09
4	30.56	64.37	–	0.53	1.43	–	3.11
5	33.47	52.95	3.03	8.00	2.08	–	0.48

Table 5.9 Microhardness comparison

	Top surface	*Microhardness, HM penetrating zone*	*Bulk alloy*
Hei-Qi-Gu patina	336	318	290
Hg–Sn treated mirror	340	321	297

than that of the bulk. By chance this is similar to the hardness of a *Hei-Qi-Gu* mirror surface which is patinated (Table 5.9).

An etching test has been conducted on the surface layer of the simulated mirror. Even if the mirror is placed in a 100°C heat-treatment furnace with acidic oxidizing atmosphere for a fortnight, the mirror surface still remains bright and shiny.

Archaeological excavations have provided material showing that tin-plating was known from at least as early as the late Shang dynasty in China. For example, a copper helmet plated with a thick tin layer was excavated from a site of this period (Guo 1963), although we are not sure that this object was tinned by the amalgam method. Other evidence for early tinning is provide by *Shi Jing* (The Book of Song, compiled during the Western Zhou dynasty and the Spring and Autumn Period) which mentions that chariot fittings and weapons were plated with tin to make them appear to be white.

From the historical texts, tin amalgam-plating was the most likely main method of tinning in ancient China. As for bronze mirrors, both high-tin and low-tin mirrors were tinned by *Xuan Xi* (tin amalgam method). In ancient China, mirror polishing was an important handicraft, some people went from house to house, street to street with *Xuan Xi* for polishing purposes. According to the historical records, this method was very convenient and effective.

Mechanism of formation of *Hei-Qi-Gu*

The mechanism of *Hei-Qi-Gu* (black surface patina) formation is very complicated. Here, we shall put forward several hypotheses rather than conclusions. As previously mentioned the patina can often be divided into three regions, the top surface is a colourless transparent film, the underlying interim and penetrating zones are black.

1. *Formation of transparent surface film*
The surfaces of the mirrors are transparent and this film is formed through the oxidation of the high-tin layer which is produced by the application of *Xuan Xi*. If an imitation of a tin-plated mirror is exposed in air for one year, a very thin film is formed, the main composition of which is SnO_2. Other investigators have examined the top film of bronze mirrors in recent years and have obtained very similar results. For example, the Structure Analysis Laboratory of the University of Science and Technology of China (1988) analysed the surface layer of a *Hei-Qi-Gu* mirror of the Han dynasty with SEM. The composition of this layer is Cu 6.2%, Sn 73.3%, Pb 7.6%, Si 6.6%, O 6.3% (no analysis for mercury was made), while the composition of the bulk alloy is Cu 74.6%, Sn 22.8%, Pb 2.6%. Mossbauer spectroscopy studies showed that in the surface film tin exists mainly in the form of tin

oxide with a valency of four (i.e. SnO_2). Wang *et al.* (1989) examined 23 ancient mirrors by X-ray diffractometer and proved the existence of crystalline SnO_2 in the surface layers of Han mirrors.

Wu (1988) examined several *Shui-Yin-Qin* mirrors (Han to Tang dynasties) and conducted some imitative experiments. These silvery mirrors also had a thin transparent surface layer. This transparent surface contained 40-60% Sn, with the chief crystalline component being SnO_2. The surface layer also contained small amounts of calcium, potassium, aluminium, chlorine, carbon, nitrogen and sulphur, etc. Some of these elements might be brought in by *Xuan Xi* while others come from the soil.

2. *Regional ocurrence of* Hei-Qi-Gu *(black) and* Shui-Yin-Qin *(bright metal) mirrors: the formation of patina*

As mentioned above, the *Hei-Qi-Gu* (black), *Lu-Qi-Gu* (green) and *Qian-Bei* (grey) mirrors mainly appeared in south China while the *Shui-Yin-Qin* (bright metal) mirrors were mostly found in north China. This is primarily because of the *burial* conditions rather than the composition of the mirror itself. One key factor is that soil tends to be acidic in south China and basic in north China. The pH values of the soils were determined several years ago as follows:

south China	pH 5–6 (a few samples reached pH 3)
north China	pH 8–9 (a few samples reached pH 10)

This is probably why in general the mirrors have different surface colouration when found in these different areas. A very good example is given by two Tang dynasty mirrors with the same relief of two birds, two animals and exquisite flowers on the back. These two mirrors were almost certainly made by the same craftsman. The one excavated in south China and now preserved in Shanghai Museum is a typical *Hei-Qi-Gu* (black), while the other unearthed in Luoyang, Henan province recently is a typical *Shui-Yin-Qin* (bright metal).

In general, other tin-plated bronzes are the same colour as mirrors found in the same tombs. However, local burial environments may be very complex; even mirrors excavated from the same tomb may be different colours, and different patinas often appear on the same mirror. For example, a mirror of the Tang dynasty excavated at Echeng, Hubei is a typical *Lu-Qi-Gu* mirror (with green patina) but there is a round area (about 2 cm in diameter) of silvery white patina (typical *Shui-Yin-Qin*) on its surface. Perhaps a round-shaped object became stuck on the mirror in the burial and thus retarded the oxidation. The surface composition of the two regions is quite different:

SEM EDX analyses normalized (no oxygen data)

	Composition wt%					
	Cu	Sn	Pb	Si	Fe	Al
Lu-Qi-Gu green region	31.0	57.2	5.9	2.9	1.7	0.2
Shui-Yin-Qin silver region	63.3	31.5	2.9	1.4	0.8	0.0

The burial environments in some cases must have been complicated, and the degree of corrosion varied very locally.

Conclusions

In ancient China, bronze mirrors were rubbed with *Xuan Xi*, leaving a thin film of tin on the surface to make them bright and white. The main components of *Xuan Xi* were tin, mercury, alum and deerhorn ash. Depending on the different burial conditions, the mirrors exhibit shiny jet black, emerald green, dark grey or silver-white surfaces. In general, the latter mainly appeared in north China, the others in the south. Soil tends to be basic in the north and acidic in the south, perhaps this is the main reason for the mirrors having different surface colourations.

Some other ancient bronzes also have shiny surfaces with different colours. Examination has shown that these bronzes have the same surface coating as the mirrors.

After examining a large number of mirrors and other bronze wares, we have come to the conclusion that the black patina was formed naturally during the long period of oxidation during burial, as were the green, grey and white patinas. They were not artificially patinated.

References

All references are in Chinese, and are given here in translation.

Chen Yuyun, Huang Yunlan, Yang Yongning and Chen Hao (1987) An experimental imitation of the *Hei-Qi-Gu* bronze mirror. *Kaogu (Archaeology)*, **2**, 175–178

Guo Baojun (1963) *The Bronze Age of China*, Beijing, p.47

Han Rubin, Ma Zhaozeng, Wang Zengjuan and Ke Jun (1983) Studies related to the black passive oxide film on bronze arrowheads unearthed with the terra-cotta warriors near the tomb of the first emperor of Qin (210 BC). *Studies in the History of Natural Sciences*, **4**, 295–302

Ma Zhaozeng and Han Rubin (1987) Further studies related to the black passive oxide film on bronze arrowheads unearthed with the terra-cotta warriors near the tomb of the first emperor of Qin (210 BC). *Studies in the History of Natural Sciences*, **4**, 351–357

Museum of Zibo City, Shandong Province (1985) The funerary pits round the princely tomb of Qi Kingdom of the Western Han Dynasty. *Kaogu Xuebao (Acta Archaeogica Sinica)*, **2**, 223–266

Structure Analysis Laboratory of the University of Science and Technology of China and the Laboratory of the Institute of Archaeology (1988) The components and structure of Han bronze mirrors. *Kaogu (Archaeology)*, **4**, 371–376

Wang Changsui, Xu Li, Wang Shengjun and Li Huhou (1989) Analysis of the structure and components of the ancient bronze mirrors. *Kaogu (Archaeology)*, **5**, 476–480

Wu Laiming (1988) Some problems concerning the research on ancient bronze mirrors. *The First Symposium on Laboratory Archaeology*, Nanking

6

Patination phenomena on Roman and Chinese high-tin bronze mirrors and other artefacts

Nigel Meeks

Abstract

High-tin bronze is a silver coloured, hard metal that takes a fine polish. These characteristics made the metal particularly suitable for the manufacture of various cast items of jewellery and belt fittings in antiquity; it was also used for Chinese mirrors of various periods and for some types of Roman mirror. Some of these items are found with their original polished silver coloured surfaces, while others have part or extensive patination of various shades of grey and green through to black, or a combination of these.

Of particular interest in recent years, have been the black Chinese mirrors that have been the subject of modern analytical research in both the East and the West. From the literature it is generally agreed that the surface of these mirrors is no longer metallic but mainly tin oxide(s) which give a 'tin' enriched surface and associated severe copper depletion with respect to the body metal composition. However, the route to the oxidized state, whether by deliberate patination in antiquity or as a natural corrosion process, is neither obvious nor agreed.

This chapter presents the comparative results of the examination of some black surfaced Chinese mirrors, Roman mirrors and other high-tin bronze artefacts showing their microstructural and analytical characterization. Generally there are common features between these objects showing pseudomorphic mineralization of the normally corrosion resistant intermetallic compound, and the ingress of various mineral elements derived from the environment. No added surface layers are seen.

Nature certainly does blacken high-tin bronze artefacts as is shown by the black surface continuing over ancient broken surfaces and worn areas.

Introduction

Bronze with 20% or more tin in the alloy is silver coloured, hard and takes a fine polish, so it is ideally suited to the manufacture of mirrors. High tin-bronze mirrors were being made in China from as early as the Warring States period and from references in the ancient texts they clearly held almost magical significance to the Chinese. A large number of Chinese mirrors have found their way into collections in the West (Watson 1962, Zhu 1986) and many of these have a beautiful, uniform black polished surface which reflects colour and flesh tones like a conventional silver coloured mirror (Plate 6.1). Chase and Franklin (1979) report one mirror having both silver colour and black patination which appears to follow the cast-in pattern. In ancient Chinese texts mirrors are variously and subjectively described as 'burning', 'magic', etc. (Needham 1962) which tends to give credibility to the idea that the *black* mirrors might have had special significance in ancient times. However, there appears to be no mention in the ancient texts of any surface blackening treatment (Zhu 1990, personal communication), whereas 'brightness' and 'whiteness' of mirrors and similar descriptions are commonly found (Needham 1962). Furthermore, not all the mirrors are black: some are silver coloured while others have corroded to shades of grey and green or a combination of these. Unfortunately there is little or no information about their burial conditions, particularly of mirrors in museum collections.

The nature and origin of the surfaces of the black Chinese mirrors have long been a subject of study (Collins 1934, Gettens 1969, Chase 1977, Chase and Franklin 1979) and more recently have been the subject of analytical research in both the East and the West (Yao Chuan and Wang Kuang 1987, He Tankun 1985, Han Rubin and Ma Zhaozeng 1983, Wang Changsui *et al.* 1989, Han Rubin *et al.* 1983, Li Shuzhen 1984, Tan Derui 1985, Ma Zhaozeng and Han Rubin 1986, Sun Shuyun *et al.* forthcoming, Zhu Shoukang see Chapter 5). It is generally agreed that the black surfaces of these high-tin bronze mirrors are no longer metallic but are mainly tin oxide(s) which give a 'tin' enriched surface. In addition, relatively low concentrations of mineral type elements, silicon, aluminium, iron, phosphorous and potassium, are found in this surface. It has been suggested that the presence of a few per cent of chromium oxide found in the black surface patination of a 12% tin bronze spearhead was important, although similar artefacts did not contain significant chromium (Han Rubin *et al.* 1983). The combination of these and various ideas has led to work on unusual projects to fuse oxidizing mixtures of nitrates and chromates onto high-tin bronze at high temperature (Yao Chuan and Wang Kuang 1987, He Tankun 1985), which were apparently successful in producing a surface layer of appropriate composition, but no micrographs were published to show whether the surface appearance matched that of the ancient mirrors.

A treatment of mirror surfaces, which does appear in the ancient Chinese texts (Needham 1962), is the application of mercury and tin. This could either mean fire tinning (see p. 79), or the use of a polishing compound, *Xuan Xi*, containing mercury and tin amongst other ingredients (Barnard 1961, Zhu see Chapter 5). Mercury has only rarely been found analytically on mirror surfaces. Fabrizi (1988) and Kerr (1990) do report the presence of mercury traces, detected by X-ray fluorescence analysis, on Chinese high-tin and low-tin bronze mirrors in the collections of the Victoria and Albert Museum.

The Romans used both high-tin and low-tin bronze in the manufacture of mirrors (Meeks 1988a, in press, Craddock 1988). Pliny discusses the tinning of low-tin bronze mirrors to produce the colour of the expensive silver metal mirrors so that even the serving maids could possess a 'luxury' item (Rackham 1952, Book XXXIV chapter XLVIII). The Roman high-tin bronze mirrors are now either silver coloured or patinated grey-black, like the Chinese mirrors, but there is no suggestion by Pliny that the mirrors were anything other than silver coloured when they were made and used.

Over the years a number of papers have been published on the characterization of black Chinese mirrors, but they generally have missed vital evidence regarding the microstructural characteristics of the mirror surfaces.

The approach of this study was to compare the Chinese mirror surfaces with high-tin bronze artefacts with silver coloured and black surfaces from other cultures. It is the purpose of this chapter to try to clarify the contradictory data regarding the black patina on high-tin bronze by characterizing not only the black surfaces of Chinese mirrors, but also Roman mirrors and related high-tin bronze material. Microstructural and analytical characterization shows important details that go beyond what has been reported previously on Chinese mirrors.

Physical properties of high-tin bronze

High-tin bronze is the term generally applied to antiquities made from bronze having a composition of between 15% and 27% tin (often with about 5% lead present) (Barnard 1961, Caley *et al.* 1979, Masaaki Sawada 1979). The physical properties of bronze change rapidly with increasing tin content above about 10% (the limit of solid solubility of tin in copper, i.e. low-tin bronze) although the colour changes less rapidly and does not become particularly silver coloured until well over 20% tin. These properties are due to the increasing presence of the hard, brittle and silver-white coloured δ intermetallic compound ($Cu_{31}Sn_8$ containing 32.6%Sn) which appears in the characteristic α+δ eutectoid microstructural form (Hanson and Pell-Walpole 1951 and this volume pp. 252–53). This eutectoid microstructure dominates the matrix of a cast high-tin bronze and hence these objects are hard, brittle and silver coloured. When polished the metal gives a brilliant reflecting surface, which is normally corrosion resistant but benefits from occasional repolishing (Needham 1974). The colour is only truly silver-white when the intermetallic compound is present without any α phase, which adds a shade of copper colour to the alloy. These properties of high-tin bronze were recognized by the ancients who also realized the impossibility of cold working this brittle metal, but found, when alloyed with a little lead, that it was a particularly suitable material with which to *cast* items that would benefit from the durable, silver-like polished finish that could be achieved. It is found extensively used for Dark Age belt fittings, buckles and brooches and items of similar size such as finger rings. The Chinese from the Han period onwards and the Romans both used this alloy to great advantage for certain types of mirrors. Mirrors of this alloy are still made in a few craft workshops in Kerala, India (Glover, personal communication. Sudhir 1991).

Thus the original colour of any cast, polished high-tin bronze object would be silver coloured, not black. The term 'speculum metal' is often used to describe this material (although specifically it has been used more recently for hyper-eutectoid composition alloys of around 30–40% tin which therefore contain δ and ϵ intermetallic compounds, Hedges 1964). It is described in ancient Chinese texts as 'truly white and reflects without tinning or silvering' (Needham 1962). Even the inscriptions on some of the Chinese mirrors describe them as being 'bright', and their purpose was to provide light in the afterworld (Collins 1934). However, items of all these types can be found to have surfaces which may be anything from silver colour through patinas having shades of grey, dark green through to black. Some items have various combinations of these shades and may also have superficial deposits of green and blue crystalline copper corrosion products as well. Some of the patinated items and some Chinese mirrors, as previously mentioned, are now completely black and are very impressive (Plate 6.1).

Bronze of similar high-tin content, but no lead, was hot worked and quenched to produce the ancient bowls (Chapter 21, Figure 21.24) and gongs found in south-east Asia, e.g. Thailand, Korea, China, the Philippines and the Nilgiri hills in south India, where the earliest date to around 200 BC–AD 500. The metal of these objects is a golden colour and they have a completely different microstructure to the cast bronzes examined here (Seeley and Rajpitak 1979, Goodway and Conklin 1987). They corrode to produce a shiny, transparent, golden-brown patina, containing tin oxide and silicon etc., similar in composition to the black patinas on Chinese mirrors.

The results of the examination of the structures of cast high-tin bronze antiquities

The techniques used to examine the high-tin bronze material were primarily scanning electron microscopy (SEM) with quantitative energy dispersive X-ray analysis (EDX) and digital elemental X-ray mapping, optical microscopy, X-ray fluorescence analysis (XRF), X-ray diffraction (XRD), FT infra-red analysis (FTIR) and optical spectrography.

The antiquities examined were generally small enough to insert into the SEM for direct observation of their surfaces, and required no surface preparation apart from dusting. Polished cross-sections were made of typical black surfaced mirrors that had been broken in antiquity, and ancient fracture surfaces were also examined directly. All of the material was

Table 6.1 Analyses of high-tin bronze antiquities and patinas. EDX analyses except where indicated

Wt %	*Cu%*	*Sn%*	*Pb%*	*Al%*	*Si%*	*P%*	*Cl%*	*K%*	*Ca%*	*Fe%*	*Ni%*	*Sb%*	*As%*	*Ag%*	*Hg%*	*Total%*
Chinese mirror																
OA 1937,4-16,210																
Body metal	69.4	25.6	4.8							0.2						100.0
Black patina	12.0	53.1	7.3		0.8	0.3	0.8	0.4	0.2	0.3						76.8
XRF trace analysis-																
Elements found	Cu	Sn	Pb							Fe	Ni	Sb	As	Ag	None	+Co
Chinese mirror																
OA 1973,7-256,66																
Black patina	6.4	42.1	5.9	0.3	2.3	0.4	0.2	0.1	0.6	0.3						61.0
XRF trace analysis-																
Elements found	Cu	Sn	Pb								Ni	Sb	As	Ag	None	
Chinese mirror																
OA 1950,11-17,230																
Black patina	7.4	54.4	6.6		2.1	0.4	0.2	0.4	0.3	0.6	0.2					
XRF trace analysis-																
Elements found	Cu	Sn	Pb							Fe	Ni	Sb	As	Ag	None	
Chinese mirror																
OA 193,11-18,264																
Black patina	8.1	52.9	6.2	0.2	2.6	0.8	0.2	0.5	1.4	0.3						74.0
XRF trace analysis-																
Elements found	Cu	Sn	Pb							Fe	Ni	Sb	As	Ag	None	
Chinese mirror																
OA 1947,7-12,339																
Black patina	7.7	46.8	7.5		2.8		0.3	0.6	0.8	1.1						69.1
Chinese mirror																
OA 1936,11-18,90																
Black patina	9.7	48.0	5.6		0.7	1.1	0.2	1.0	0.5	1.8						70.6
Chinese mirror																
OA 1920,2-18,1																
Surface of silver-coloured mirror	58.3	26.3	1.0		0.2		0.3	0.3								87.0

Chinese mirror OA 1955,7-14,2 Surface of silver-coloured mirror XRF trace analysis-Elements found	Cu	Sn	Pb							Fe	Ni	Sb	As	Ag	None	
Chinese mirror from Changsha, 1990																
Body metal	73.5	22.2	4.0								0.3					100.0
Black patina	7.3	50.2	8.9	0.3	1.8	0.1	0.2		0.2	0.7						69.7
Roman mirror GR 1814,7-4,1063																
Body metal	69.5	22.8	6.8							0.3		Zn0.6				100.0
Black patina	10.1	55.6	6.9	0.2	2.8	0.2	0.2	0.6	1.2	1.1						79.8
Ptolemaic ring GR 1930,7-15,3																
Body metal, atomic absorption analysis	69.2	24.6	4.5							0.081	0.032	0.02	0.15	0.022	Co0.03	98.5
Black patina	11.0	55.4	5.0	0.4	3.1		1.2	0.7	0.7	1.1						79.8
Black fracture	5.5	49.5	5.0	0.5	1.6		0.5	0.4	1.1	0.9						68.3
Nilgiri Hills Bowl OA 1882,10-9,18 Quenched high-tin bronze																
Body metal	76.6	22.4								0.5	0.5					100.0
Patina	8.0	58.0		0.2	1.8	0.4	0.2	0.2		0.9	0.4					70.1
Dark Age Buckle Silver colour																
Body metal	69.7	23.2	3.4							0.15		Zn1.9				100.0
Alpha corrosion	29.8	33.8	3.6	0.2	0.9	0.4	0.4			3.50		Zn1.2				74.7

subjected to detailed microstructural and analytical examination in the SEM. Digital X-ray mapping was used on cross-sections of both silver coloured and black surfaced bronzes to determine the distribution and relative concentrations, by colour intensity, of not only tin, copper and lead from the original metal but also of the mineral elements within the mineralized surface zones.

XRF analysis was used to search for the presence of mercury on Chinese mirror surfaces (detection limit 0.05%). When no mercury was found by XRF, this was supplemented by the more sensitive spectrographic analysis (detection limit 100 ppm). XRD and FTIR were applied to the black mineralized surfaces of objects to determine their crystallographic structure.

Collins (1934) was unable to see microstructure in the patinated mirror surfaces by optical observations, and Gettens (1969) predicted that advanced techniques such as SEM may hold the key to understanding the nature of these mirror surfaces, but until now the technique has not been fully effective in its application to the critical areas that hold the key information. However, the use of backscattered electron imaging in this study to provide compositional contrast information in the SEM shows important microstructural detail within the surface oxide layers.

The antiquities chosen for study were, firstly, conventional *silver* coloured high-tin bronze Roman mirrors, Chinese mirrors and Dark Age belt fittings, in order to determine the normal structures that occur on these cast high-tin bronze surfaces. Secondly a variety of *black* surfaced high-tin bronze Roman mirrors, Dark Age belt fittings and Ptolemaic rings, which acquired their patina during burial. Thirdly, *black Chinese* mirrors including a sample from a newly excavated black mirror from Changsha, China. This group were examined from the viewpoint that they may have been deliberately patinated in antiquity, and therefore the composition and microstructure of these mirrors were compared to the naturally patinated material to see if any differences could be found.

For microstructural purposes, the non-Chinese objects can be grouped together irrespective of type, provenance or date because they have fundamentally the same cast eutectoid microstructure and similar composition (Table 6.1). Variation in structure between objects is primarily the result of their relative compositions and is determined by the phase diagram in the usual manner (Hanson and Pell-Walpole 1951).

The results are discussed below under the following headings:

Roman and Dark Age high-tin bronze objects
- (a) Uncorroded body metal
- (b) Silver coloured surfaces
- (c) Black surfaces

Chinese mirrors
- (d) Uncorroded body metal
- (e) Silver coloured surfaces
- (f) Black surfaces

Mercury tinning
- (g) Formation of mercury/tin amalgam
- (h) Mercury (fire) tinning
- (i) Cold mercury amalgam tinning.

Roman and Dark Age: uncorroded body metal

The typical body microstructure of these cast high-tin bronze objects is exemplified by the Roman mirror from St Albans, Hertfordshire shown in Figure 6.1 at low magnification and Figure 6.2 at higher magnification (Craddock *et al.* 1989). The microstructure and optical colour of the sectioned body metal are dominated by the silver coloured δ intermetallic compound, $Cu_{31}Sn_8$ containing 32.6% tin, which manifests itself in the $\alpha+\delta$ eutectoid (average composition 27% tin) and occurs by way of a solid state transformation as the cast metal cools through the eutectoid isothermal at 520°C. A small amount of excess copper-rich α phase (solid solution) is seen as laths in this mirror structure, and the white globules are lead, which always occurs in this form. The melting point of this typical alloy, 22.2% Sn, 6% Pb, 71.8% Cu, is about 800°C.

This is the classic high-tin bronze structure, and is silver in colour. After millennia of burial the surfaces of some objects stay silver coloured and some develop green, grey or black patination. This St Albans mirror has a polished black surface.

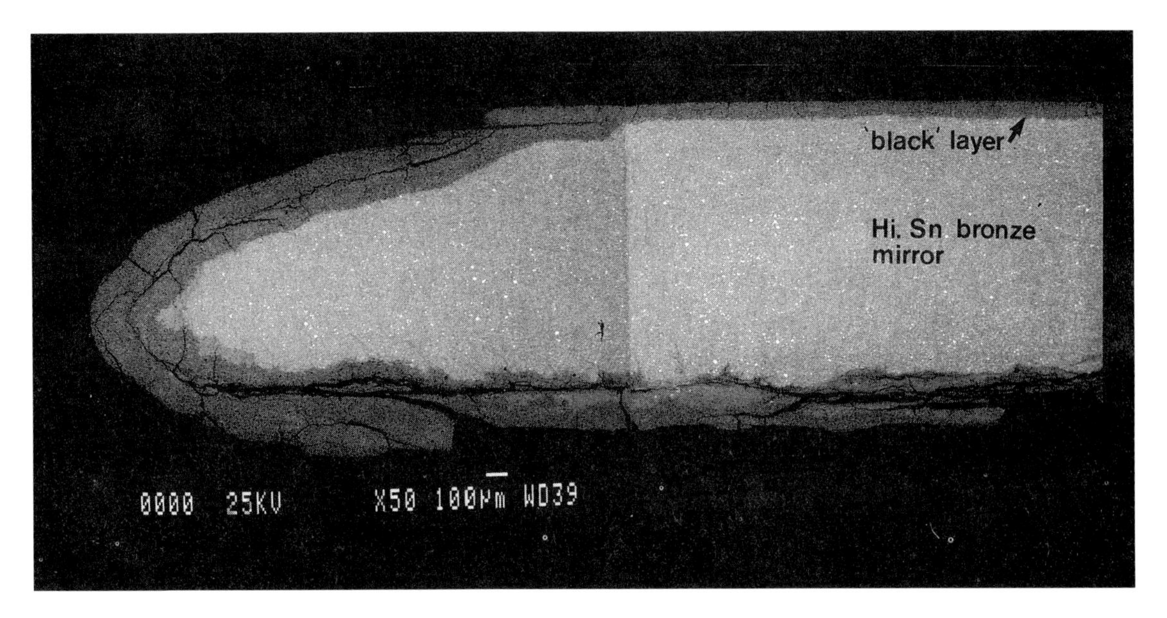

Figure 6.1. *Cross-section of the black Roman mirror from St Albans showing the uniform thickness of patination on the polished front and the deeper patinatination elsewhere. SEM low magnification. The white scale bar represents 100 microns.*

Roman and Dark Age: silver coloured surfaces

It has been previously shown (Meeks 1986, 1988a and b, see also Chapter 21) that with a silver coloured high-tin bronze surface, corrosion usually attacks the anodic α solid solution within the eutectoid and any excess α laths, as well as lead globules to a depth of typically 25–100 microns. The δ intermetallic compound even at the very surface, is uncorroded and therefore still imparts a silver colour to the object because it is the major component present. These characteristics are illustrated on the surface of a Roman high-tin bronze mirror shown in Chapter 21 (Figure 21.21), and similarly in the cross-section of a typical Dark Age buckle (Figure 21.22), of the type shown in Chapter 21, Figure 21.2. The microstructures are classic for the surface of cast high-tin bronze antiquities.

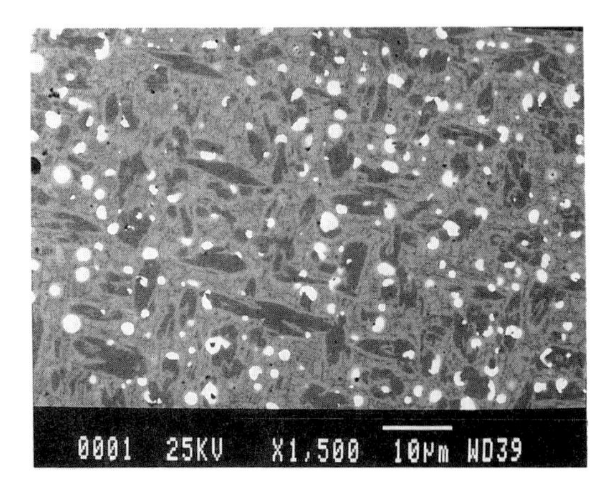

Figure 6.2. *Cross-section of the Roman mirror from St Albans showing the characteristic microstructure of the uncorroded leaded high-tin bronze body metal. SEM backscattered electron image. The white scale bar represents 10 microns.*

Microanalysis shows that corrosion has severely depleted copper from the α solid solution leaving in its place mainly tin oxide, interdiffused with some residual copper and lead oxides. Cuprite is not a major corrosion product because the copper is mostly lost to the environment, although some small round cuprite inclusions are seen in this and other objects, deposited in some of the porosity left by the leached lead globules. There is an ingress of mineral type elements such as silicon, aluminium, iron, etc. into the tin oxide-replaced α interstices, and the typical composition of α phase corrosion of a silver coloured Dark Age buckle is shown in Table 6.1. Digital X-ray mapping through the cross-section clearly showed the severe loss of copper and the relative increase in tin content as well as the distribution of mineral type elements in the corroded α phase in the silver-coloured surface. The matrix differences between the metallic body with high copper and the mineralized α at the surface with low copper accentuate the relative tin enrichment effect although there is no contribution of tin from the environment.

It is important to note that although this is in a silver coloured surfaced high-tin bronze, it shows the same compositional gradient of relatively increased tin at the surface within the corroded regions that is seen in the mineralized surfaces of black Chinese mirrors and other material for which the assumption is made that excess tin must have been deliberately added to account for the tin oxide concentration in the surface. This does not appear to be the case.

Roman and Dark Age: black surfaces

By comparison, the black surfaced high-tin bronze Dark Age buckles and Roman mirrors show in cross-section the same depth of α corrosion (forming the 'altered zone') as the silver coloured high-tin bronze objects, but there is one important difference. At the very surface the δ intermetallic compound is also mineralized but only to a depth of a few microns. In some cases the mineralization of the δ is much thinner but the surface is still black. Tylecote notes that some corrosive archaeological environments allow the degradation of the intermetallic compound in bronze even before the α phase is much affected (Tylecote 1979, Chase and Franklin 1979). Figure 6.3 shows the typical microstructure of a cross-section through the black Roman high-tin bronze mirror, illustrated in Figure 6.4, which shows both uncorroded body metal and and a clear 30 micron thick 'altered zone' (Chase and Franklin 1979) in which the δ intermetallic compound is largely unaffected, although the α phase and lead globules within this zone are mineralized (as occurs with the silver coloured high-tin bronze). Notice that in

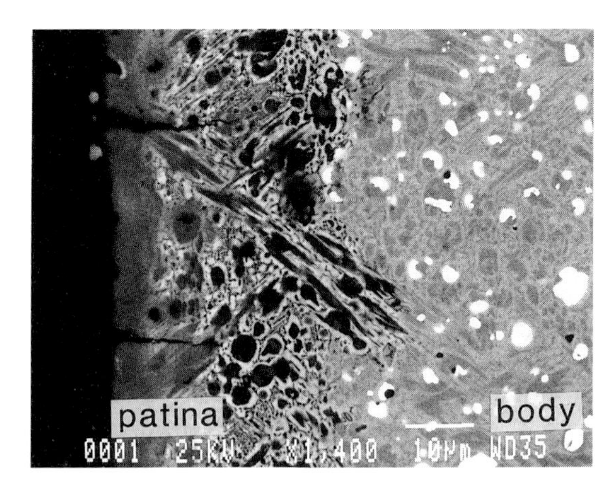

Figure 6.3. *Cross-section through the black surfaced Roman high-tin bronze mirror showing uncorroded body metal, α phase corrosion forming the partly mineralized 'altered zone', and the mineralized surface patina, on the left, with corrosion of the δ intermetallic compound. Note the residual ghost pseudomorphic eutectoid microstructure reaching the surface. SEM backscattered electron image. Width of field = 88 microns.*

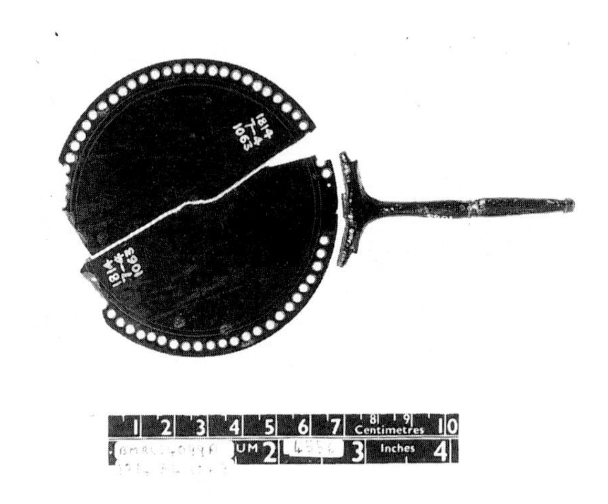

Figure 6.4. *Roman high-tin bronze mirror with black patinated surface. British Museum GR. 1814, 7-4, 1063. Diameter 85 mm.*

this case the top 10 microns of the surface the δ compound is seen to change from metallic to mineral form, and most importantly the eutectoid microstructure is retained as a ghost pseudomorphic structure in the mineralized surface. Cracks run through the mineralized surface and into the altered zone. Viewing the mirror surface directly in the SEM clearly shows the extent of the ghost structures across the crazed mineralized surface (Figures 6.5 and 6.6). The crazing can be seen optically but the micro-cracks propagate a little under the influence of the electron beam showing that the surface may be hydrated.

The composition of the black mineralized surface of this mirror is shown in Table 6.1 and illustrates the severe depletion of copper, the relative concentration of tin and the uniform distribution of lead within the layer derived from the corrosion of lead globules. Digital X-ray mapping across the polished section clearly shows these elemental distributions and the mineral type elements which have diffused in from the environment during burial. The quantitative analytical totals of the mineralized surface come to around 60%–80% indicating the non-metallic nature of the surface. The presence of the low concentrations of silicon, aluminium, iron, phosphorus, chlorine, potassium and calcium (oxides) in the mineralized surface can be accounted for by exchange and absorption with soluble ions from the environment over a long period of time (Scott 1985, Gettens 1969, Tylecote 1979, Werner 1972, Zwicker *et al.* 1979). Clearly, the tin-oxide rich surface layer must be semi-permeable to have allowed corrosion of the α to a depth tens of microns below the surface. This has also allowed the exchange of mineral elements, in the same way as it does with the silver coloured objects. However, once the tin oxide layer forms, the rate of permeation of corrosive elements and gases through the oxide layer is very low (Soto *et al.* 1983). Hence the often-stated 'corrosion resistance' of patinated Chinese mirrors (Soto *et al.* 1983, Yao Chuan and Wang Kuang 1987).

XRD analysis of the mineralized layer on the black Roman mirror shows diffuse patterns of tin oxide, SnO_2. Other naturally blackened high-tin bronzes show similar diffuse patterns and include SnO as well. Soto *et al.* (1983)

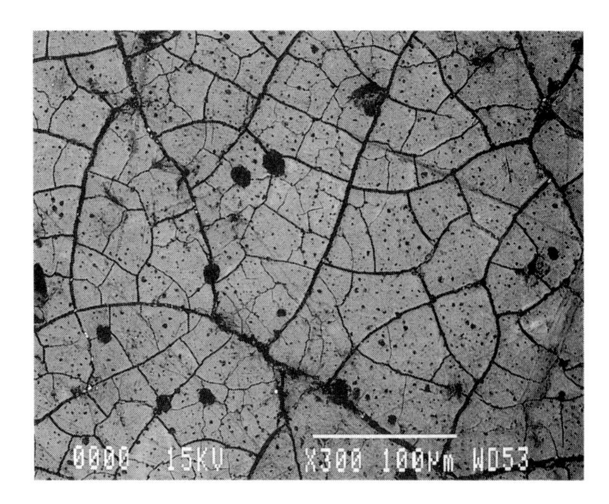

Figure 6.5. *Surface of the black Roman mirror showing the mineralized ghost pseudomorphic eutectoid microstructure, the lead globule porosity and surface crazing. SEM backscattered electron image. The white scale bar represents 100 microns.*

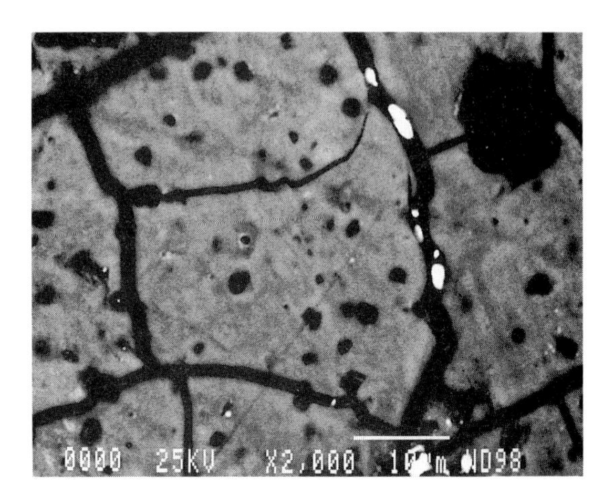

Figure 6.6. *Detail of Figure 6.5 showing ghost structures in the surface. The white scale bar represents 10 microns. Compare with surface in cross-section (Figure 6.3).*

notes that stannous and stannic oxides have similar stability in a wide range of commonly encountered environments and either or both may be formed over time.

Thus the black mineralized surface is now a mixture of partly crystalline tin oxides with some residual mineralized copper. Most of the original lead has been substituted within the oxide matrix, with additional diffusion into the mineralized surface of low concentrations of

Figure 6.7. *Ptolemaic high-tin bronze ring with black patina covering its surface, including the ancient fracture. British Museum GR. 1930, 7-15, 3. Height about 30 mm.*

various mineral elements commonly found in the soil.

This is the classic microstructure and composition of naturally blackened, patinated high-tin bronze (Figure 6.3). The only difference between the objects in this category is the degree or depth of degradation of the δ intermetallic compound, but the oxide layer can be quite thin and still have a black colour.

Not only are original surfaces of objects found to be black, ancient fracture surfaces are also black. The Ptolemaic ring shown in Figure 6.7 has an ancient catastrophic fracture that rendered the ring unusable. The ring has a uniform black patina that runs over the entire surface including the fracture. This ring therefore shows both in section and on the surface, all the microstructural attributes (Figures 6.8 and 6.9) and composition (Table 6.1) as the patinated Roman mirror illustrated in Figure 6.3. The ring would have been silver in colour when worn in antiquity and the surface has patinated black during burial.

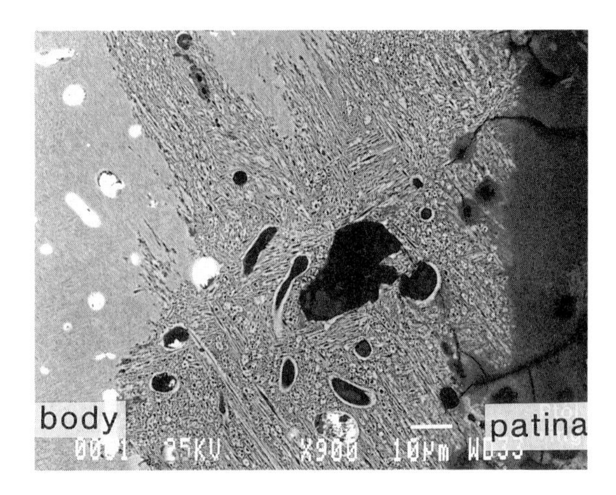

Figure 6.8. *Cross-section of the Ptolemaic ring showing the mineralized patina, on the right, and other similar features to the Roman mirror in Figure 6.3. SEM backscattered electron image. The white scale bar represents 10 microns.*

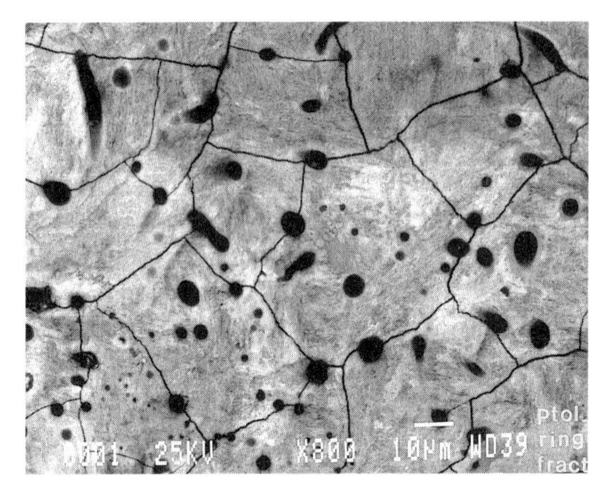

Figure 6.9. *Surface of Ptolemaic ring showing pseudomorphic, mineralized eutectoid microstructure. SEM. Width of field = 150 microns.*

Chinese mirrors: uncorroded body metal

Chinese mirrors are exemplified by the Han mirrors shown in Figure 6.10(a), (b) and Figure 6.12, and have a typical composition of 24% Sn, 71% Cu, 5% Pb. Barnard (1961) gives a comprehensive list of Chinese mirror analyses. The homogeneous uncorroded body microstructure (Figure 6.11) is again dominated by

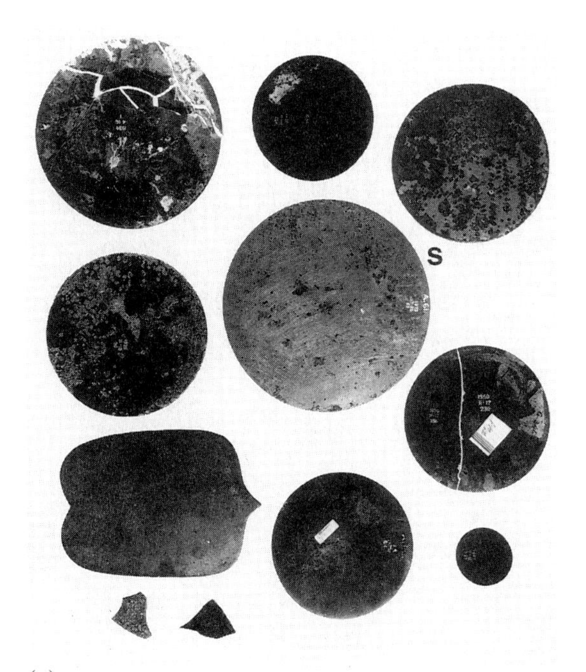

(a)

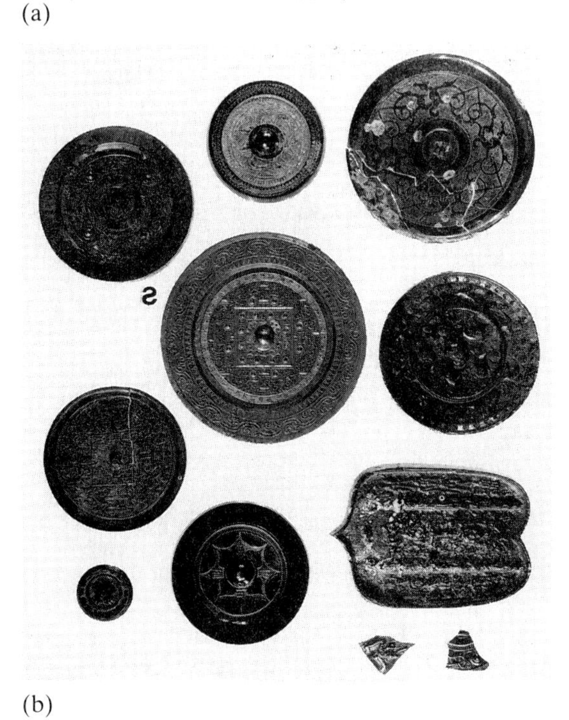

(b)

Figure 6.10 (a) and (b). *Selection of the Chinese mirrors showing both the polished sides (10a) and cast decorative sides (10b). Central mirror 157 mm diameter. s = two adjacent silver coloured mirrors, all other mirrors are black.*

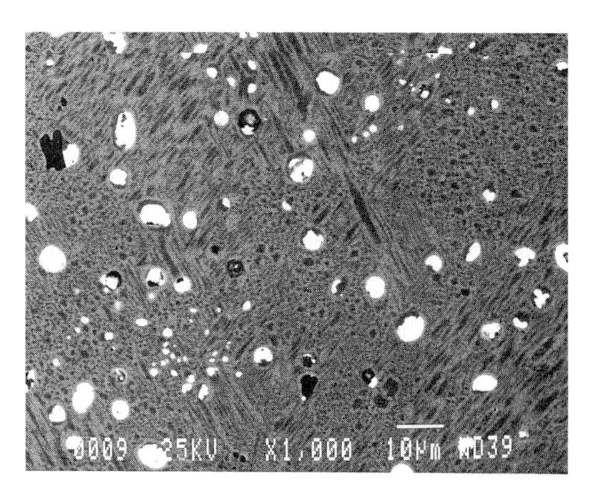

Figure 6.11. *Cross-section of a fragment of a black Chinese mirror showing the characteristic microstructure of the uncorroded, leaded, high-tin bronze body metal. SEM backscattered electron image. The white scale bar represents 10 microns.*

the eutectoid with fine lead globules. There is little evidence of casting defects, segregation, gas porosity or entrapped non-metallic inclusions, showing advanced practical skills in metal purification and casting techniques (Hanson and Pell-Walpole 1951, Barnard 1961). The cooling rate of the cast metal in the clay moulds (Barnard 1961) would have been slow enough to avoid stress-induced cracks being introduced into the variable thickness mirrors; particularly important with this brittle metal.

The structure is not quenched, it does not have the martensitic β structure common to the hot worked high-tin bronze bowls and gongs of the Far East mentioned earlier (see also Chapter 21 Figure 21.24).

Chinese mirrors: silver coloured surfaces

The surfaces of four fine examples of the silver coloured Chinese mirrors were examined, two are shown in Figure 6.12. They have similar eutectoid microstructures with lead porosity (Figure 6.13), and hence similar compositions, to the surfaces of the silver coloured Roman and Dark Age material. These cast mirrors are homogeneous throughout the body metal, and in common with Roman and Dark Age objects have also lost the α phase and lead through

Figure 6.12. *Chinese high-tin bronze mirrors with silver coloured surfaces. British Museum OA. 1955, 7-14, 2 (fluted edge mirror), and OA. 1963, 2-11, 2. Diameters are respectively 198 mm and 196 mm.*

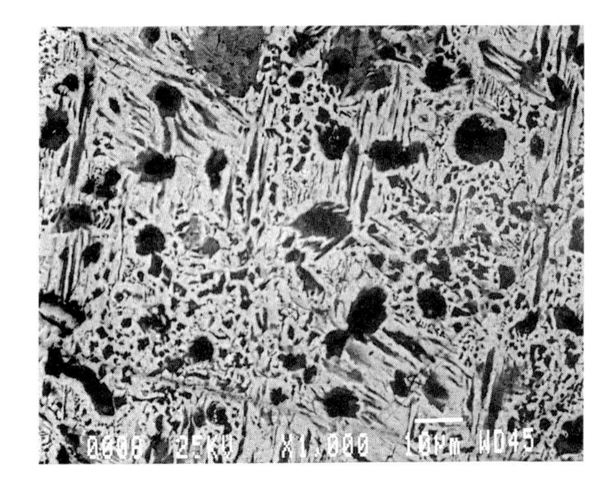

Figure 6.13. *Surface of silver coloured Chinese high-tin bronze mirror. SEM. Width of field = 115 microns.*

surface leaching, while the δ compound at the surface is uncorroded. The surfaces of the mirrors were not hot-tinned. There is no evidence of ϵ intermetallic compound layers, that are characteristic of hot tinning (see Chapter 21). The eutectoid structure is clear and unobscured.

Mirrors such as these have often been thought of as mercury tinned by polishing with *Xuan Xi* (see Chapter 5) to account for the silver coloured surface. The silver colour on these mirrors is from the original δ compound of the cast eutectoid structure. No mercury was detected by XRF analysis, but there is no doubt that the mirrors were mechanically polished to a fine surface and accurate convex curvature as part of the manufacturing process. If Chinese mirrors were polished with a compound containing mercury and tin, no evidence of this survives on these particular mirrors.

One form of corrosion that is found on some of these Chinese and Roman silver coloured high-tin bronzes is a localized eruption of the surface from below resulting in the raising and star-shaped cracking or terracing of the metal around the central corrosion pit (Chase and Franklin 1979). Similar physical effects are also found on some of the black Chinese mirrors. This type of corrosion would result from surface defects in the metal allowing direct access of moisture from burial to the core. Cuprite, malachite and redeposited copper have been seen associated with these corrosion pits, and their formation within the body metal may account for the physical eruption of the surface. McDonnell *et al.* (1992, in press) have examined this type of corrosion pit on Roman mirrors in the collections of the Museum Kam, Nijmegen (Lloyd-Morgan 1981).

Chinese mirrrors: black surfaces

A selection of nine black high-tin bronze Chinese mirrors of similar type to those published previously (Gettens 1969, Chase 1977, Chase and Franklin 1979, Barnard 1961) were chosen from the Han to Tang periods (Figures 6.10(a) and (b)). Some of the mirrors were truly black with shiny reflecting surfaces (although it is unknown whether they have been polished since discovery), others were black and pitted, and one was black/grey/green with pitting on the polished side while the reverse side was black with mercury gilding following the cast-in patterns. Two large mirror fragments were available for polishing taper sections and a third fragment, recently excavated at Changsha in China, was provided by Professor Zhu from Beijing for inclusion in this paper. The four silver coloured Chinese high-tin bronze mirrors, described above, were used as the unpatinated reference material for comparison with the microstructures on the surface of black patinated mirrors.

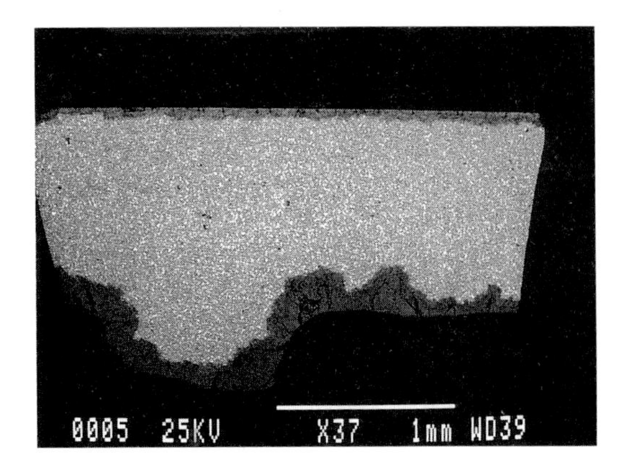

Figure 6.14. *Low magnification of mirror section showing thinner patina on polished side (similar to the Roman mirror in Figure 6.1). SEM. The white scale bar represents 1 mm.*

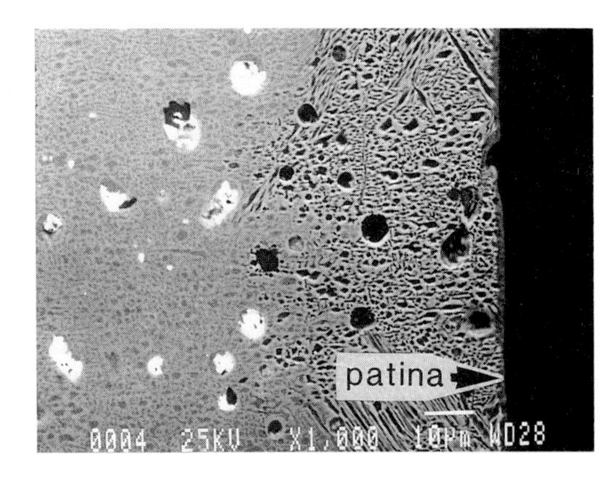

Figure 6.16. *Cross-section through a fragment of a black surfaced Chinese mirror showing α and lead corrosion and a thin surface patina. SEM. Width of field = 115 microns.*

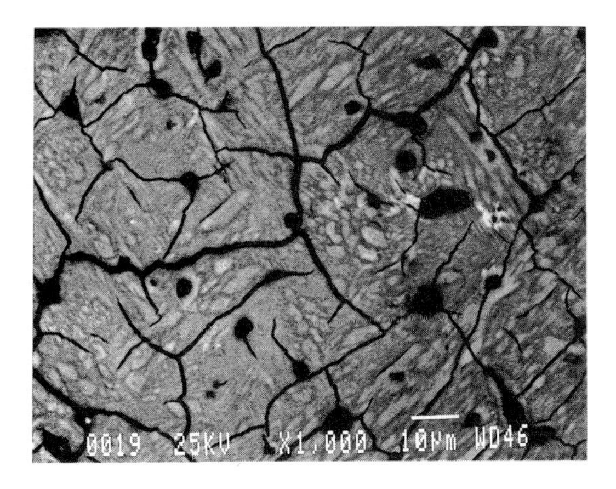

Figure 6.15. *Surface of a black Chinese mirror showing the ghost pseudomorphic mineralized eutectoid microstructure, the lead globule porosity and the crazing of the mineralized surface. SEM backscattered electron image. The white scale bar represents 10 microns.*

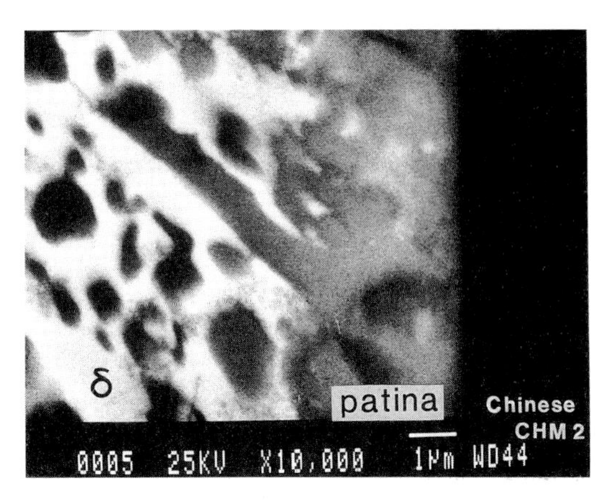

Figure 6.17. *Detail of the section through the black surface patina of the mirror in Figure 6.16, showing the pseudomorphic mineralized 2 micron thick layer. SEM, 10,000× magnification. The white scale bar represents 1 micron.*

The typical body microstructure of these black high-tin bronze mirrors is shown in Figure 6.11 from the polished section of a black surfaced mirror fragment of composition 25% tin, 5% lead and 70% copper. In cross-section, the body of this mirror is optically silver coloured due to the α+δ eutectoid microstructure which dominates the matrix. It also contains a fine globular distribution of lead, like the Roman mirror (Figure 6.2), the silver coloured Chinese mirrors and indeed is similar to all cast leaded high-tin bronze antiquities. The tin-copper ratio is approaching the eutectoid composition in this sample and consequently little excess α phase is seen. At low magnification the region of surface mineralization can be seen to be thinner on the polished reflecting side and thicker on the reverse as-cast side (Figure 6.14). This appears to be a common feature of both Chinese and Roman patinated mirrors.

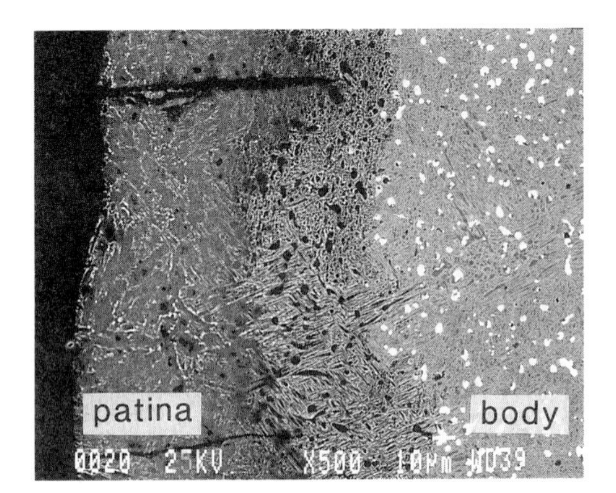

Figure 6.18. *Cross-section through the cast figure side of the mirror from Changsha showing the deeper patination zone on the mirror back. Width of field = 230 microns.*

The classic type of black Chinese mirror is illustrated by the mirror in Plate 6.1 which has a uniform black patina over both sides, with a few superficial deposits of green corrosion. The convex surface is highly polished and optically excellent, reflecting colour and flesh tones without distortion. Surface observation of this mirror in the SEM, using the compositional contrast mode, clearly shows the *ghost eutectoid structures* (Figure 6.15) with associated crazing of the mineralized surface. These ghost pseudomorphic structures are in the original polished surface. There is *no* observable layer applied to the mirror surface over the eutectoid structures. This fact is emphasized by the round porosity in the surface which is where the original lead globules resided before mineralization. All of the black mirrors which were examined show the same characteristics, some being more clear than others.

Examination of the cross-section through the surface of one of the broken mirror fragments, at moderate and high magnification (Figures 6.16 and 6.17), shows a similar microstructure to the black Roman mirror in Figure 6.3. They both have a mineralized surface, an 'altered zone' of α mineralization and uncorroded body metal (with a little less α phase in the Chinese mirror). The eutectoid is clearly mineralized to a depth of about 2 microns and unquestionably the original $\alpha+\delta$ eutectoid microstructure reaches the mirror surface in mineralized form. The mirror would have originally been silver in colour.

Similarly the surface and cross-section of the recently excavated mirror from Changsha, China, shows exactly the same characteristic microstructures. Figure 6.18 shows in cross-section the back, cast-figure side of the mirror, with deeper patina. Figures 6.19 and 6.20 show two magnifications of the section through the

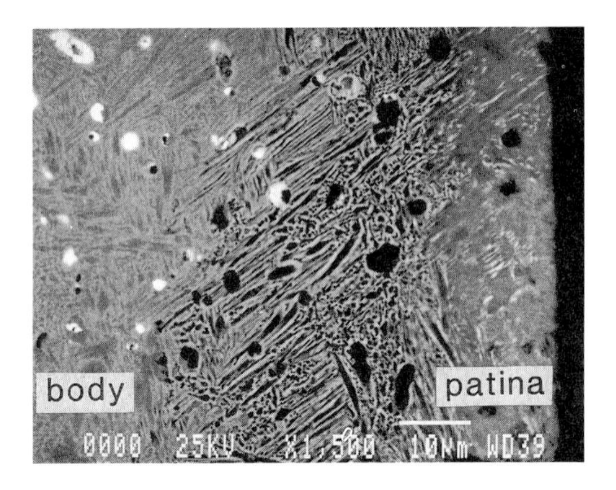

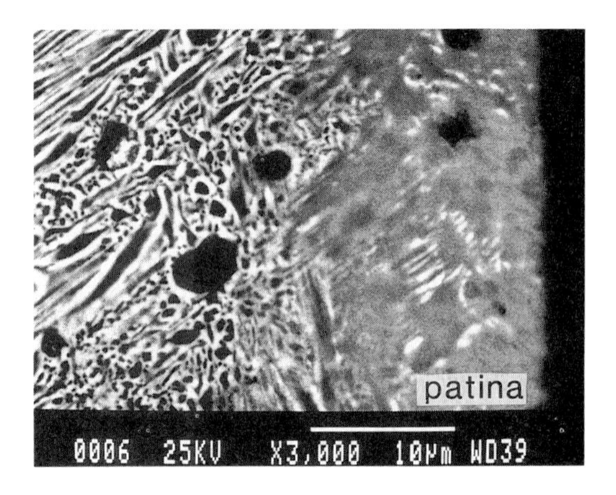

Figures 6.19 and 6.20. *Two magnifications of the cross-section through the polished surface of newly excavated black Chinese mirror from Changsha, showing the body metal, the corroded α 'altered zone', and the mineralized tin oxide-rich surface with the mineralized δ phase, on the right. Note the pseudomorphic eutectoid microstructure reaching the surface. SEM backscattered electron image. The white scale bars represent 10 microns.*

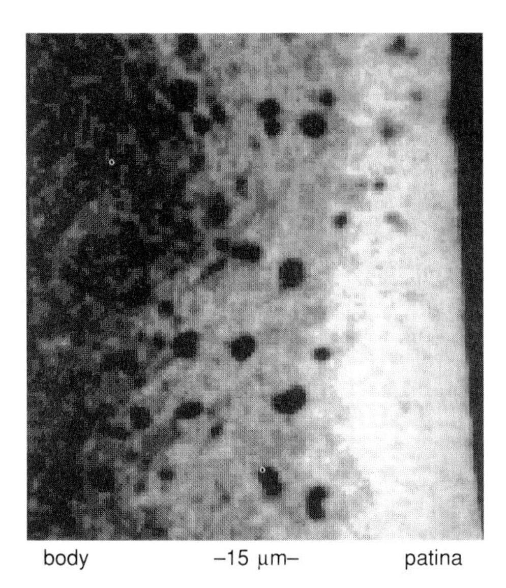

Figure 6.21. *Digital X-ray distribution map of tin in the cross-section of the black Chinese mirror from Changsha showing the relative enrichment of tin in the mineralized surface.*

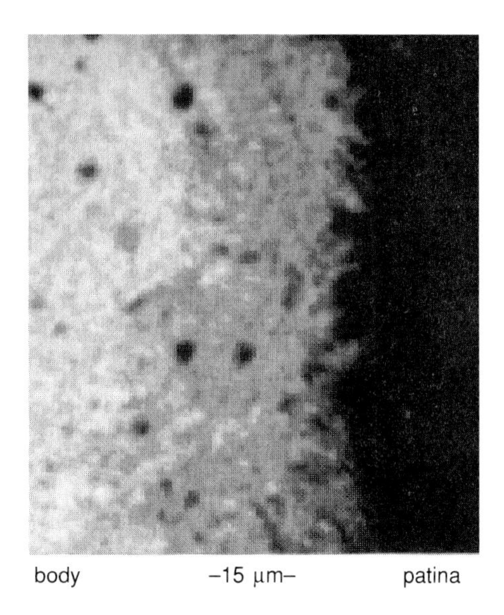

Figure 6.22. *Digital X-ray distribution map of copper in the cross-section of the black Chinese mirror from Changsha showing severe copper depletion to the environment.*

body and mineralized surface on the polished side of this mirror, which has a composition of 23% Sn, 72.5% Cu and 4.5% Pb. Digital X-ray mapping across the section shows, at the mineralized surface, the relative enrichment in tin (Figure 6.21), the depletion of copper (Figure 6.22), the distribution of mineralized lead (Figure 6.23) and the ingress of silicon from the environment (Figure 6.24). Again this is seen to be identical to the naturally patinated material.

Most of the black mirrors had small chips on their edges, reaching into the core metal, where the brittle mirrors had been knocked during use in antiquity, and these were also black. Similarly those mirrors which were either fragments or were broken into a few pieces in antiquity had black fracture surfaces that were identical to their black mirror surfaces. Clearly these damaged areas can only have been blackened by the passage of time during burial. Similarly, where deep corrosion pitting has occurred, the exposed sub-surface metal is black.

The black mirror fragment with fire gilded decoration has worn areas from ancient use, exposing the eutectoid microstructure of the bronze underlying the gilding. It is now black in colour and mineralized (Figure 6.25). The gilding adjacent to the now exposed bronze is seen to be diffused into the δ intermetallic compound of the body metal so the eutectoid microstructure is rich in gold (Figure 6.26). This illustrates that the fire gilding must have been carried out on the original silver coloured metallic mirror surface because no fusion could occur on a mineralized surface, and indeed a patinated surface would be severely damaged by heating during gilding. Hence the black patination can only have developed after the gilding had worn through. The reflecting side of this mirror has areas that are coloured black, green and grey. The fractured edges are black. These areas are also mineralized and have no added surface layer. It is concluded that this mirror was originally silver in colour with fire gilded decoration, the patination having occurred during burial.

The surface compositions of typical black Chinese mirrors are not significantly different to the other black high-tin bronzes, and are shown in Table 6.1. If the above micrographs and analyses of the black Chinese mirrors, the black Roman mirrors and the black Ptolemaic ring were unlabelled and compared it would

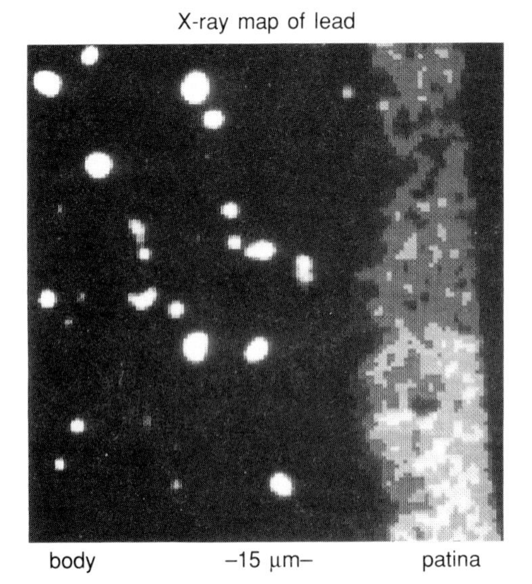

Figure 6.23. *Digital X-ray distribution map of lead in the cross-section of the black Chinese mirror from Changsha showing how the mineralized lead is retained in the patina. The black scale bar represents 15 microns.*

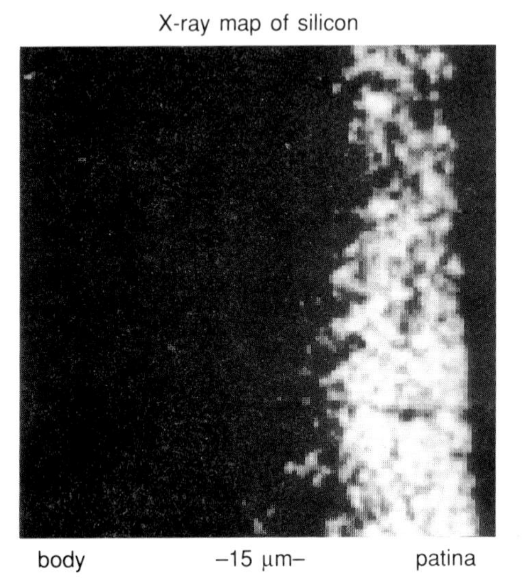

Figure 6.24. *Digital X-ray map of silicon in the cross-section of the black Chinese mirror from Changsha showing the diffusion of silicon into the mineralized surface from the environment. The black scale bar represents 15 microns.*

not be possible to distinguish between them with any certainty. There are no significant differences in microstructure or composition.

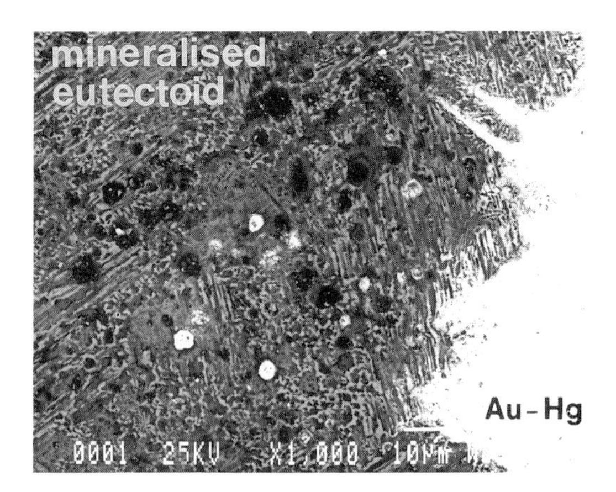

Figure 6.25. *Surface of the mercury gilded high-tin bronze black Chinese mirror showing areas where the gilding has worn away to expose the bronze that has subsequently patinated to a black surface. SEM backscattered electron image. Width of field = 115 microns.*

Figure 6.26. *Detail of Figure 6.25 showing the gold diffusion bond with the δ intermetallic compound of the eutectoid. SEM. The white scale bar represents 10 microns.*

Mercury tinning: formation of mercury/tin amalgam

The ancient texts mention mercury in the surface treatment of Chinese mirrors during the final stages of manufacture (Needham 1962, 1974) but the residual evidence is missing on all of the mirrors examined here and many elsewhere (Fabrizi 1987, Chen Yu-Yun *et al.* 1987).

The first question to be asked is whether 'quicksilver colour', etc. in ancient texts (Yetts 1931) has one or more meanings. Does it mean that a mirror simply has the colour of mercury (Collins 1934), or does it refer to true use of mercury amalgam tinning of high-tin bronze, low-tin bronze or both, either as a direct process or as a combined process by the application of the mixture *Xuan Xi* for the final polishing of mirrors during manufacture (see Chapter 5). Does mercury have any significance in the formation of the black patina? The simplest way to answer the problem is to examine experimentally what structures form on a mercury tinned bronze, consider what residual structures might be found in an ancient patinated mirror surface and then compare with what is found on the ancient mirrors.

The mercury/tin phase diagram (Figure 6.27) shows that at room temperature mercury and tin are almost mutually insoluble. When a mixture of 50/50wt% mercury and tin is shaken together there is no obvious sign of amalgam formation, unlike mercury and gold which react instantly. However, by heating the mixture until the tin melts (232°C) a homogeneous liquid forms, and on cooling produces a friable pellet of interlocking crystals of γ compound, $HgSn_6$ (*c.* 19% Hg) which acts like a sponge holding the excess mercury (Figure 6.28).

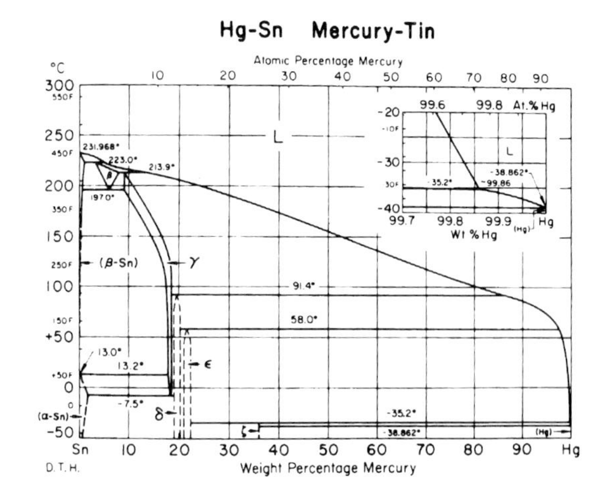

Figure 6.27. *Mercury/tin phase diagram showing the γ compound, $HgSn_6$ and the low melting point.*

Figure 6.28. *Experimental mercury/tin amalgam showing cube shaped crystals of $HgSn_6$ in excess mercury. The white scale bar represents 10 microns.*

Mercury (fire-) tinning

During mercury (fire-) tinning, the γ phase of the amalgam decomposes at 214°C and mercury is therefore easily driven off by heating above its boiling point of 357°C, releasing tin onto the bronze surface. At this temperature the molten tin fuses with the bronze surface in exactly the same manner as conventional tinning.

The microstructures associated with fire-tinning are shown in Figure 6.29, which is the cross-section through an experimentally fire-tinned copper sample in which a thick amalgam layer was put on the surface and heated to drive off the mercury. A thick layer of ϵ intermetallic compound is formed by diffusion of tin, which was released from the amalgam, into the substrate. Between this and the copper base metal is a layer of eutectoid which signifies that, in this case, the heating reached above 520° C. while driving off the mercury. At the very surface is a thin layer of η intermetallic compound. Heating to 357°C would have produced the same structure but without the eutectoid layer. Final polishing would produce a silver-white coloured ϵ compound surface. With a high-tin bronze substrate the ϵ and η intermetallic compound layers will also form but would be thinner

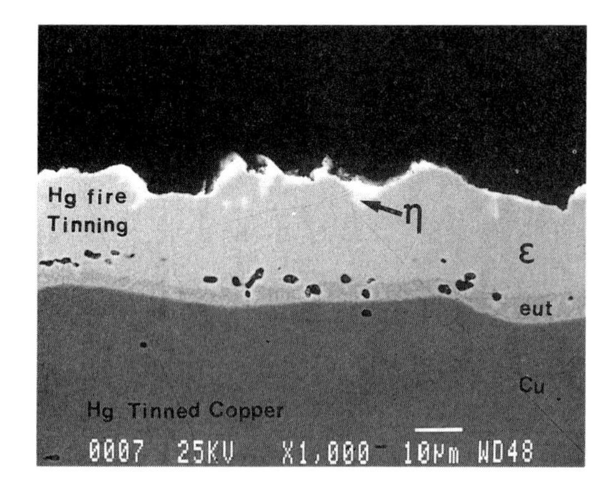

Figure 6.29. *Experimental mercury fire-tinning on copper showing the same sequence of tin-copper intermetallic compounds as overheated, conventional wipe-tinning. The white scale bar represents 10 microns.*

because less copper is available for interdiffusion. XRF analysis found no mercury in the newly prepared layers of the sample (detection limit 0.05%). Fire-gilding often shows significant residual mercury (Northover *et al.* in press). The reason that mercury is retained in gilding is because mercury and gold form the intermetallic compound Au_3Hg, which is stable up to 420°C. Above this temperature mercury still forms a solid solution with gold to 1000°C. even though mercury boils at 357°C. It is therefore difficult to drive off all of the mercury. By comparison both tin and mercury are fluid at above 232°C. and it is far easier to drive off the mercury from this mixture.

However, even if there is no detectable mercury on an ancient mirror that had been mercury (fire-) tinned one would expect to find some evidence of residual or mineralized ϵ, Cu_3Sn, compound layer which would cover the eutectoid structure of the underlying bronze body metal.

Such a layer was not found on any of the high-tin bronze black (or silver coloured) surfaced mirrors examined in this paper; the pseudomorphic eutectoid structures reach the surface and there were no resolvable applied layers. Mercury was not found, even in trace quantities, on those black mirrors subjected to spectrographic analysis (detection limit 100 ppm), although as explained above this might not be surprising. However, microstructure is the best guide to whether the original surface was mercury (fire-) tinned or not.

The question remains as to why it would be necessary to use mercury amalgam tinning on ancient mirrors when the straight application of tin by conventional melting onto the surface is simpler. Indeed, was tinning really necessary, because by definition the high-tin bronze already 'reflects without tinning or silvering'? The answer probably lies in the use of mercury and tin along with other ingredients as a polishing compound *Xuan Xi* described by Professor Zhu, Chapter 5, rather than its use solely as a tinning process. Thus, if the mirrors were not heated after application of the mixture, no tin-copper intermetallic compounds would be formed. This could explain the absence of any tangible evidence for 'mercury tinning' in the mirrors examined here. Hence the following experiments with cold amalgam tinning.

Cold mercury amalgam tinning

Experiments were carried out to test the 'cold' mercury amalgam tinning hypothesis. A previously polished 25% high-tin bronze (hypo-eutectoid and therefore a shade coppery in colour) was rubbed with a small amount of crushed amalgam pellet on a cloth. After 2 minutes polishing, a noticeably more silver coloured surface had appeared in the amalgam treated area compared with the original polished surface. Polishing a second sample for 8 minutes produced no further optical improvements to the silver coloured surface. No heating was necessary to deposit the silvery layer.

Under the optical microscope the silver coloured coating obscured all but a faint impression of the underlying eutectoid microstructure. The layer was continuous and extremely thin.

SEM examination of the coated 'mirror' surface confirmed the thinness of the layer by being visible only at low accelerating voltage (<2 kV) with corresponding low electron penetration (Figure 6.30). At 25 kV the underlying eutectoid microstructure appeared clearly as if untinned, but with a hint of surface streaking seen particularly over the α phase

Figure 6.30. *Cold mercury/tin amalgam tinning/polishing. SEM 1.0 kV showing the thin surface coating, white, (and porosity from the surface structure). Width of field = 4 mm.*

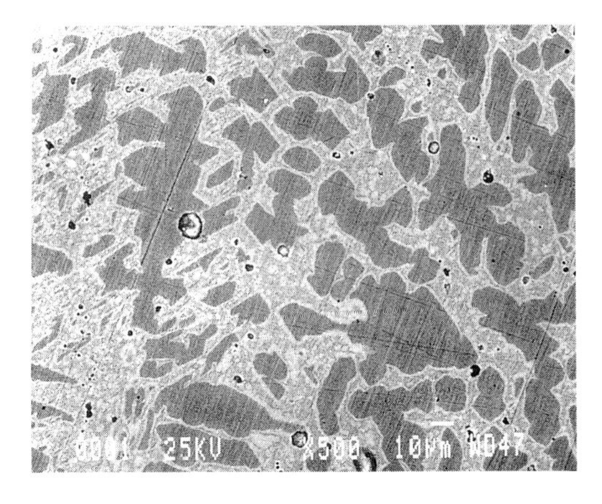

Figure 6.31. *Same sample as in Figure 6.30 but at high kV in the SEM allowing beam penetration of the very thin tinning and thus showing the underlying eutectoid. Width of field = 230 microns.*

regions (Figure 6.31). The thickness of the layer is estimated to be in the range of some tens to perhaps a hundred nanometres.

The composition of the layer is difficult to quantify by EDX analysis because it is so thin. However, it appears to be predominantly tin with some mercury. The formation of the layer is thought to result from friction during rubbing causing local point of contact heat and breakdown of the γ compound into its constituent parts, allowing the tin to bond to the surface.

The Chinese mirrors examined in this paper do not show residual evidence of cold mercury tinning; the eutectoid structures of the cast bronze are clearly seen optically and no mercury was detected by XRF analysis. If they were polished in this way in antiquity, there is no evidence of it now. Professor Zhu (private communication) notes that when an experimentally prepared sample polished with *Xuan Xi*, containing mercury and tin, was left for a few years, evaporation had left no detectable mercury (see Chapter 5).

Does mercury play a direct part in the black oxidation process? Comparison with the Roman and other naturally blackened tin oxide rich high-tin bronzes shows that mercury itself is not important in the process and also it is not necessary to have added tin, by way of the amalgam, to produce the tin oxide enriched surface. The extra tin provided by the extremely thin, cold mercury amalgam tinning is insignificant compared with the microns of thickness of the black tin oxide rich mirror surfaces derived from mineralization of the δ bronze compound.

Conclusions

This study presents conclusive evidence that the black Chinese mirror surfaces are not analytically or structurally different from the black surfaces either on high-tin bronzes of other cultures or on fracture surfaces, both of which are unarguably the result of corrosion over long periods of burial.

It is clear why the patina of Chinese mirrors is corrosion resistant: it *is* the stable product of a corrosion mechanism, passivation by tin oxide(s). The mechanisms by which the δ compound mineralizes in the black mirrors when it does not on the silver coloured mirrors is not yet fully explained. The answer must lie in the burial environment, but unfortunately most of the Chinese mirrors in collections in the West are from unknown archaeological contexts. However, Barnard (1961, Chapter 7) gives a useful table of loess soil analyses in his chapter on patinas. Collins (1934) describes one Han dynasty mirror that was half-covered in loess from burial. The uncoated half was silver in colour while removal of the loess from the other half revealed a 'blue-black' patina which suggests the black patina originated in the burial conditions (see also Chapter 5).

Can this mineralization be accelerated artificially? A recent paper by Sun *et al.* (1992) has made an important contribution to the discussion of patination mechanisms with experiments using humic acids to patinate high-tin bronzes. They found that the experimental patinas are 'similar to that of *Hei-qui-gu* ancient mirror surface both in visual appearance and in composition and structure'. They conclude that 'the formation of the tin-rich shiny black surface was the product of soil corrosion of the mirrors'. Further experimental work, with reference to modern surface treatments (Schwartz 1982), may clarify whether accelerated patination is a possibility. Additionally, the examination of numbers of ancient mirrors from known burial conditions is essential to complete the record. It is important to repeat that so far there is no direct evidence from the known ancient texts that the blackening was a part of the manufacturing process.

The pseudomorphic ghost structures at the very surface of the mineralization prove that no significant layers were applied to the original bronze surface. This eliminates the addition of silicacious polishes, fusion of oxide layers and hot mercury tinning (fire tinning). Polishing with *Xuan Xi* may leave no traces detectable by present methods: none were found in this study. However, even if it was used it cannot account for the thickness of the black mineralized layer on the mirrors or the amount of tin oxide present. This arises solely from the natural corrosion of the δ intermetallic compound of the surface metal.

The evidence from this investigation clearly shows that the black surfaces found on high-tin bronze mirrors can be accounted for by natural patination mechanisms of the original silver coloured mirrors which were buried to provide 'light' in the afterworld where 'the visions of imortality meet with the scenes of daily life in the tombs of the Han dynasty' (Rawson, 1980).

Acknowledgements

Paul Craddock of The British Museum Research Laboratory is thanked for his important contributions to this paper. Jessica Rawson, Keeper of Oriental Antiquities in the British Museum, is thanked for providing the selection of key Chinese mirrors used for this investigation. Dafydd Kidd, Assistant Keeper of medieval collections in the Department of Medieval and Later Antiquities, is thanked for providing Dark Age material. Professor Zhu from the Institute for Non-Ferrous Metals, Beijing, is thanked for his discussions on the subject and for providing the freshly excavated Chinese mirror sample for inclusion in this paper. Tom Chase of the Smithsonian Institution and Professor Michael Notis of Lehigh University are thanked for their contributions to the discussions on related Chinese material. Jimping Zhang from the Museum of Chinese Revolution, Beijing is thanked for providing translations of Chinese texts.

References

Barnard, N. (1961) *Bronze casting and bronze alloys in ancient China. Monumenta Serica Monograph XIV*, published jointly by the Australian National University and Monumenta Serica, Canberra

Caley, E.R., In Soon Moon Chang, Nilufer Paranyi Woods (1979) Gravimetric and spectrographic analysis of some ancient Chinese copper alloys. *Ars Orientalis*, **11**, 183–193

Chase, W.T. (1977) What is the smooth lustrous black surface on ancient bronze mirrors? In *Corrosion and Metal Artefacts*, a dialogue between Conservators, Archaeologists and Corrosion Scientists, National Bureau of Standards Special Publication No. 479, pp. 191–203, Washington.

Chase, W.T., and Franklin, U.M. (1979) Early Chinese black mirrors and pattern-etched weapons. *Ars Orientalis*, **11**, 215–258

Chen Yu-Yun, Huang Yun-Lan, Yang Yong-Ning and Chen Hao (1987) *The study of ancient bronze mirror* 'Hei-Qi-Gu' *and its imitation*, Research Laboratory of the University of Science and Technology, Beijing

Collins, W.F. (1934) The mirror-black and 'quicksilver' patinas of certain Chinese bronzes. *Journal of the Royal Anthropological Institute*, **LXIV**, 69–79

Craddock, P.T. (1988) Copper alloys of the Hellenistic and Roman World. In *Aspects of Ancient Mining and Metallurgy, ACTA of The British School at Athens Centenary Conference*, Bangor 1986 (ed. J. Ellis-Jones) pp. 55–65

Craddock, P.T., Hook D.R., Meeks, N.D. (1989) Appendix B. Technical report on Roman Mirrors. In *Verulamium: The King Harry Lane Site* (eds I.M. Stead and V. Rigby) English Heritage report 12, pp. 271–272

Fabrizi, M. (1988) *Later Chinese Copper Alloys*. An internal report on the objects in the collection of the Victoria and Albert Museum

Gettens, R.J. (1969) *The Freer Chinese Bronzes in the Smithsonian Institution, Freer Gallery of Art, Oriental Studies, No.7, Volume II*, Technical Studies, Washington, pp. 121–139 and 171–195

Goodway, M. and Conklin, H.C. (1987) Quenched high-tin bronzes from the Philippines. *Archaeomaterials*, **2**(1), pp. 1–27

Han Rubin, Ma Zhaozeng, Wang Zengjun, T,Ko (1983) Studies of the black passive oxide film on bronze arrowheads unearthed with the terra-cotta warriors near the Qiu Tomb (210 BC) at Lintong, Xian. *Studies in the History of Natural Sciences*, **2**(4), 295–302

Han Rubin and Ma Zhaozeng (1983) Studying oxide coating of unearthed bronze from Qinshihuang Mausoleum site No.26

Hanson, D. and Pell-Walpole, W.T. (1951) *Chill-cast Tin Bronzes*, Edward Arnold, London

Hedges, E.S. (1964) *Tin in Social and Economic History*, Edward Arnold, London

He Tankun (1985) A scientific analysis of the transparent coating on the surface of ancient mirrors. *Studies in the History of Natural Sciences*, **4**(3), 251–257. (in Chinese)

Kerr, R. (1990) *Later Chinese Bronzes*, Victoria and Albert Museum Far Eastern Series, Oxford University Press with Bamboo Publishing, Victoria and Albert Museum, London

Li, Shuzhen (1984) Composition of the coating of the Chinese antique bronze mirror and sword. *Chengdu Keji Daxue Xuebao*, **3**, 153–158 (in Chinese)

Lloyd-Morgan, G. (1981) *Description of the Collections in the Rijksmuseum, G.M. Kam at Nijmegen, IX The Mirrors*, Ministry of Culture, Recreation and Social Welfare, Amsterdam

McDonnell, R.D, Meijers, H.J.M. (in press) A study of the composition and microstructure of six fragments of Roman mirrors from Nijmegen, The Netherlands. In *Proceedings of the 12th International Congress on Ancient Bronze, Nijmegen, 1–4 June 1992*, in conjunction with the Provincial Museum G.M. Kam and the Katholieke Universiteit Nijmegen.

Ma Zhaozeng and Han Rubin (1986) Gu tong-qi biao-mian hua-xue chu-li di yan-jiu. The research on the chemical treatment of the surface of ancient bronze vessels. *Hua-xue tongbao*, **8**, 59–61 (in Chinese)

Masaaki Sawada (1979) Non-destructive X-ray fluorescence analysis of Ancient Bronze Mirrors excavated in Japan. *Ars Orientalis*, **11**, 195–213

Meeks, N.D. (1986) Tin-rich surfaces on bronze – some experimental and archaeological considerations. *Archaeometry*, **28**(2), 133–162

Meeks, N.D. (1988a) A technical study of Roman bronze mirrors. In *Aspects of Ancient Mining and Metallurgy, ACTA of The British School at Athens Centenary Conference, Bangor 1986* (ed. J. Ellis-Jones) pp. 66–79

Meeks, N.D. (1988b) Surface studies of Roman bronze mirrors, comparative high-tin bronze Dark Age material and black Chinese mirrors. In *Proceedings of the 26th International Archaeometry Symposium, University of Toronto, Canada 1988* (eds R.M. Farquhar, R.G.V. Hancock, L.A. Pavlish) pp. 124-127

Meeks, N.D. (in press) A technical study of Roman bronze mirrors. In *Acta of the 12th International Congress on Ancient Bronzes*, Nijmegen 1–4 June 1992 (ed. A. Gerhartl-Witteveen) in conjunction with the Provincial Museum, G.M. Kam and the Katholieke Universiteit, Nijmegen

Needham, J. (1962) *Science and Civilisation in China, Volume 4, Part 1: Physics*, Cambridge University Press

Needham, J. (1974) *Science and Civilisation in China, Volume 5, part 2*, p. 249, Figure 1326, Cambridge University Press

Northover, P., Salter, C.J., Chase, W.T., and Bunker, E.C. (in press) Early mercury gilding and silvering in China. *Proceedings of Symposium International Outils et Ateliers D'orferes, Saint-Germain-en-Laye, Paris*, January 1991

Rackham, H. (1952) (translator). *Pliny Historia Naturalis* Vol. IX, Book XXXIV, Ch XLVIII, pp. 243–245, William Heinemann, London

Rawson, J. (1980) *Ancient China, Art and Archaeology*, British Museum Publications Ltd, London

Schwartz, W. (1982) Blackening of metals. *Plating and surface finishing*, **69**(6), 26–29

Scott, D. (1985) Periodic corrosion phenomena in bronze antiquities. *Studies in Conservation*, **30**(2), 49–57

Seeley, N.J. and Rajpitak, W. (1979) The bowls from Ban Don Ta Phet, Thailand: an enigma of prehistoric metallurgy. *World Archaeology*, **11**(1), 26–32

Soto, L., Franey, J.P., Graedel, T.E., and Kammlott, G.W. (1983) On the corrosion resistance of certain ancient Chinese bronze artefacts. *Corrosion Science*, **23**(3), 241–250

Sudhir, T.S. (1991) Metallic Marvels. In the *Indian Sunday Times*, an article on the metal mirror craft from Kerla at an exhibition at Pragati Maidan, Delhi, (about) September 1991

Sun Shuyun, Ma Zhaogeng, Jin Lianji, Han Rubin and T.Ko (forthcoming) The formation of Black patina on Bronze Mirrors. *Proceedings of the Archaeometry Conference*, Los Angeles, 1992

Tan Derui (after 1985) Study of coating technique of Eastern Han bronze mirror. *Shanghai Museum periodical*, **4**, 405–427 (in Chinese)

Tylecote, R.F. (1979) The effect of soil conditions on the long term corrosion of tin-bronze and copper. *Journal of Archaeological Science*, **6**, 345–368

Wang Changsui, Cheng Tingzhu, Zhang Jinjguo and Huang Yanlan (1989) X-ray quantitative analysis for the surface layer of ancient bronze mirrors. *Wenwubaohu Yu Kaogen kexue*, **1**(2), 28–31 (Chinese with English summary)

Watson, W. (1962) *Ancient Chinese bronzes*, pp. 81–103, plates 90–104, Faber and Faber, London

Werner, O. (1972) *Spektralanalytische und Metallurgische Untersuchungen an Indischen Bronzen*, E.J. Brill, Leiden

Yao Chuan and Wang Kuang (1987) A study of the black corrosion- resistant surface layer of ancient Chinese bronze mirrors and its formation. *Corrosion Australasia*, **12**(5), 5–11

Yetts, W.P. (1931) Problems of Chinese Bronzes. *Journal of the Royal Central Asian Society*, **XVIII, Pt.3**, 1–4

Zhu, Shoukang (1986) Ancient metallurgy of non-ferrous metals in China. In *Proceedings of Symposium on the Early Metallurgy in Japan and the Surrounding Area,* October 1986, special issue Vol. 11, Bulletin of the Metals Museum, pp.1–13

Zwicker, U., Nigge, K., and Urbon, B. (1979) Distribution of metallic elements in patina layers. *Mikrochem Acta, Supplement*, **8**, 393–419

7

Japanese traditional alloys

Ryu Murakami

Abstract

Colour is an important feature of Japanese traditional metal art and craft. *Shakudo* is a copper-gold alloy and has a purplish-black surface, *shibuichi* is a copper-silver alloy and has a greyish-brown surface. These colours are achieved by *nikomi-chakushoku*, which is a traditional colouring treatment in Japan. In this study each surface was characterized by many scientific methods and the colouring mechanism was considered.

Introduction

Colour is one of the most important attributes of Japanese traditional artefacts and they are famous for their sophisticated techniques and range of colour. The main representatives of Japanese traditional alloys are *shakudo* and *shibuichi*. These two alloys are widely known, because visiting researchers introduced them to the western world in the late 19th century (Pumpelly 1866, Gowland 1910 and Roberts-Austin 1893). They were used mainly for sword ornaments, for example *tsuba*. Since the wearing of swords was prohibited in 1876 after the Meiji Restoration, these alloys have been applied to a wider range of metal artefacts (Murakami 1989a).

Japanese traditional alloys are characterized by a special process, by which they are coloured without any pigment, paint or metal plating. The colour obtained has a close relationship to both the metallographic structure and the elements in each alloy matrix so these colouring processes could be defined as matrix colouring. *Nikomi-chakushoku* is the matrix colouring process which turns *shakudo's* surface purplish-black and *shibuichi's* greyish-brown (Murakami 1989b). *Shakudo*, in particular, has been much studied (Savage and Smith 1979, Notis 1979 and Oguchi 1983). Other alloys with a black surface from outside Japan have been reported, e.g. Indian bidri (La Niece and Martin 1987, and Stronge this volume, Chapter 11), Chinese high tin bronze mirrors (Chase and Franklin 1979, Meeks this volume Chapter 6 and Zhu, Chapter 5) and Corinthian bronze (Craddock 1982 and this volume Chapter 9). Although *shakudo* is well known, the details of the colouring mechanism have not been perfectly understood (Miyazawa 1917, Uno and Aachen 1929 and Oguchi 1983).

In this study, the coloured layer of *shibuichi* and *shakudo* were characterized by various scientific techniques and the colouring mechanism of each alloy was considered. The knowledge of Japanese traditional alloys contributes to the thinking on the history of colouring metals.

Nikomi-chakushoku: traditional colouring process

The colouring process, *nikomi-chakushoku*, has a history of at least 600 years. Some recipes have been passed down by tradition like the recipe below, which was developed in Tokyo National University of Fine Arts & Music and can be considered as a modern standard (Murakami and Niiyama 1985).

Before colouring, all alloys have to be polished to a mirror finish and well-degreased. Traditionally, charcoal was used for the final polishing and Japanese radish for degreasing. The metal is boiled in an aqueous colouring solution containing small amounts of verdigris, copper sulphate and alum. The colouring solution for *shakudo*, for example, contains 1.9 g natural verdigris, 1.2 g copper sulphate and 0.2 g alum in one litre of water. The pH of the colouring solution is about 5.6, the standard treatment time is 30 minutes.

The characteristic colour of each alloy does not appear until it is boiled in the colouring solution. Both *shibuichi* and *shakudo* are copper alloys; *shibuichi* is a copper-silver alloy and *shakudo* is a copper-gold alloy. Before the colouring process they look very like pure copper, but on boiling in the colouring solution the surface of *shibuichi* becomes greyish or whitish-brown and the surface of *shakudo* becomes purplish-black. Pure copper and copper-zinc are also coloured by *nikomi-chakushoku*; pure copper becomes reddish-brown and an alloy of copper-zinc becomes yellowish-brown on the surface. The alloys coloured by *nikomi-chakushoku* are sometimes called *irogane*, which means 'colour alloy'. Plate 7.1 shows a cigarette case, the work of Shoumin Unno (1844–1915), who was typical of the metalworkers in the 19th century (the Meiji period), famous for superb workmanship and frequent use of *irogane*. The black horse is made of *shakudo*, the white horse is made of *shibuichi* and the reddish-brown background is pure copper, all colours obtained by *nikomi-chakushoku*.

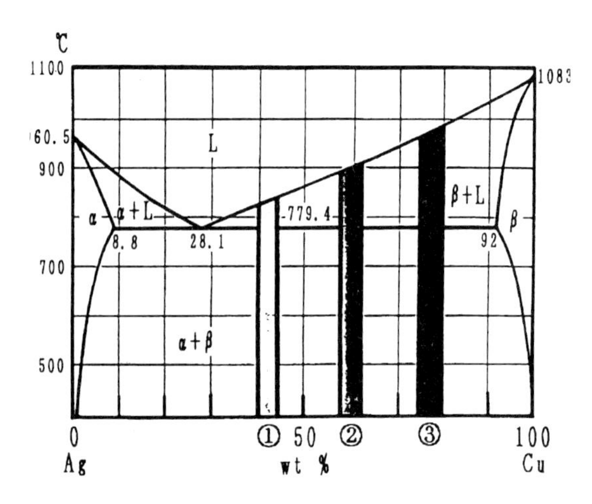

Figure 7.1. *Various* Shibuichis *illustrated in the copper-silver phase diagram. (1)* Shiro-shibuichi *(white); (2)* jou-shibuichi *(upper grade); (3)* nami-shibuichi *(ordinary).*

Surface characterization of Japanese traditional alloys

1 Shibuichi

Shibuichi, 四分一 in Japanese means a quarter. Indeed, *shibuichi* is basically a copper alloy containing 25% silver; but there are some variations in silver content, as indicated in Figure 7.1. Copper–silver is a typical eutectic system and all *shibuichis* have a hypereutectic composition. The colour of *shibuichi* changes from brownish-grey to whitish-grey with increased silver. In *shiro-shibuichi shiro* means white.

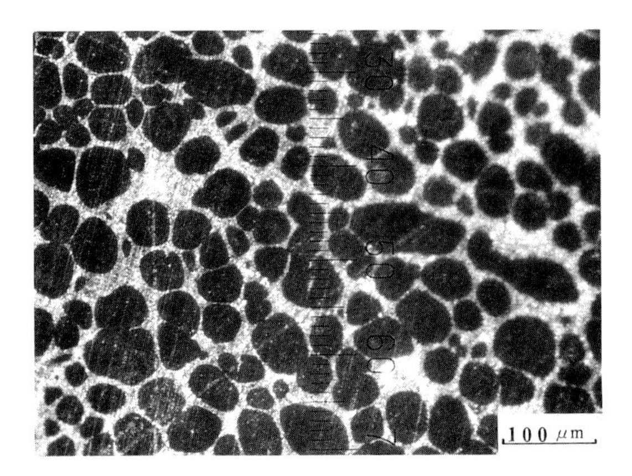

Figure 7.2. *Microphotograph of experimental* shibuichi *(annealed).*

Plate 1.1. *Section through a typical naturally corroded bronze. Initially the copper oxidizes to form the red cuprite layer. This in turn is converted to the familiar green patina. The nature of this depends on the environment.*

Plate 1.2. *Small Chinese bronze sleeve weight, British Museum OA 1894.1-8.8, made in the Song Period, but imitating the style of a thousand years before and with an 'original' 12th century AD fake patina. The patches of green patina are ground-up malachite stuck to the surface.*

Plate 1.3. *Japanese* tsuba, *British Museum OA 3432. The superb velvet black patina was produced by the chemical patination of the very carefully prepared surface of the* shakudo *alloy, which contains copper with small quantities of gold and sometimes a little silver and arsenic.*

Plate 6.1. *Black Han dynasty Chinese high-tin bronze mirror showing the reflecting properties of the black patinated surface. British Museum OA 1936, 11-18, 264. Diameter 91mm.*

Plate 7.1. *Cigarette case by Shoumin Unno. The black horse is made of* shakudo, *the white horse is made of* shibuichi *and the reddish-brown background is pure copper. (Museum of Tokyo National University of Fine Arts & Music).*

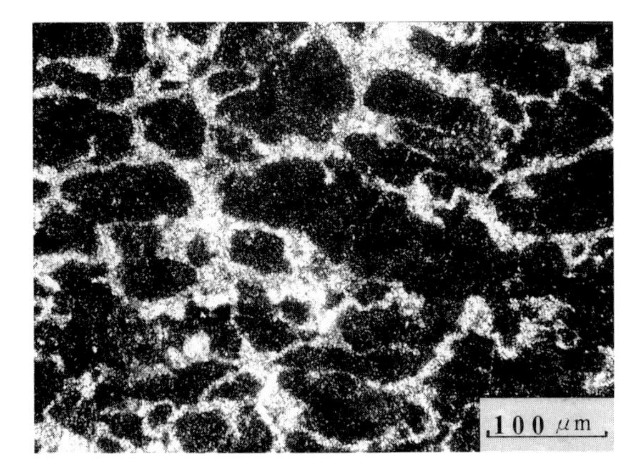

Figure 7.3. *'Kingfisher' by Shoumin Unno. (Museum of Tokyo National University of Fine Arts & Music.) (a) A microphotograph of* shibuichi *used for the background of 'kingfisher'. (b) A general view (a plate chiselled and inlayed).*

To examine the coloured layer scientifically, *shibuichi* containing 20% silver was prepared experimentally. Figure 7.2 is a micrograph of a sample of experimental *shibuichi*, annealed before colouring to emphasize the typical structure. The white network is a silver-rich eutectic region; the dark part is a copper-rich region. Figure 7.3(a) is a micrograph of *shibuichi* used in a piece by Shoumin Unno, for the background to the kingfisher, shown in Figure 7.3(b) (Murakami and Niiyama 1985). This too is a typical eutectic structure, with the silver-rich eutectic network surrounding the copper-rich region. This *shibuichi* plate was slightly cold worked.

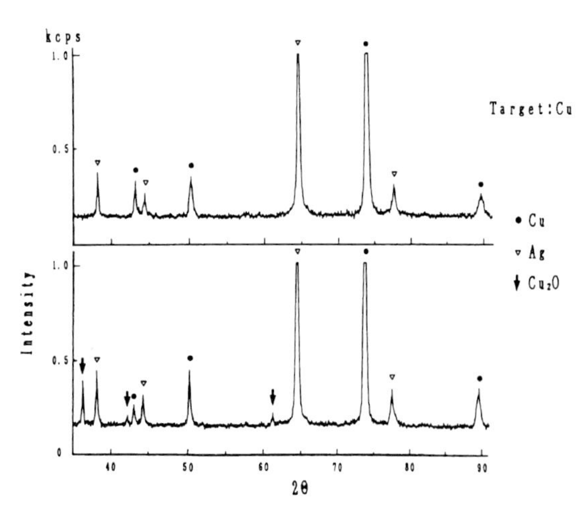

Figure 7.4. *X-ray diffraction scan of experimental* shibuichi. *Upper: before colouring. Lower: after colouring*

Figure 7.4 gives the results of X-ray diffraction analyses of experimental *shibuichi* before and after colouring. Only the peaks of copper and silver were present before colouring and they are attributable to the alloy composition. After colouring, however, cuprite (Cu_20) was detected in addition to copper and silver. The same result, shown in Figure 7.5(b), was obtained for the *shibuichi* used in the piece by Shoumin Unno (Figure 7.3(b)). New X-ray diffraction apparatus has recently been developed in Japan, which is able to obtain information directly from an artefact without sampling. The result in Figure 7.5(b) was gained with this apparatus, illustrated in Figure 7.5(a).

A scanning electron microscope revealed the details of the surface of the experimental *shibuichi* (Murakami and Niiyama 1985). In Figure 7.6 it was observed that only the copper-rich region was etched by the colouring process and the silver-rich network remained proud. These two phases create a micro-galvanic cell in contact with the colouring solution of *nikomi-chakushoku* and a thin cuprite layer is formed on the surface of the copper-rich region. The overall colour, from brownish-grey to whitish-grey, can be controlled by the amount of silver contained in the alloy (Murakami 1989a). Figure 7.7

(a)

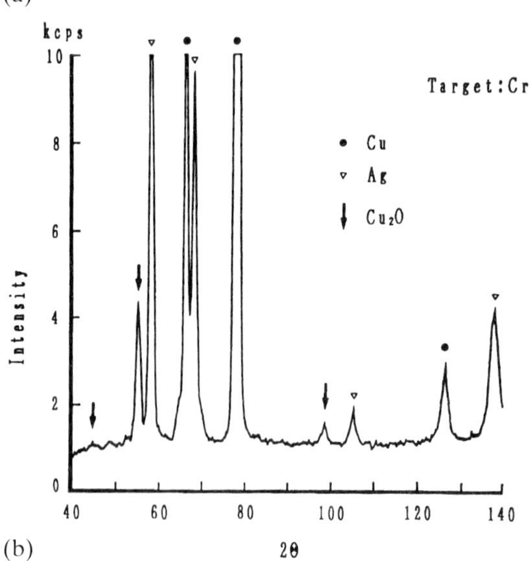

(b)

Figure 7.5. *X-ray diffraction analysis of* shibuichi *used for 'kingfisher'. (a) A general view of the newly-developed apparatus for cultural property to analyse the 'kingfisher'. (b) X-ray diffraction scan.*

demonstrates the microstructure of low silver and high silver *shibuichi*. Colour difference can also be easily influenced by the degree of working, which changes the shape and size of the silver-rich networks. The fine irregularity of the structure and division of colours disperses the light reflected from the surface giving the matt effect typical of *shibuichi*. The colouring process of *shibuichi* makes the best use of the intrinsic metallographic structure.

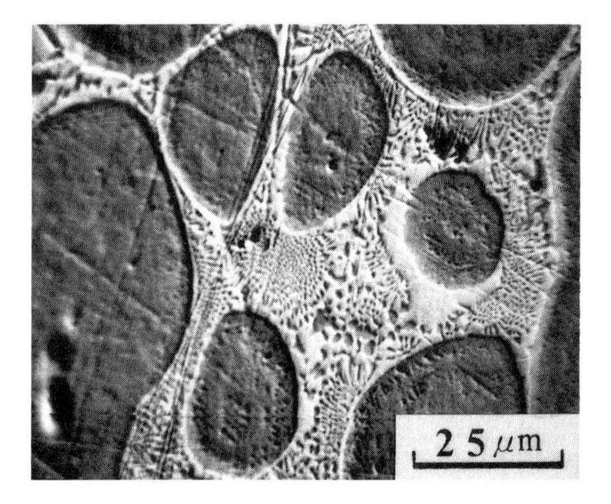

Figure 7.6. *Scanning electron microphotograph of experimental* shibuichi.

2 Shakudo

Shakudo is a curious name. The *kanji* character 赤 standing for *shaku* is usually pronounced *aka*, which means red and *do*, 銅 means copper, so *shakudo* seemingly indicates red copper, but it has a black surface. However, another *kanji* character 烏金 is sometimes applied to *shakudo*. This translates as crow gold, that is, black gold. Standard *shakudo* is a copper alloy containing 3–5% gold and with this composition the alloy can be considered a solid solution. The old type of *shakudo*, mainly used in the Edo period, sometimes contained small amounts of silver. The copper alloy containing arsenic is called *kuromido* 黒味銅, translating as black taste copper. It also has a black surface after colouring and it was often substituted for *shakudo* except for high-grade articles. A copper alloy containing over 10% gold is called *shi-kin* or *murasaki-kin* 紫金, which means purple gold.

To characterize the black surface of *shakudo*, copper-gold alloys were prepared containing 0.25%, 0.5%, 1%, 3%, 6%, 10% gold. The copper alloy containing 3% gold was used for most of the following experiments.

Before being coloured, copper alloys containing 0.25–10% gold are almost the same colour as pure copper. The drastic change in colour arises from boiling in the colouring solution of *nikomi-chakushoku*. The influence

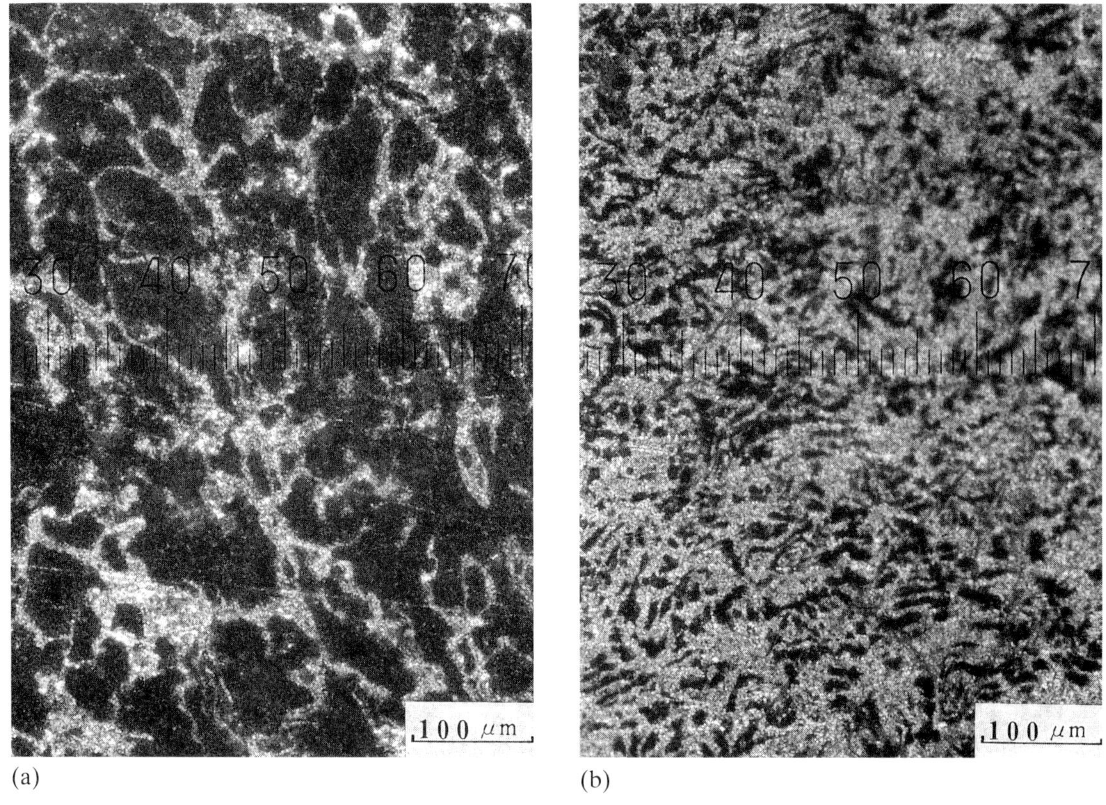

(a) (b)

Figure 7.7. *Microphotographs of* shibuichis. *(a) Low silver content* shibuichi (nami-shibuichi). *(b) High silver content* shibuichi (shiro-shibuichi).

of gold content on the colour was measured with a spectrophotometer (Murakami *et al.* 1987). The result is shown in Figure 7.8. After colouring, pure copper without gold became a reddish-brown colour. As the gold content increased, the red region (600–750 nm) of the surface reflection spectrum of the alloys decreased. At 3% gold content a strong absorption occurred over the visible region of the spectrum, and *shakudo* showed only a black colour. The substance formed on the surface of both pure copper and *shakudo* by *nikomi-chakushoku* has already been identified as cuprite by X-ray diffraction (Oguchi 1983 and Notis 1988), by X-ray diffraction, infra-red spectroscopy and X-ray photospectroscopy (Murakami *et al.* 1986 and 1987).

The *shakudo* used for a piece made by Shoumin Unno was examined by the X-ray diffraction equipment developed for cultural property (Murakami *et al.* 1987). The piece

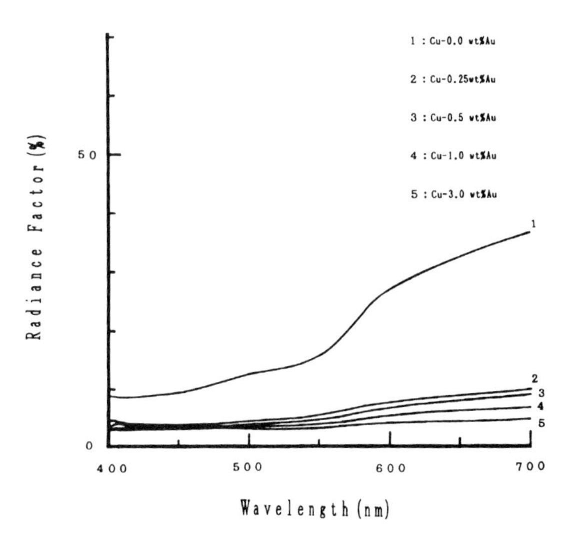

Figure 7.8. *Influence of gold content on the surface colour of copper gold alloy after colouring process.*

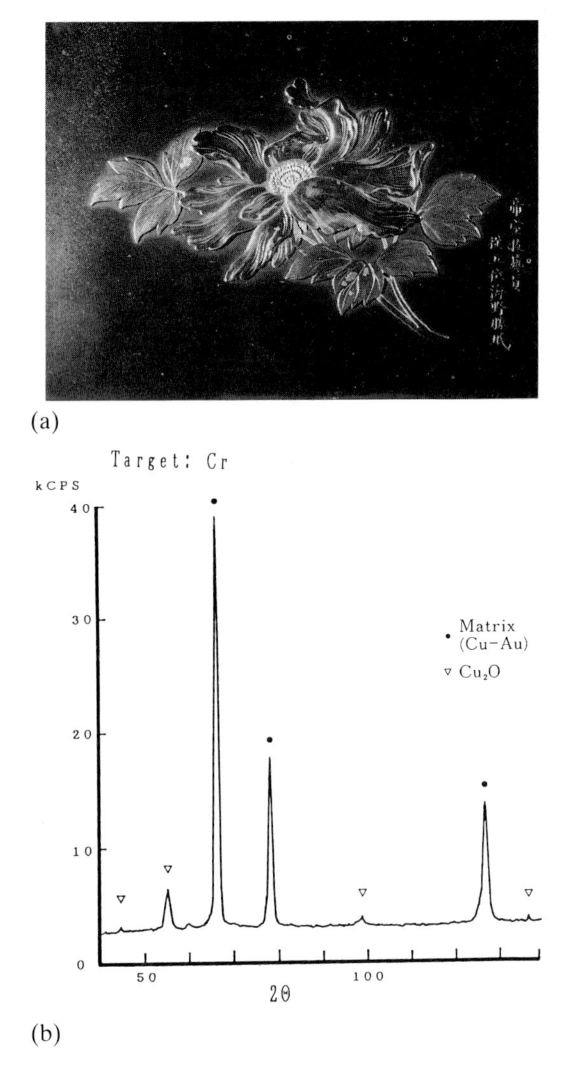

Figure 7.9. *X-ray diffraction analysis of* shakudo *used for 'peony flower' by Shoumin Unno. (a) A general view (a plate chiselled and inlayed). (b) X-ray diffraction scan gained with newly developed X-ray diffraction apparatus for cultural property.*

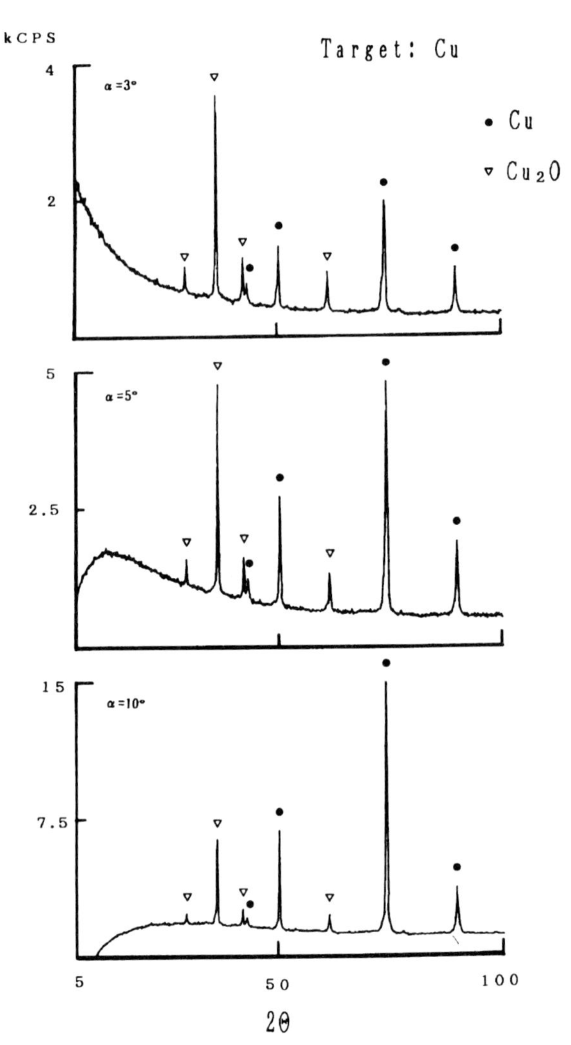

Figure 7.10. *X-ray diffraction scans of the surface of experimental* shakudo *gained with thin film X-ray diffraction apparatus.*

analysed and the result are shown in Figure 7.9. Only cuprite was identified, as has been found by other analytical methods.

The surface of the experimental *shakudo* was also examined by thin film X-ray diffraction (Murakami *et al.* 1987). The result is shown in Figure 7.10. The α in this figure indicates the incident angle of the X-ray beam on the surface of a specimen. The smaller the angle α, the thinner the layer analysed by X-rays becomes. Only cuprite was detected throughout the coloured layer formed on the surface of the experimental *shakudo*. The influence of gold content on the lattice constant of cuprite formed on the experimental *shakudo* is shown in Figure 7.11 (Murakami *et al.* 1987). As the gold content increased, the lattice constant of the matrix increased linearly, following Vegard's law; on the other hand, the lattice constants of the coloured

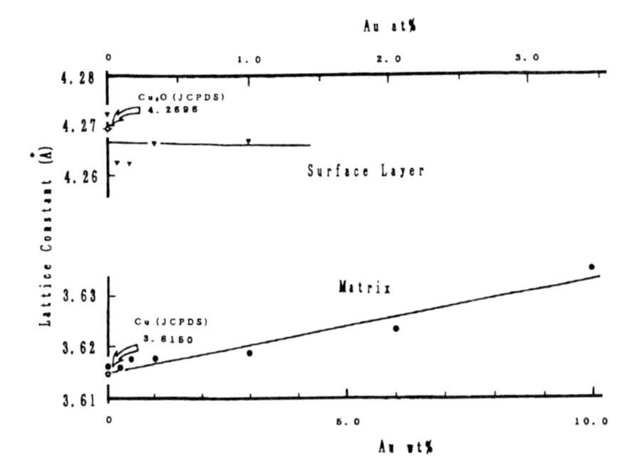

Figure 7.11. *Influence of gold content on the lattice constant of both the coloured layer and the matrix of copper-gold alloy.*

surface layers were thought to be influenced sensitively by the colouring condition, but seemed to be stable. From this alone, however, it is difficult to say that the same substance was formed in the coloured layer of both pure copper and *shakudo*.

What is the difference between the surface cuprite layer *shakudo* and that on pure copper? The micromorphology of the surface layers of both pure copper and experimental *shakudo* was observed with a scanning electron microscope (Murakami *et al.* 1986, 1988a and 1988b). The coloured layer of the pure copper was fragile and scaled off easily, whereas that of the *shakudo* was flexible and adherent. As shown in Figure 7.12(a), the coloured layer of pure copper was 2–3 nm in thickness and was the aggregate of fine particles of cuprite. A particle of cuprite was 20–50 nm in size and tended to grow in the perpendicular direction to the interface of the matrix and the coloured layer. This unidirectional development of cuprite particles seems to result in fragility. Figure 7.12(a) shows a cleavage fracture which occurred on grain boundaries under the interface. It is presumed that the oxygen diffusing into the matrix under the interface reduces the grain boundary binding

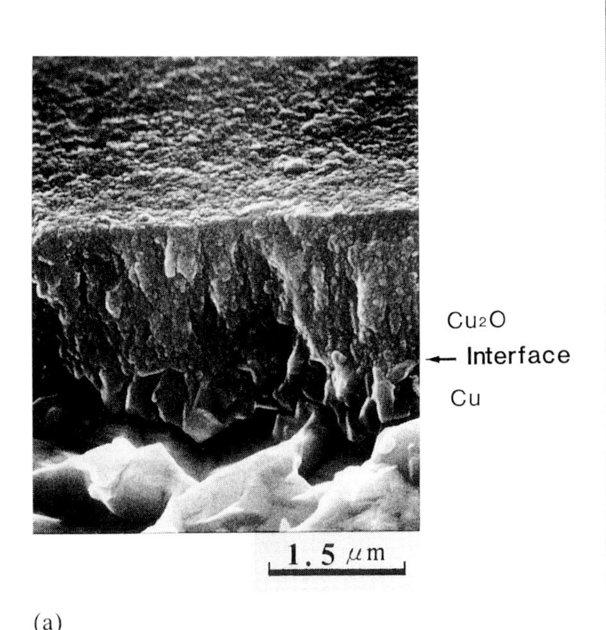

(a)

(b)

Figure 7.12. *Scanning electron microphotographs of the coloured layers. (a) Pure copper. (b) Experimental* shakudo.

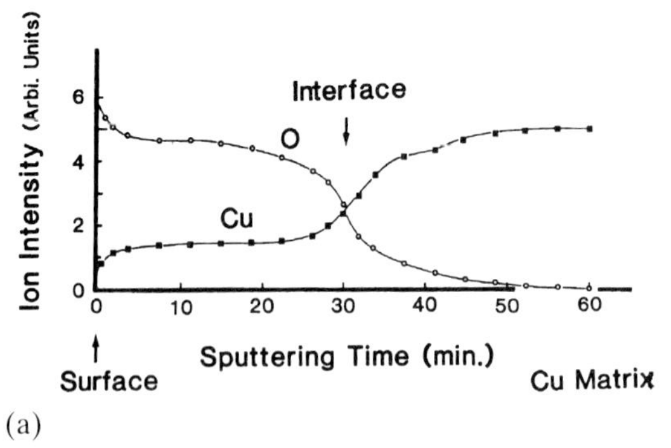

(a)

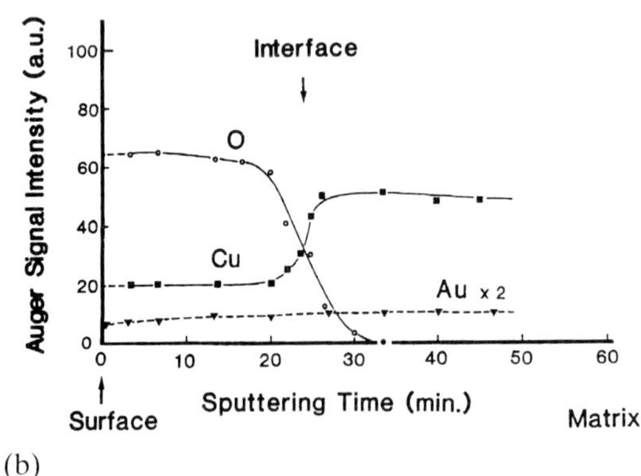

(b)

Figure 7.13. *Distribution of elements in the coloured layers. (a) Pure copper (by SIMS). (b) Experimental* shakudo *(by AES).*

energy and cleavage fracture occurs. The structural detail of the coloured layer on the experimental *shakudo* is shown in Figure 7.12(b). Though the thickness of the coloured layer was almost the same as that on the pure copper, the cuprite particles grew isotropically. It is thought that the isotropic development of cuprite particles gives the flexibility and adhesion of the coloured layer of the experimental *shakudo*, which make it difficult to peel off.

The distribution of the main elements in the coloured layer of both pure copper and experimental *shakudo* were surveyed (Murakami *et al.* 1986, 1988a, and 1988b). The oxygen and copper in the coloured layer of pure copper were studied by secondary ion mass spectroscopy (SIMS). Figure 7.13(a) shows the result. Near the surface, the copper-enriched thin layer $Cu_{2-x}O_{1+x}(0<x\leq0.5)$ was formed. The region between this thin layer and the interface consisted of Cu_2O. Both the oxygen and copper concentrations gradually changed near the interface. Oxygen diffused into the matrix under the interface and formed a Cu(O) region with a long slope, which is thought to be the intermediate stage in cuprite formation. In this way, the diffusion of oxygen from the surface develops the coloured layer on pure copper.

The copper, oxygen and gold in the coloured layer of the experimental *shakudo* were examined by Auger electron spectroscopy (AES). As shown in Figure 7.13(b), the copper decreased more drastically at the interface than on the pure copper sample above. The concentration of gold in the coloured layer did not differ greatly from that in the metal matrix containing 3% gold; hence it is suggested that gold is present in the coloured layer of *shakudo*.

To obtain more detailed information about the coloured layer of *shakudo*, transmission electron microscope (TEM) was applied (Murakami *et al.* 1988a). The electron diffraction pattern of the coloured layer from the experimental *shakudo* and its analysis are given in Figure 7.14. This result suggests that gold atoms exist as fine particles dispersed in it.

Modern chemistry has revealed the close relationship between fine particles and colour (Turkevich 1985). The hypothesis proposed is that the fine gold particles formed in the coloured layer of *shakudo* may be the critical size to absorb the red part of the visible region of the incident spectrum, causing *shakudo* to appear black, as shown in Figure 7.8 (Murakami *et al.* 1988a). On this point further investigation is required to give a full account of the phenomenon.

Conclusion

Colouring is one of the most important finishing processes for metal arts and crafts. Many historical metal artefacts must have been treated by some colouring process, but the traces disappeared from the surfaces during burial. Another reason for not recognizing the evidence of colouring is that many of the ancient techniques have gone out of use. In Japan, unlike most other parts of the world,

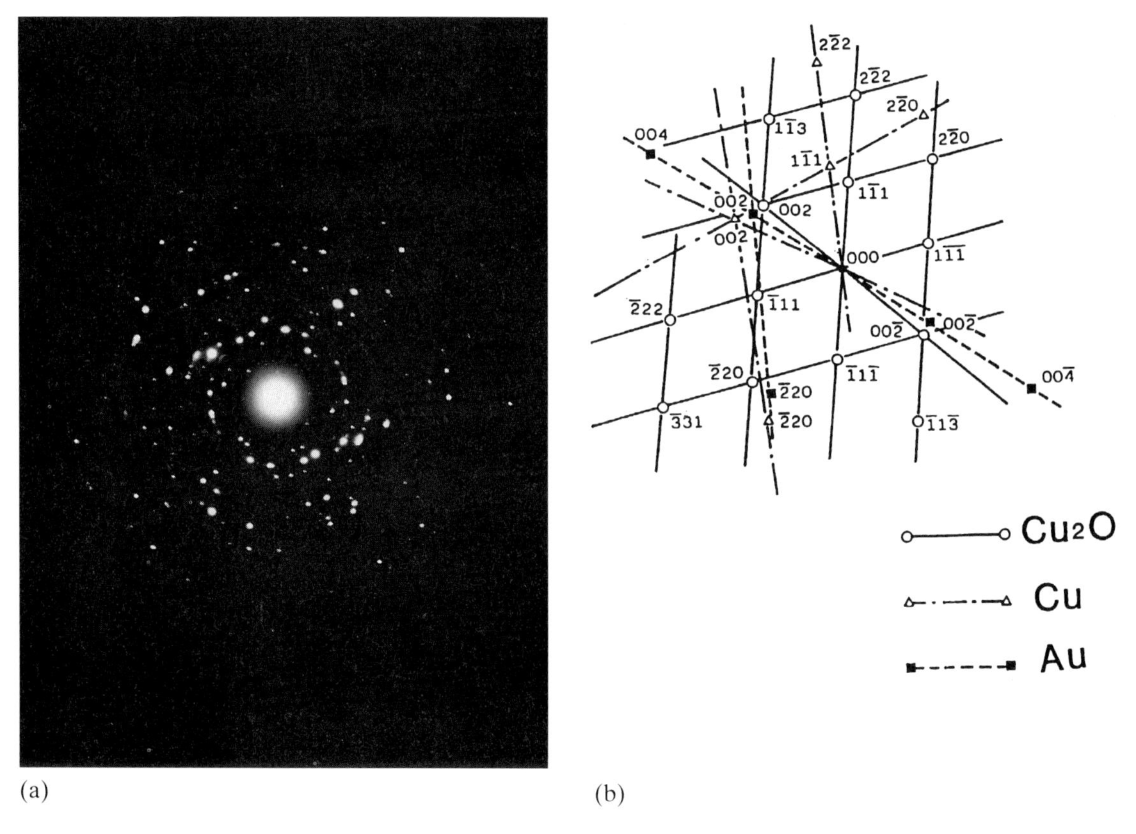

Figure 7.14. *Electron diffraction analysis of the coloured layer of experimental* shakudo *with TEM. (a) Electron diffraction pattern. (b) An example of indexing.*

the traditional colouring techniques have been practised continuously since the Middle Ages and metal artefacts coloured by the traditional methods have been passed down through many generations. Because many of the recipes have been preserved, the investigation of Japanese traditional alloys gives fundamental information about the colouring of historical metal artefacts generally.

In this research, *shibuichi* and *shakudo* were studied. The substances formed on the surfaces of these alloys were scientifically detected and micro-morphologically observed. The colouring of Japanese traditional alloys has been empirically developed by making the best use of their metallographic structure. Through excavated objects, the origins of Japanese traditional alloys need to be archaeologically pursued in Asia, especially the connection between *shakudo* and other alloys with a black surface.

Acknowledgements

The author is grateful to Professor Sakae Niiyama of Tokyo National University of Fine Arts & Music, who gave me much useful advice on the techniques of Japanese traditional alloys, and Dr Masahiro Kitada of Hitachi Central Laboratory Co. Ltd who helped with the analytical work. The author also thanks the Museum of Tokyo National University of Fine Arts & Music for permission to use the collections for the research.

References

Chase, W.T. and Franklin, U.M. (1979) Early Chinese black mirror and pattern-etched weapons. *Ars Orientalis*, **11**, 215–258

Craddock, P.T. (1982) Gold in antique copper alloys. *Gold Bulletin*, **15**, 69–72

Gowland, W. (1910) The art of working metals in Japan. *Journal of the Institute of Metals*, **2**, 4–41

La Niece, S. and Martin, G. (1987) The technical examination of bidri ware. *Studies in Conservation*, **32**, 97–101

Miyazawa, S. (1917) On metal colouring. *The Journal of Chemical Industry*, **20**, 109–135 (in Japanese)

Murakami, R. and Niiyama S. (1985) Observation of colouring process of irogane alloy. *Scientific Papers on Japanese Antiques and Art Crafts*, **30**, 1–10 (in Japanese)

Murakami, R. Niiyama, S. and Kitada, M. (1986) Observation of colouring layer of copper by Japanese traditional treatment. *Scientific Papers on Japanese Antiques and Art Crafts*, **31**, 11–17 (in Japanese)

Murakami, R. Niiyama, S. and Kitada, M. (1987) The characterization of black surface of shakudo(I). *Scientific Papers on Japanese Antiques and Art Crafts*, **32**, 31–39 (in Japanese)

Murakami, R. Niiyama, S. and Kitada, M. (1988a) The characterization of black surface of shakudo(II). *Scientific Papers on Japanese Antiques and Art Crafts*, **33**, 24–32 (in Japanese)

Murakami, R. Niiyama, S. and Kitada, M. (1988b) Characterization of the black surface layer on a copper alloy coloured by traditional Japanese surface treatment. In *The Conservation of Far Eastern Art* (eds. J.S. Mills, P. Smith and Kazuo Yamasaki) Preprints of the IIC Congress, Kyoto. IIC, London, pp. 133–136 (in English)

Murakami, R. (1989a) Reconstruction of Japanese metal craft developed in the *Muromachi* and *Edo* period. *Journal of the Society of Instrument and Control Engineers*, **28**, 693–696 (in Japanese)

Murakami, R. (1989b) Surface colouring of metal cultural artifacts. *Journal of the Japan Colour Material*, **62**, 658–665 (in Japanese)

Notis, M.R. (1979) Study of Japanese *mokume* techniques by electron microprobe analysis. *MASCA Journal*, **1**, 67–69

Notis, M.R. (1988) The Japanese alloy *shakudo*: its history and its patination. In *The Beginning of the Use of Metal and Alloys* (ed. R. Maddin) MIT, London, pp. 315–327

Oguchi, H. (1983) Japanese *shakudo*. Gold Bulletin, **16**, 125-132

Pumpelly, R. (1866) Notes on Japanese alloys. *American Journal of Science*, **42**, 43–45

Roberts-Austin, W.C. (1893) Application of alloys in metalwork. *Journal of the Society of Arts*, **41**, 1022–1043

Savage, E.I. and Smith, C.S. (1979) The techniques of the Japanese *tsuba*-maker. *Ars Orientalis*, **11**, 291–328

Turkevich, J. (1985) Colloidal Gold (II). *Gold Bulletin*, **18**, 125–130

Uno, D. and Aachen, Z. (1929) Kœnstliche Korrosion von japanischen Speziallegierungen. *Korrosion und Metallschutz*, **5**, 121–130.

8

Why *shakudo*?

Victor Harris

Abstract

This chapter considers the origins of *shakudo* and related alloys in Japan. It is suggested that *shakudo* treatments were carried out originally to produce surfaces on metals similar to those on the lacquered fittings on the earliest swords.

The subsequent development of the technique is described here, especially in relation to its most famous practitioners, the Goto family.

The word *shakudo* has caused some confusion in the past, since the characters with which it is usually written are those for red and copper, whereas *shakudo* is black. *Shakudo* has also been called *ukin* or *udo* (cormorant gold, or copper). The term *jukudo* has been used for highly refined copper, but the two characters for red copper which are pronounced *shakudo* have been used for copper, irrespective of the degree of purity, for centuries. Early records include, for example, the 8th century Todaiji temple inventory of the objects kept in the Shosoin repository, in which a long handled censer is among objects described as 'red copper'. It is patinated to a rich dark colour, and is made of copper, but possibly an impure form like *yamagane* (mountain metal). The confusion is further compounded since the nascent *shakudo* alloy is indistinguishable from copper to the eye. It is only recognizable as *shakudo* when it is patinated to give the characteristic rich black or purple-black surface colour.

The quality of the alloy depends upon the amount of gold used. The best quality, or *gobu zashi*, included 5% gold, while the poorer kind might have only 1%. However, other metals, including silver, might be found in *shakudo*. Before the Edo period (1603–1868) *shakudo* was generally made using a low purity copper, or *yamagane*, whose constituents varied considerably and which are described in detail elsewhere. But *shakudo* of the Edo period and later employed refined copper. Gowland, Wakayama and Sato, and others have clearly indicated the conscious use of trace elements found in *yamagane* by the addition of *kata shirome* to the refined copper used to make *shakudo* during the Edo period (Gowland 1914–15, Wakayama and Sato 1974), and confirmed by recent analyses (La Niece 1990). In addition, later *shakudo* generally contains less gold. The increasing use of *shakudo* for sword fittings during the Edo period, and the policy of the Tokugawa government to hoard gold are no doubt in some measure to blame for this. But perhaps the reason might also be found in a spiritual paucity during an era of comparative peace, when the superficial was judged equal to the essential. The difference between old and new *shakudo* is seen both in the quality of the black colour and in the quality of the workmanship. It is also different in that old *shakudo* seems to naturally revert to black even if it is damaged or somehow worn down to the red metal. Whether this is due to the natural effect of handling, whether it is an inherent quality of the metal, or whether it arises from some reaction with the atmosphere, will eventually become clear. An interesting example of such an object is an *uchigatana* type sword known to have been in the collection of the castle Schloss Ambrass in the Austrian Alps since the early 17th century, on which some of the *shakudo* components are quite worn and yet jet black. What in the Alpine air might have caused that? An example of a *kozuka* (hilt of a utility knife) in the British Museum is also black despite being very evidently worn from use (Plate 8.1). Both the *shakudo nanako* ground (see p. 98) and the gold and silver inlay are worn almost smooth.

Since it was an expensive metal *shakudo* was used only to make small objects. There were sword mounting fittings, sliding door inset handles, pouch clasps, and similar objects. The metal was highly regarded by the *shogun*, military rulers of Japan from 1185 to 1868, and thereafter by the Imperial Government. After the restoration to Imperial power in 1868 *shakudo* became used in ornamental metalwork made for export to the West. European and American craftsman adopted the Japanese concept of a black metal together with the racey Japanese approach to pictorial art which had so much influenced the Art Nouveau movement. Black pigments, niello, lacquer and other materials enjoyed a glow of popularity in the crafts world. Some Westerners even used *shakudo* itself (see Plate 14.3).

Following the Sino–Japanese war of 1894–5 the government ordered campaign medals to be struck from the bronze of captured Chinese guns, 'but they must have a black patina resembling *shakudo*' (Gowland 1914–15, p. 62). This was achieved by the use of a speise containing 60% iron and 32% arsenic mixed into the bronze. The use of arsenic is interesting since it is one of the constituents of *kata shirome* and is found in some forms of *yamagane*. The significance of this is discussed elsewhere (Chapter 9).

The sophistication of the old Japanese metalworkers is indicated in documents like the printed books *Soken Kisho*, by Inada Tsuryu (1781), and the *Bankin Sugiwai Bukuro*, by Miyaki Yarai (1732), in which details of alloys, patinating processes, and manufacturing methods are described. The 8th century Shosoin records list dozens of terms for metals, and includes a particularly concise list of materials used in casting and polishing mirrors. In the light of such metallurgical erudition it is the more remarkable that *shakudo* seems not to have been produced in Japan until the end of the Kamakura period (1185–1333), and even then that its use has been largely restricted to making metal fittings for sword mountings.

It is generally recognized that the blade of the Japanese sword far surpasses that of any other culture as a cutting weapon and even as a work of art in itself. The blades are often mounted in rich lacquered scabbards with precious metal fittings, and those for court use are inlaid with precious stones.

There are gilt and silvered mountings of swords excavated from the *Kofun* (ancient burial mounds) period (around 4th to 7th centuries AD), and especially luxurious mountings on some swords kept in Buddhist temples and Shinto shrine collections since that time. Especially notable is the 'Chinese style sword' in the Shosoin, the 8th century treasure house of the Todaiji temple in Nara, which is thought to have belonged to the Emperor Shomu. This sword is in a decorative lacquered scabbard, and the gilt and silvered metal fittings are of finely carved scrolling with *lapis lazuli* inlay. Alongside such rich mountings, throughout the ages, some of the very finest swords were mounted using the cheapest of materials and in the humblest style. They were plain black. The reason for the popularity of black *shakudo* for sword fittings must be considered in the context of the long tradition of the use of black lacquer on the metal components of the swords.

Before the Muromachi period, warfare in Japan was largely a matter of combat between mounted and armoured *samurai*, but the incessant civil wars of the 14th through to the 16th centuries saw the rise of large armies, and a change to combat on foot. The style of sword mountings changed accordingly. The mounted man had carried a *tachi*, a sword suspended loosely from the belt by cords or chains with the cutting edge downwards. The foot soldier during the Muromachi period carried an *uchigatana* thrust through the belt with the cutting edge uppermost. The materials used for black sword mountings since the Nara period were magnolia wood for the scabbard, ray fish skin for wrapping the wooden hilts, leather and wood for the *tsuba* (sword guard), lacquer blackened with charcoal, and iron and *yamagane* and its alloys.

Early examples of black mountings include the *chokuto* (straight sword), which belonged to the 8th century hero Sakanoue Tamura Maro, now held by the Kurama temple in Kyoto. Another black-mounted sword also said to have belonged to him is kept in the Kiyomizu temple of Hyogo prefecture. The 12th century National Treasure sword named *Kitsune ga Saki* (Fox Peak), considered by many to be the finest of all blades, is mounted with *yamagane* fittings and lacquered black overall. Another of the same period, called *shishi-O* (Lion King) is an *Itomaki no tachi* (cord wrapped sword), with silk braid wrapped partway along the scabbard to prevent abrasion against the wearer's armour. It is lacquered black over the original white fish skin, wooden scabbard, *yamagane* fittings, and the wood or leather *tsuba* (Harris and Ogasawara Cat. no. 8, p. 31). A 16th century sword from the collection of the Tokyo National Museum, originally kept in the Saiendo hall of the Horyuji temple, has fittings of lacquered leather, *yamagane*, and horn (Harris and Ogasawara Cat. no.38, p.71).

At some time, probably during the late Kamakura period, *shakudo* was first used for black sword mountings. Among the clearly identifiable pieces with *shakudo* mounts is a *tachi* type sword in the Sanage shrine collection which is known to date from the Kamakura period. The scabbard and hilt are lacquered black, but the fittings are of *shakudo* with birds sculpted in high relief on a *nanako* (fish roe) ground which is covered with a regular array of small hemispherical protruberences made with a hollowed punch. The overall style of the sword is identical to those with black lacquered metal fittings, indicating with little doubt that *shakudo* was introduced as a replacement for black lacquer. A number of examples of black lacquered *uchigatana* of the Muromachi period in collections such as those in the Itsukushima shrine and the Horyuji temple clearly indicate the transition from the old *kuro urushi* (black lacquer) *tachi* to the standard black lacquered *kamishimo goshirae* of the Edo period: *kamishimo* (top-bottom), referring to the formal dress of jacket and *hakama* (pleated trousers), and describing the pair of swords required for formal wear.

Whatever the tactical advantages of having weaponry coloured black, shades of black seem to have been particularly attractive to the military class in Japan. Their spiritual exercises were steeped in Zen Buddhism, and all manner of arts and crafts sponsored by the successive *shoguns* reflected this. Decorative motifs favoured by the Ashikaga *shoguns* during the 15th and 16th centuries had a strong Chinese bent reflecting the Chinese origins of Zen. Often regarded as the highest form of Japanese art, *sumie* ink painting and calligraphy exemplifies the most elegant use of black. The appreciation of subdued designs and colours, particularly black, shows in the concepts of *wabi* and *sabi*, as defined best in the Way of Tea and propounded in the Muromachi period. *Wabi* might be defined as a kind of inner richness in outer poverty, and *sabi* as the beauty of desolation. Objects associated with these principles include, among the highest regarded of wares for the Tea Ceremony, the black rusted iron kettles, and dark glazed tea caddies and drinking bowls. These concepts are evident in both the choice of materials and the decorative motifs on black sword fittings.

In the Muromachi period there grew up a connoisseurship of sword fittings as a form of art. Undoubtedly combat on foot required more robust *tsuba*, and during the Muromachi period, metal, and particularly iron, replaced the earlier wood and leather pieces. Iron *tsuba* were at first plain sheet, perhaps pierced or engraved with simple motifs expressing the religious sensibilities of the *samurai*. They were patinated black in accordance with traditional aesthetics, with the added advantage that further accidental rusting of the iron was thereby prevented. The motifs were similar to those of *sumie* ink paintings of the time, pierced in either positive or negative silhouette, and with a powerful use of space characteristic of ink painting. In the 16th century a *tsuba* maker, Kaneie of Fushimi, established the concept of high relief pictorial sculpture on iron sword guards. His themes were similar to those of *sumie*, showing a kind of cooperation between painter and sculptor which was to flower later in the peaceful *Edo* period.

Yamagane itself was appreciated for its dark brown rustic patina, and makers such as the Mino school sculpted motifs from nature on their sword fittings, predominantly insects, animals, and flowers, in sunken relief. *Shakudo*, developed by the addition of gold to *yamagane*, became the favourite metal for the sword fittings of the military elite Ashikaga clan. Early *shakudo* is in the style of Mino Province, and it was from the Mino tradition that the family whose name has become synonymous with *shakudo*, the Goto, emerged.

From the time that the Shogun Ashikaga Yoshimasa comissioned Goto Yujo (1440–1512) to make sword fittings of *shakudo*, the Goto family enjoyed a position of some prestige with successive governments, through seventeen generations over four centuries. The main family was retained by the *shogun*, and branch families worked for certain provincial lords. Goto work is known as *iebori* (Clan sculpture) since they were retained by the *samurai* clans. In this respect the Goto were in a similar position to the Kano school of painters, who also specialized in Chinese themes, and the Koma school of official lacquer artists. Their most prevalent subjects included the flowers, grasses, insects and animals favoured by the Mino school, and all the themes of the Zen-inspired arts like the Noh theatre, the Tea ceremony, Swordplay, and Chinese Buddhist subjects like *Hotei* (The Chinese priest with a sack and a stick), and legendary animals like the *kirin* (a kind of gryphon), dragons, and the *shishi* (a kind of dog-like lion). The early generations specialized in the manufacture of the so-called *midokoro mono*, or 'things of three places'. This was a set of a pair of *menuki* (ornaments fitted under the braid binding either side of the hilt), the *kozuka* (the hilt of the utility knife), and the *kogai* (a kind of bodkin used for dressing the hair, cleaning the ears, and sundry other functions). The latter two were kept in pockets in the scabbard, and drawn for use through two kidney-shaped holes in the *tsuba* (Plate 8.2).

The extent of the patronage of the Goto clan, and the prestige of *shakudo*, are demonstrated by a number of edicts made during the 17th century known as the *Buke Shohatto*. These regulated the dress and conduct of the *samurai* when on official duty in the capital, Edo (present-day Tokyo). Their swords had to have black lacquered scabbards, and Goto style fittings, although iron was acceptable for the *tsuba*, and the *kashira*, or pommel, could be of horn. Black or dark coloured braid was specified for the hilt binding. The *menuki*, and the carved or inlaid designs on the other fittings were often in the form of the *mon*, or clan badge.

Interestingly, a very slight brownish tinge to the black lacquer of the scabbard, known as *roiro* (wax colour), was admired, reflecting perhaps the traditional appreciation of russet iron *tsub*a. For the pommel and other small fittings buffalo horn was preferred since it was black. The *Bankin Sugiwai Bukuro* (Yarai 1732) tells us 'The *kashira* (pommel), *koiguchi* (mouth of the scabbard), the *kurikata* (a protrusion on the side of the scabbard which prevents it from slipping through the belt), etc., are made of buffalo horn and are lacquered black, so that even if the lacquer were to wear off it would not be noticeable. But the horn of the common ox is very ugly when spoiled, since it is somewhat white in colour.'

Goto work was predominantly in *shakudo* with a *nanako* (fish roe) ground covered overall with a regular array of small hemispherical protruberences made with a hollowed punch, although a plain polished ground was also produced. Details on the sculpted motifs were depicted in gold or silver overlay, and sometimes the whole piece was carved from solid gold. Yujo and the early generations specialized in *repousse* work, beating out the decorative overlays over blocks of pitch. This method was used for *menuki* and pieces of miniature sculpture which were to be pinned onto *shakudo* bases. The early method of inlaying gold is known as *uttori*, in which the leaf was mechanically worked onto the ground with hammers, points, and chisels. This was a job requiring considerable skill, and yet the gold might work loose on even the best pieces with passage of time. Imitators of early Goto work are known to have used lacquer to adhere the gold. During the Momoyama period, when the fashion was for copious use of gold on all kinds of objects, the most prevalent method was with the use of solder, rubbing the gold on to the ground with burnishers while keeping it hot. During the Edo period mercury gilding, or *tokin*, was common. The *Bankin Sugiwai Bukuro* (Yarai 1732) describes mercury gilding thus:

> The objects to be plated are washed and then polished with a bamboo brush and powdered charcoal. Next a little plum vinegar and mercury are applied, so that the object becomes white. Then a few pieces of gold leaf are rubbed onto the surface, which remains white. The objects are then placed on a metal grill over a fire and heated strongly until a gold sheen emerges. If the gold colour is too light it can be enriched by applying the juice of *Kihada* bark (this was also used for dyeing textiles yellow), and heating once more. If this is repeated ten times the true colour of gold is obtained. Such pieces are sometimes immediately put on sale, and sometimes they are smoked and wiped with an oily cloth to give an appearance of age.

The end of the 17th century saw a decline in the manufacture of sword blades, reflecting the decades of peace and prosperity under the Tokugawa regime. But people had become fashion conscious, and the demand for decorative sword mountings increased among the wealthier samurai and the merchant class who were entitled to wear one short sword. Many independent commercial studios known as *machibori* (Town sculpture), as opposed to *iebori* (Clan sculpture), came into being.

From around this time, *shibuichi* began to rival *shakudo* in popularity for sword fittings. *Shibuichi* (literally one part in four) is an alloy of copper with one quarter silver. It could be patinated to give a range of colours from silver through to dark greys and brownish greys. Like *shakudo*, *shibuichi* is a sombre colour, and like *shakudo*, it provides an ideal ground for applied sculpture and inlay of gold and other metals (Plate 8.3). The alloy was used extensively by the *machibori* artists from the latter part of the 17th century onwards, although the Goto clan also sometimes used it. The *machibori* artists devised a method of engraving in imitation of calligraphic brush strokes, known as *katakiri bori*. A triangular-pointed chisel was used to cut away the ground to form the outline of the design. The angle of the chisel could be varied so that lines of varying depth and thickness were obtained. The variation of angle of the cut further formed shadows to heighten the brush-like effect. *Katakiri bori* is said to have been first popularized by Yokoya Somin (1670-1733), a metalworker of the Goto school who rebelled against the strict dictates of the tradition and set up an independent studio. His work influenced many schools. Other techniques popular among the *machibori* artists of the Edo period included *shishiai bori*, or carving in sunken relief deriving in part from the old technique of the Mino school. Many artists specialized in *iroe zogan*, (coloured pictures inlay), using a number of different coloured metals to 'paint' pictures (Plate 8.4).

Although Goto work was to a great extent standard and repetitive, the *machibori* artists were provided with popular subject matter by the fashionable artists of the day. But, however elegant the design, it was the quality of the metal and the confidence of the hand holding the chisel which made decorative sword fittings serious works of art. Like brush painting or calligraphy, it is the inner dynamism of the artist which brings life to the composition.

The intimate relation between painters and metalworkers survived the restoration of imperial rule in Japan, after which they co-operated on ornamental pieces for export to the West. Among the greatest artists of the time, whose life spanned both the feudal and modern democratic eras, were the painter and lacquerer Shibata Zeshin, and the metalworker Kano Natsuo, whose joint works earned international acclaim a hundred years ago. Zeshin perfected a means of imitating old corroded material, which he called *sabi age*, in a conscious statement of the ideals of understated beauty expressed so well by *shakudo*.

Zeshin has left a document indicating the quiet aura of *wabi* and *sabi* in a lacquered wood copy of a 16th century iron *tsuba* by Nobuie (Plate 8.5), made during the lifetime of a generation who had carried two swords in black mountings in the immediate past as naturally as they wore dark suits, black hats and carried umbrellas in the Japan of the Meiji era.

References

Gowland, W. (1914–15) Metals and metalworking of old Japan. *Transactions of the Japan Society of London*, **13**, 19–100

Harris, V. and Ogasawara, N. (1990) *Swords of the Samurai*, British Museum Publications, London

La Niece, S. (1990) Japanese poychrome metalwork. In *Archaeometry '90* (eds A. Hauptmann, E. Pernicka and G. Wagner) Birkhauser Verlag, Basel, pp. 87–94

Tsuryu, I. (1781) *Soken Kisho* (in Japanese)

Wakayama, H. and Sato, K. (1974) *Toso Kodogu Koza Yusankaku*, Tokyo (in Japanese)

Yarai, M. (1732) *Bankin Sugiwai Bukuro* (A Bagful of Ten Thousand Metal Industries), Kyoto (in Japanese)

9

Ḥśmn-Km, Corinthian bronze, *shakudo*: black-patinated bronze in the ancient world

Paul Craddock and Alessandra Giumlia-Mair

Abstract

Distinctive and highly prized black-patinated bronzes inlaid with precious metals are known from a number of early cultures. Investigation of these has shown that they are technically very similar, an alloy of copper or bronze with small amounts of gold, silver and sometimes arsenic, inlaid and then patinated with hot aqueous solutions. They include the Egyptian *ḥśmn-km*, the black panels on the Mycenaean daggers from the shaft graves, the Corinthian bronze of Roman fame, culminating more recently in the Chinese *wu tong* and the Japanese *shakudo*.

Introduction

Lost civilizations, lost cities, materials that survive only as a name have a fascination all their own. The archaeologist can uncover the former, but it sometimes can take the combined skills of antiquarian and scientist to identify the famed products of the past from the meagre evidence that has come down to us from antiquity.

The subject of this chapter is the identification of the Egyptian material . *Ḥsmn-Km*, literally black copper, and the classical *Corinthium aes*, Corinthian bronze. It is suggested here that the two were very closely related copper alloys containing small quantities of gold, silver and arsenic, treated to produce a dense matt purple-black surface which was inlaid with gold and silver.

This is of course a concise description of the more recent Japanese *shakudo*, one of the group of patinated metals collectively known as *irogane*. The possibility that knowledge of this process originated in the West and travelled over half the world to the Far East is discussed in the final part of this chapter together with the implications for the diffusion of technology generally.

The study has been as diverse as its subject and relies on evidence of many kinds – curators' observations, ancient documentary sources, scientific examination, etc. Sometimes the evidence was no more than a clue, perhaps an inference that only became significant as more information was collected. Striking parallels were found in descriptions and materials separated by millennia and from around the world which together built up the identity of the material and the basic continuity of the process through many different cultures and traditions. First the present day Japanese process will be described.

Japanese *Irogane*

The history and technology of these alloys are described in detail elsewhere in this volume and thus will only be given in outline here.

When Japan was opened up to the outside world in the mid-19th century many of the traditional processes were noted with interest by technical advisers called in from the West. One of these was William Gowland who worked at the Imperial mint, but who was keenly interested in all matters relating to metallurgy from mining to the decorative processes of *irogane*. He recorded in some detail the alloys and treatments and obtained specimens for analysis (Gowland 1894, 1915). Slightly later the eminent metallurgist William Roberts-Austen also took an interest in these alloys (Roberts-Austen 1888, 1892, 1898), and undertook a long series of experiments creating specialized patination effects on copper and other alloys, some of them in collaboration with the sculptor Alfred Gilbert (Dorment 1985).

Since at least the 14th century the Japanese have been producing patinated copper alloys with a wide range of very attractive colours and textures known collectively as *irogane*. The patinated metals were invariably inlaid with other metals, notably gold and silver, but also copper and brass to obtain a range of colours and contrasts (see Plates 7.1, 8.1–8.5). The colour of the patinated metal depends on both the treatment, *nikomi-chakushoku*, and the composition of the alloy. The most familiar alloys are *shakudo*, *shibuichi* and *shirome*. These have a wide range of composition in themselves but basically *shakudo* (which literally means red copper, or in another variant black gold, see Murakami p. 88, and Harris, this volume p. 95) is an alloy of copper with a small amount of gold, typically between 1% and 5%, usually some silver of the order of 1%. Arsenic has also been found in some of the pieces analysed by La Niece (1990). The arsenic was probably present as an impurity in the copper, but nevertheless it is significant in the context of this paper as many of the earlier Egyptian and Roman pieces also have enhanced levels of arsenic. A Japanese recipe quoted by Lyman (1890–1) states that 4% of gold should be added to a *shirome* copper which itself normally contained about 1% of arsenic (see below). The gold content varied greatly but seems to have been typically around 4–5%. Above 5%, the colour is said to be increasingly purple, and experimental pieces with more than 10% appeared to have a bluish tinge (Notis in conference), which may also have some significance in relation to the patination on some ancient bronzes as well as to the comments of some of the ancient writers on blue-coloured copper alloys.

Shibuichi is an alloy of copper with about 25% of silver, although in the past the content could vary between 10 and 50% depending on the surface colour required.

Shirome seems to have originally been a by-product of the liquation and refining of copper and thus could have a variable composition, but the consistent and important constituent was about 10 to 20% of arsenic (Gowland 1894). This metal could be added to ordinary bronzes in small controlled proportions to promote the formation of a fine grey patina. The metallic *shirome* by itself had a pleasing soft grey coloration.

These alloys were patinated by the *nikomi chakushoku* process, of which the best description in English of the traditional recipes was given by Gowland (1894). The metal was first boiled in a lye prepared by lixiviating wood ashes, that is a caustic solution of pearl ash and potassium hydroxide, which would act as a very effective degreasing agent. After this, if necessary, the metal was carefully polished with charcoal powder. This part is very important, the final effect obtained in many surface treatments is heavily dependent on the degree of polishing of the original surface. It was then placed in a saline solution of plum vinegar, washed in dilute lye to neutralize the vinegar and then washed in a tub of water to remove every trace of the chemicals. After this treatment it was then digested in a boiling solution of blue vitriol (copper sulphate) and verdigris (copper acetate), to which nitre (potassium nitrate) was sometimes added, until the desired patination had developed. Roberts-Austen (1892 p. 51) recorded actual quantities of the materials for the final patinating solutions: 292 drachms of blue vitriol and 438 drachms of verdigris to a gallon of water, equivalent to about four and six grams per litre respectively. This is not dissimilar to the figures given by Murakami (p. 86) in his description of the modern process, and shows that the actual 'pickle' which produced the coloured patina did not have to be very strong at all. This is of relevance later in this paper when the role of the water from the fountain of Peirene in Corinth is considered (p. 112), and also in the modern hot water repatination treatments applied to Egyptian bronzes after conservation where a dark bronze was required (p. 117).

A great deal of research has been carried out to establish the nature of the patination in the *shakudo* alloy (Notis 1988, Murakami *et al.* 1988, and in this volume). This work shows that the principal component in the patina is copper oxide, cuprite.

Figure 9.1. *Detail of Osiris figure, EA 64477, showing composite necklace of black bronze and electrum inlaid into the body of the figure.*

Egyptian *Ḥśmn-km*

Many Egyptian bronzes have a black patina and for at least a century scholars have speculated on its nature and significance, of which the most perceptive and detailed study so far was carried out by Cooney (1966 and 1968), although apparently without the help of any scientific study or analysis. Cooney stated that from early in the 18th Dynasty on there are repeated references to a specific material [hieroglyphs], *ḥśmn-km*, literally black copper, in inscriptions recording donations to temples etc. He pointed out that the black bronzes were always described as inlaid with gold or electrum, indeed he states that the purpose of his 1966 paper was to show that 'black bronze always indicated an inlaid bronze'. Inlaid black bronzes are indeed known from the 18th Dynasty on, and include statuettes, especially depicting Osiris, and a number of prestige artefacts often associated with temple ritual such as the superb adze handle from the tomb of Tutankhamen (Carter Handlist no. 402) with the dedicatory inscription picked out in gold against the black bronze. On other items, notably statuettes the black bronze could itself be used as an inlay, often in conjunction with gold or electrum inlays, as, for example on a statuette of Osiris, now in the British Museum, (EA 64477), where the complex necklace is made up of black bronze inlaid with electrum, and itself inlaid into the neck of the bronze (Figure 9.1).

The surviving artefacts tend to be small, but the inscriptions mention bronze doors and other architectural features which would have necessitated large castings.

Very naturally those who wrote about the black bronze speculated on the nature of the black surfaces. It was recognized as being the product of a surface treatment, with Cooney emphasizing again that the whole process to produce *ḥśmn-km* was threefold: casting the metal, patinating the surface and inlaying. Cooney's work and judgement on this subject was generally excellent, he established the significance of *ḥśmn-km* and correctly identified it amongst surviving bronzes, and his description of the black patina, investigated using a low power microscope, are meticulous and quite accurate. In only one area was he in error, and that was in the actual chemical composition of the patinated layer and how it was formed. He apparently carried out no scientific tests but was quite adamant that

> Producing this black surface was so very easy that the process could have been known at an early date but not utilised. A sulphide must have been the basic agent in any process used. Its simplest production would be the dark film made on silver by contact with an egg and it is possible that some acute Egyptian craftsman observed this homely fact and applied it to his craft. Probably the Egyptians heated the bronze which was to be inlaid and then applied a sulphide in a colloidal, almost certainly a resin. Or the stain could be obtained merely by applying a very ripe egg. (Cooney 1966)

He gave no authority for these incorrect assertions and neither Lucas (1962) or Partington (1935) mention black bronze. Garland and Bannister (1927 p. 82) in their excellent book on Egyptian metallurgy mention an artificial blue patina on Egyptian bronzes and even make the cryptic comment 'the bronze itself, not improbably containing gold,' suggesting they may well have been on to something but they pursued the matter no further and there is certainly no mention of a patinating process.

However, a sulphidic patination process had been postulated long ago for a superb statuette of Ptah of the late Dynastic period, that has now very recently entered the collections of the Ashmolean in Oxford (Whitehouse 1987). The god's facial features, cap, beard, jewellery and the regalia he holds are all patinated black and inlaid with gold and silver. Over a century ago the self-same statuette was exhibited at the Society of Antiquaries of London, and Professor Middleton produced a detailed and in some ways prescient note for the *Proceedings* of 1887. In this he states that 'The method employed to colour the bronze was probably much the same as that used by modern Japanese bronze workers.' So far so good, but then he continues

> who darken their metal by creating an artificial patina, not by applying any surface lacquer or enamel. This is done in Japan by applying sulphur in some form in a pasty state, laid carefully on the parts where the dark patina is wanted. The metal is then heated, its surface chemically absorbs some of the sulphur and thus a thin coating of sulphuret of copper is produced on the places where the sulphurous paste has been applied

The source of Middleton's apparently detailed knowledge of Japanese metalwork is not given, but irrespective of whether or not his information was correct he seems to have been ignorant of the *irogane* processes. At this time Gowland was actually in Japan, but from the 1890s he became a frequent participant at the meetings of the Society of Antiquaries; if he had been in the audience at the 1887 meeting he could and surely would have supplied Middleton with the information to make the correct analogy; instead over a century had to pass before the true connection between Egyptian and Japanese metalworking practice was realized.

Black-patinated bronzes from Egypt

Recent work carried out by Shearman (1988) on the black inlays on the faces of statuettes of Horus (EA 11528) and of Montu-Re (EA 606339), showed that copper oxides were present, notably cuprite, but no trace of sulphides. In the article Shearman suggested that the blackening may have been due to arsenic concentrated on the surface, but she has since abandoned this idea (personal communication). Analysis of the body metal of the *Horus* figure showed it to be an ordinary bronze with about 10% of tin and the usual trace metals, but both gold and silver fell below the detection limit of the atomic absorption analytical method. X-ray fluorescence analysis (XRF) of the patinated inlay suggested it was of copper with some arsenic, and traces of silver were detected in one area analysed. Although the detection limits of this analytical method are fairly high, it does seem that the patinated areas did not contain significant amounts of either gold or silver.

However, encouraged by the work of Cooney, and the identification of cuprite in the patination of the pieces examined by Shearman, the authors of this paper examined more pieces of black inlaid Egyptian bronzes (Figure 9.2). The composition of the bronzes is given in Tables 9.1 and 9.2, showing two of them contain gold and silver. X-ray diffraction analysis (XRD) of the surviving genuine patinas showed them to be principally of cuprite (Table 9.3). The results are discussed in detail below (p. 116)

Some other Egyptian bronzes were already known with this distinctive composition, although they are very rare, as the high status of *ḥśmn-km* would suggest. Cowell's (1986) analysis of about 300 Egyptian axes of all periods failed to produce any with high silver or gold content, but as these were mainly utilitarian tools or weapons, and none were inlaid, this is hardly surprising. However, in 1929 the Sumerian Copper Committee published the

Figure 9.2. *Group of six bronze Egyptian figures, selected at random but all having black surfaces, and heavily inlaid. From left to right the figures are Osiris EA 64477, Osiris EA 24718, Ptah EA 27363, Aegis of Ra 16037, Amun-Re EA 63581, and Montu-Re EA 60342.*

analysis of an otherwise unidentified 'copper chisel of early Dynastic period' which contained 93.21% copper, 2.51% silver, 4.14% gold, 0.05% lead, and 0.06% of arsenic (Bannister 1929), given to Bannister by Lucas who later stated that it came from Nubia (Lucas 1962 p. 200). They comment that the gold and silver content indicate a native copper, but this is incorrect, such an alloy is never encountered in metals occurring in the natural state. Indeed it does seem possible that this could be the earliest known example of the *ḥśmn-km* metal, but unfortunately there was no mention of patina or inlay and, as no provenance or registration number was given, the piece cannot now be identified for further study and checking, although the analyst, Professor Bannister was a reliable chemist. The extensive programme of analyses carried out by Riederer (1978, 1982, 1983) on several hundred Egyptian statuettes now in German museums are of rather limited use because gold was not reported, and in the bronzes from the Munich Museum (Riederer 1982) arsenic was also omitted. A very few of the bronzes had moderately high silver contents in the range 0.5 to a maximum of 1.5%, but nothing else to indicate that these may have been *ḥśmn-km*.

Fishman and Fleming (1980) examined a bronze statuette of Tutankhamen and found it to contain 88.7% of copper, 4.7% of gold, 1.1% of arsenic, and 0.75% of silver together with 4.6% of tin, 1.57% of iron, 0.25% of lead and 0.038% of antimony. The statuette is inlaid with gold and silver, and apparently has 'an odd dark patina', but this was established to be recent. The authors speculate the patina 'might even have been made to crudely simulate the appearance of "black copper (*hmty km*)"'. The statuette now stands revealed as a good example of *ḥśmn-km*, and as it is quite easy to

Table 9.1 Atomic absorption body metal analyses of black-patinated bronzes

Description	*Registration no.*	*Main patina*	Composition (wt%)											
			Cu	*Pb*	*Sn*	*Au*	*Ag*	*As*	*Fe*	*Sb*	*Ni*	*Co*	*Bi*	*Zn*
Egyptian figures														
Ptah	EA 27363	Cu_2O	90.5	0.12	2.15	2.71	0.45	0.57	0.2	<0.04	0.032	<0.006	<0.02	0.01
Amun-Re	EA 63581	–	90.9	2.18	5.11	0.008	0.02	0.11	0.46	0.04	0.01	0.02	0.03	0.01
Montu-Re	EA 60342	–	82.1	8.93	6.63	0.013	0.06	0.45	0.13	0.13	0.04	0.005	<0.016	0.01
Osiris	EA 24718	–	89.7	0.34	6.09	0.064	0.05	0.4	0.09	<0.04	0.04	<0.006	<0.019	0.007
Aegis	EA 60637	–	84.8	7.4	5.2	0.005	0.05	0.7	0.2	0.5	0.08	0.007	<0.02	0.02
Classical pieces														
Plaque	GR 1979. 12–13.1	Cu_2O	92.0	1.9	1.4	0.6	1.2	1.1	<0.01	0.3	0.3	<0.005	0.005	0.05
Body of ink pot	GR 1853. 2–18.7	Cu_2O	92.8	0.06	5.58	<0.008	0.03	<0.08	0.08	0.08	0.014	<0.008	<0.028	0.12

The figures have a precision of ± 5–10% for the alloying elements and ± 30% for the minor and trace metals.

Table 9.2 EDX/SEM and XRF: surface analyses of black patinated bronzes

Description	*Registration no.*	*Main patina*	*Cu*	*Pb*	*Sn*	*Au*	*Ag*	*As*	*Fe*	*Sb*	*Zn*
			Composition (wt%)								
Egyptian figure											
Black necklace inlaid into Osiris	EA 64477	Cu_2O	68.1	3.5	10.6	3.8	2.0	(1.2)	nd	nd	nd
Classical pieces											
Top of ink pot	GR 1853.42–18.7	Cu_2O	93	nd	nd	(0.8)	2.3	nd	0.15	(0.3)	nd
Top of ink pot (cleaned)	GR 1853.42–18.7	Cu_2O	90	(0.9)	nd	(0.9)	4.0	(0.7)	0.2	nd	nd
Side of ink pot	GR 1853.42–18.7	Cu_2O	43	1.9	35.5	nd	0.5	nd	0.5	nd	nd
Base of ink pot	GR 1853.42–18.7	–	33	nd	36	nd	nd	nd	0.7	nd	nd
Black band inlaid into forceps	GR 1814.7–4.969	Cu_2O	52	3.75	0.8	3.1	3.9	nd	nd	nd	1.6
Green surface of body metal of forceps	GR 1814.7–4.969	–	37	3.1	11.0	nd	0.5	nd	nd	0.8	8.4
Black-patinated inset on bronze couch end	GR 1784.1–31.4	Cu_2O	83.5	1.1	13.8	<0.05	1.25	<0.3	0.1	0.04	<0.14

Each analysis is the average of three separate determinations, where the figures are in parenthesis, the element was only detected once. EDX/SEM detection limits are 0.6% for Pb, and As, 0.5% for Au, 0.3% for Pb, Ag, Sb and Sn, and 0.2% for Cu and Zn.
Note: nd = not detected

Table 9.3 XRD determinations of the principal crystalline compounds in the black patination

Description	*Registration no.*	*Principal compounds identified*
Egyptian figures		
Ptah	EA 27363	Cuprite, Cu_2O
Osiris (neck band)	EA 64477	Cuprite, Cu_2O
Amun-Re	EA 63581	no ancient patina
Montu-Re	EA 60342	no ancient patina
Osiris	EA 24718	no ancient patina
Aegis of Re	EA 60637	no ancient patina
Classical pieces		
Plaque	GR 1979.12–13.1	Cuprite Cu_2O
Top of ink pot	GR 1853.2–18.6	Cuprite Cu_2O + silver chloride AgCl + Ag
Black patina from draperies	GR 1853.2–18.6	Cuprite Cu_2O + tenorite CuO
Forceps	GR 1814.7–4.969	Cuprite Cu_2O
Couch end	GR 1784.1–31.4	Cuprite Cu_2O

produce the black patina on a bronze of correct composition the recent repatination probably inadvertently reproduces quite well the black patina that the statuette originally possessed. Fishman and Fleming considered the possibility that the gold could be a deliberate addition to colour the metal, having picked up the comment of Plutarch on Egyptian 'blue bronze' from Garland and Bannister (although in fact the bronzes concerned were at Delphi, see below for more on Plutarch and patinated bronzes) but missing the point of a patinated bronze, rejected it, favouring instead an accidental mix of recycled metal.

Finally in this survey of analysed Egyptian metalwork it is necessary to consider the recent report (El Gayar and Jones 1989) of gold in copper from Old Kingdom copper smelting sites at Buhen in southern Egypt (Emery 1963). It is claimed that gold was found irregularly distributed within both the malachite copper ore and the droplets of smelted copper. Gold is perfectly miscible in all proportions with copper, and it is difficult to understand how the gold could have come through the smelting process without dissolving in the molten copper. Until more evidence is presented it would seem advisable to set these results aside.

Summary

The previous analytical work on Egyptian bronzes would seem to indicate that the *ḥśmn-km* alloy is rare, but it has been observed in at least one other inlaid bronze, namely the Tutankhamen discussed above, even though it had lost its original patina. Black bronzes have been noted frequently and speculation on the nature of the patina has usually assumed it was sulphidic. The important work of Shearman showed that it was in fact an oxide, and now at last it has been established that at least some of the black patinated bronzes are of a special alloy containing gold and to link these bronzes to the *ḥśmn-km* of the ancient inscriptions.

The prestige of ancient Egypt was felt around the eastern Mediterranean, and nowhere more so than in Greece. Indeed legend and even Herodotus claimed the Greeks originated from the land of the Nile, and certainly the cultural as well as technical influence of the already old and established Egyptian Empire on the nascent Greek states must have been enormous (Bernal 1987). It is now necessary to consider patinated and inlaid bronzes in the Greek and Roman world, and the link with the 'blue bronzes' of Plutarch.

The Greek World: κύανος, *Kyanos* and Corinthian Bronze

The literary evidence

There are references to black patinated copper alloys inlaid with gold or silver from Homer onwards. The most familiar reference is in the *Iliad* (Murray 1963) where the Shield that Hephaistos made for Achilles is described in book 18. This was supposed to have been made of copper, tin, gold and silver, all melted together (474–5), and profusely decorated with inlaid scenes. The detailed description of these include references to black bronze such as 'the field was black and was like the ploughed earth although it was of gold, this was the great marvel of the work' (548–9); and again at 561–3, where a vineyard was depicted with bunches of grapes of gold, and on it there were black grapes, supported by poles of silver. The inference here is of a mixed bronze–gold–silver alloy that was itself inlaid with silver and gold, almost exactly paralleling the inlaid Mycenaean dagger examined recently (see below p. 114).

At this early period the material apparently did not have a specific name, unless the word *kyanos* is the blue-black patinated copper alloy. The term is found quite widely in Greek literature starting with the *Iliad* where it occurs at Book 11, 24 and 35, on Agamemnon's cuirass and shield, and also on the Shield of Achilles, book 18, 564, and described as being black. *Kyanos* is a general term for blue or blue-black but could have more specific meanings. In these contexts it has previously been interpreted as tempered steel (Blümner 1922 2238–2242), niello (Higgins 1974 pp. 140–141), or even as enamel but always with meanings related to colour. Thus for example the early Christian author, Clement of Alexandria (Butterworth 1960) refers to *kyanos* in his description of the famous statue of Serapis which stood in Alexandria. He described how the statue was made from filings of gold, silver, copper, iron, lead and tin together with Egyptian stones, lapis lazuli, haematite, emerald and topaz. The statue was first polished, then inlaid (αναμίγνυμι, *anamignymi*), then treated by or with *kyanos* (apparently referring either to the process or to the actual chemicals), as a result of which the statue was nearly black. Note that the English Loeb edition ignores the dative and totally alters the sense to 'stained the mixture dark blue, on account of which the statue is nearly black'.

In reality this is probably to be interpreted as a polychrome statue, rather reminiscent of the Volubilis fragment (see below), but here the actual body metal was black and inlaid with various stones and probably some of the metal as well. Note the order of operations is basically correct with the patinating or *kyanos* being the final process.

The name *Corinthium aes* only occurs in Roman times and there is no firm evidence that this was the name used by Greeks themselves before they became a Roman province.

The literary references to the Corinthian bronze are legion, and are discussed in greater detail in Giumlia-Mair and Craddock (1993), from these it is clear that Corinthian bronze was a familiar and distinctive material that required no description and thus the modern reader is left to infer both the appearance and the composition from these references that are oblique, even opaque. Engels (1990 p. 37) suggests that Corinthian bronze was a forgery, with no precious metal content at all whereas Jacobson and Weitzman (1992) claim Corinthian bronze was no more than a surface-enhanced base gold or silver alloy, exemplified by the debased *denarii* of the 2nd and 3rd centuries AD. Surely the highly valued, eagerly sought Corinthian bronze of the classical authors merits a more prestigious material than the worthless *denarii* of the economic collapse of the Roman economy. Many of the later references have been collected by Murphy-O'Connor (1983), and the general impression is of a distinctive prestige material, but more specific clues as to its nature can be gleaned. Thus the standard English translation of the *Deipnosophistae* of Athenaeus (Gulick 1927) is rather difficult to to interpret, but a very literal translation suggests that Corinthian bronze was inlaid (XI 199e), and even gives information on how this was done (XI 488c). Seneca in the *De Brevitate Vitae* (Bassore 1932 II 2) spoke disparagingly of people collecting little pieces of rusty sheet metal, from which it can be inferred that the material was patinated.

The Hebrew and Aramaic sources have recently been listed and discussed by Jacobson and Weitzman (1992), dealing especially with the gates of Temple of Jerusalem, Josephus describing how the great gate leading to the sanctuary of Herod's Temple was of Corinthian bronze 'and far exceeded in value those doors plated with silver and set with gold'.

Other references continued to be made by authors such as Petronius in the *Satyricon* (Heseltine 1913, 50) and by others through the Roman period, with Paulus Orosius noting in the early 5th century AD that the material was called *Corinthium aes* to the present day. This was copied by Isidorus of Seville, who compiled his *Etymologiae* in the early 7th century AD (Arevalo 1878), suggesting the material was still familar right up to the end of classical antiquity, and as we shall see — beyond. One point that emerges very clearly from these references and also from those of Pliny was the problem of discerning genuine Corinthian bronze from imitations, a problem reflected in the compositions of the surviving pieces reported here.

The two longest and most revealing descriptions of Corinthian bronze are given in Plutarch's *Moralia* and Pliny's *Natural History*.

The full text of the *Moralia* reference is given elsewhere in this volume on p. 34, here we just quote the specific reference to Corinthian bronze, although it should be noted that it is introduced quite naturally in a discussion on patinated bronze generally. The Greek text used is that of the Loeb edition (Babbitt 1936, V 395) although the very literal translation is our own.

"Ἆρ' οὖν," ἔφη, "κρᾶσίς τις ἦν καὶ φάρμαξις τῶν πάλαι τεχνιτῶν περὶ τὸν χαλκόν, ὥσπερ ἡ λεγομένη τῶν ξιφῶν στόμωσις ἧς ἐκλειπούσης ἐκεχειρίαν ἔσχεν ἔργων πολεμικῶν ὁ χαλκός; τὸν[1] μὲν γὰρ Κορίνθιον οὐ τέχνῃ φασὶν[2] ἀλλὰ συντυχίᾳ τῆς χρόας λαβεῖν τὸ κάλλος, ἐπινειμαμένου πυρὸς οἰκίαν ἔχουσάν τι χρυσοῦ καὶ ἀργύρου, πλεῖστον δὲ χαλκὸν ἀποκείμενον, ὧν συγχυθέντων καὶ συντακέντων, ὄνομα τοῦ χαλκοῦ τῷ μείζονι[3] τὸ πλῆθος παρέσχεν."

Ὁ δὲ Θέων ὑπολαβών, "ἄλλον," ἔφη, "λόγον ἡμεῖς ἀκηκόαμεν πανουργέστερον, ὡς ἀνὴρ ἐν Κορίνθῳ χαλκοτύπος ἐπιτυχὼν θήκῃ χρυσίον ἐχούσῃ πολὺ καὶ δεδοικὼς φανερὸς γενέσθαι κατὰ μικρὸν ἀποκόπτων καὶ ὑπομειγνὺς ἀτρέμα τῷ χαλκῷ θαυμαστὴν λαμβάνοντι κρᾶσιν ἐπίπρασκε πολλοῦ διὰ τὴν χρόαν καὶ τὸ κάλλος ἀγαπώμενον. ἀλλὰ καὶ ταῦτα κἀκεῖνα μῦθός ἐστιν· ἦν δέ τις ὡς ἔοικε μεῖξις καὶ ἄρτυσις, ὥς που καὶ νῦν ἀνακεραννύντες ἀργύρῳ χρυσὸν ἰδίαν τινὰ καὶ περιττὴν ἐμοὶ δὲ φαινομένην νοσώδη χλωρότητα καὶ φθορὰν ἀκαλλῆ παρέχουσι."

'Wasn't there' he said 'some alloy and a (colouring) treatment used on bronze by the ancients, like the so-called tempering of swords, which being abandoned, bronze had a respite from employment in war? And actually they say that Corinthian ware acquired the beauty of colour not through art, but through accident, fire destroyed a house that had some gold and silver and a lot of stored copper (or bronze), these metals being melted and fused together, the large amount of copper gave the name because it was the major quantity.'

And Theon beginning to speak said 'But we have heard a more complicated story, that a bronze worker at Corinth, having found a casket full of gold, afraid of being detected, divided it a little at a time, quietly adding it to the bronze, which acquired a singular alloy, he gained a lot because (its) colour and beauty were appreciated.

But both this and that story are a myth: however, it seems there was some mixing and treatment (seasoning), because in some way now they also alloy gold with silver and produce some extraordinary, and for me excessive and wondrous sickly paleness, and an ugly shading of the colours.'

From the text we learn that copper was the major metal with smaller amounts of gold and silver, and the clear implication is that the metal owed its appearance to some treatment.

It is important to note the different verbs used here to describe the union of gold and silver with copper in the two apocryphal accounts of the discovery of Corinthian bronze. In the first the metal ran from the burning house and mixed or amalgamated, in this instance Plutarch uses the verb συγχψηέω, *syncheo*, but in the second account the gold was mixed by the bronzeworker in the sense of added to or incorporated and the verb ὑπομείγνυμι, *hypomeignymi* is used, that is still retaining a separate identity. This is the sense of the word that is used by Pliny (see below) when an inlay is meant. This meaning also makes much more sense here. If the bronzesmith just put small quantities of gold and silver into the bronze alloy without further treatment it would, to all intents and purposes, be lost, but if the meaning is that the smith was disposing of the gold in small quantities as *inlays* in the Corinthian bronze then all becomes comprehensible, once again the sense of both alloying and inlaying precious metals.

Pliny's *Natural History* has by far the longest and most detailed, and through no fault of his own, misunderstood descriptions of Corinthian bronze. The first reference is significantly in a section on purple dyeing at 9. 139 (Rackham 1940), where the implication is that Corinthian bronze was made by adding gold and silver to copper or bronze. The main discussion is appropriately in book 34 (Rackham 1952) devoted to copper and its alloy, where, because of its prestige and value, Corinthian bronze commences the work. Its value lay 'before silver and almost before gold' (34.1). Continuing at 34.5, Pliny describes how the bronze was first alloyed (*confundo*) and then inlaid (*misceo*) with gold and silver. Note *confundo* has the sense of the intimate or inseparable union, such as a solution, or a molten alloy (also *temperatura*), whereas *misceo* has the sense of a mixture where the components retain their separate identity. A good example of Pliny's use of *misceo* is given at 33.132, where he describes debased coins made of iron and silver and uses *misceo* to describe their union. Recent scientific examination described elsewhere in this volume (Zwicker *et al.* p. 223) has conclusively shown that these were of an iron core plated with a silver–copper alloy. That is, the iron and silver were united in such a way as to retain their separate identities (although at 34.6 *misceo* is used to describe the addition of one part *aes* to silver in a context which should mean alloying). As with the analogous Greek words used by Plutarch above, Pliny was using the terms for mixing and adding with care and precision, which must be strictly followed if the nature of these complex materials is to be elucidated.

At 34.6 and 7 Pliny continues, giving the familiar apocryphal account of the accidental discovery of Corinthian bronze, and then at section 8 describes three varieties

> eius aeris tria genera: candidum argento nitore quam proxime accedens, in quo illa mixtura praevaluit; alterum, in quo auri fulva natura; tertium, in quo aequalis omnium temperies fuit. praeter haec est cuius ratio non potest reddi, quamquam hominis manu est, at fortuna temperatur in simulacris signisque illud suo colore pretiosum ad iocineris imaginem vergens, quod ideo hepatizon appellant, procul a Corinthio, longe tamen ante Aegineticum atque Deliacum, quae diu optinuere principatum.

> There are three kinds of this bronze: one of light colour coming very near to silver in brightness, in which that assembly (*mixtura*) predominated; the second one in which there was the yellow nature of gold, the third one in which there was an equal proportion (*temperies*) of all. Besides these there is (another one) whose formula (*ratio*) cannot be given even though it is made by man, but it is prepared (*tempero*) in the right quantity by luck — in portraits and in statues, this (bronze) is valued because of its colour, which turns into a liverish appearance and is therefore called liverish (*hepatizon*), far from Corinthian bronze, nevertheless a long way ahead of the Aeginetian and the Delian, which for a long time occupied the first rank.

The silver and golden varieties refer primarily to their **inlays**, *mixtura* is the word used implying a distinct and separable union. Thus the first, that is very close to silver in brilliance, should be interpreted as Corinthian bronze heavily inlaid with silver such as on the cylindrical bodies of the pair of inkpots (G R 1853.2-18.6) which, perhaps significantly, are included in the catalogue of Greek and Roman silver (Walters 1907 Cat 90) (Plate 9.1). The second, in which the yellow nature of gold is present, should be understood as being inlaid with gold, for example on the tops of the above-mentioned inkpots. In the description of the third variety *temperatura* is not used, but *temperies*, a much more general, looser term meaning proper balance, conveying the sense of equal proportions, suggesting perhaps a more intimate form of union, either an alloy or possibly a form in which the black of the Corinthian bronze and gold and silver inlays were interwoven such as on the small Roman plaque, GR 1979.12-13.1 (Plate 9.2).

If Pliny was primarily referring to inlays then this explains why the difference was so noticeable, but one does wonder if he might also have been referring to different compositions in the actual alloy as well – the wording of the text suggests something a little more fundamental or complex than merely an inlay. Possibly the composition of the alloy was reflected in the inlay. This hypothesis is supported by the few classical pieces to be analysed so far. Thus the tops of the inkwells (Detail, Plate 9.1) are inlaid with gold and the alloy contains about 0.5% of gold, in contrast to the silver inlaid bodies which have no gold or silver. The couch end, inlaid with silver and copper has 1.2% of silver but no gold (Plate 9.3), and similarly the silver inlaid black bronze on the drapery from Volubilis (see below p. 115) has about 4% of silver, but no gold was reported. The Mycenaean dagger blade reported above and the small plaque have small amounts of both silver and gold and are inlaid with gold, silver and electrum.

This apparent association of the metal inlay with the body metal may seem inherently unlikely, and taken over such a small number of objects can be no more than a suggestion, but it does have a strong parallel in another inlay material from the classical world, namely niello. La Niece, in her 1983 study, showed that Roman niellos were normally of single sulphides: silver artifacts tended to be inlaid with the silver sulphide, acanthite, whereas copper alloys tended to be inlaid with copper sulphides, and on one gold artefact a complex gold sulphide had been used.

Finally there was another kind, *hepatizon*, which was an alloy of unknown composition, used in portraits, statues and was of a liverish appearance, that is a purple colour, but far from the true Corinthian bronze. This would seem to be something else entirely, linked only by colour. The word *temperatura* is used which implies a true inseparable union such as a solution or an alloy. Possibly *hepatizon* is to be equated with the alloy referred to at 34.98 'the addition of lead to Cyprus copper produces the purple colour seen in the bordered robes of statues', although Pliny states here that the composition of *hepatizon* is not known, and could also be equated with the *molybdochalkos* of the alchemists.

There is one specific reference to the manufacture of Corinthian bronze in that city. Pausanias in his *Description of Greece* (II:3.3, Jones 1918 p. 261) states:

> *κεκόσμηται δὲ ἡ πηγὴ λίθῳ λευκῷ, καὶ πεποιημένα ἐστὶν οἰκήματα σπηλαίοις κατὰ ταὐτά, ἐξ ὧν τὸ ὕδωρ ἐς κρήνην ὕπαιθρον ῥεῖ πιεῖν τε ἡδὺ καὶ τὸν Κορίνθιον χαλκὸν διάπυρον καὶ θερμὸν ὄντα ὑπὸ ὕδατος τούτου βάπτεσθαι λέγουσιν, ἐπεὶ χαλκός . . . γε οὐκ ἔστι Κορινθίοις.*
>
> The spring is embellished with white marble and chambers like caves have been made, from which the water flows into the fountain in the open, and not only is it pleasant to drink but also they say that Corinthian bronze made red hot and incandescent is 'dyed' [*bapto* dip, immerse, plunge, dye] by means of its water since bronze [copper] ..... the Corinthians have not.

This is the only direct reference to the process, based on the visit of Pausanias to the city in AD 165. It is unfortunate that a lacuna occurs in the last sentence and also that the Loeb edition translates *bapto* as tempered without comment or giving alternative meanings. It is not possible to temper ordinary bronze alloys and this has led some commentators to dismiss the passage altogether. Caley (1941) produced a major study on the corroded bronze of Corinth, and discussed Pausanias' statements on the fountain and Corinthian bronze in great detail. He was apparently not aware that Pliny and Plutarch specified that true Corinthian bronze contained precious metal, much less that it could itself have been a patinated material even though he quotes from the relevant passage of the *Moralia*. He noted that *bapto* was used quite frequently in the context of the dyeing of cloth, 'That βάπτεσθαι, means that the bronze received a color in dipping,... seems at first glance very likely from the ways in which the verb βάπτω is often used by ancient authors other than Pausanias', although he ultimately rejected this interpretation because he could not perceive how the bronze could become coloured, and instead came down in favour of tempering, possibly of a high tin bronze. In the *Leyden Papyrus* X, recipes 19 and 20 (Halleux 1981 pp. 89–91) the closely related verb καταβάπτω is used to describe the dipping of a base silver alloy in a corrosive liquid to change the surface appearance, that is a process not dissimilar to that which we propose for the Corinthian bronze. Similarly Plutarch uses the word *baphe* when describing the colouring of bronze (Moralia V 395B).

The fountains of Peirene were quite famous in antiquity and Athenaeus also mentions them in his description of the waters of Greece. In the *Deipnosophistae* II 43 he states:

> Σταθμήσας τὸ ἀπὸ τῆς ἐν Κορίνθῳ Πειρήνης καλουμένης ὕδωρ κουφότερον πάντων εὗρον τῶν κατὰ τὴν Ἑλλάδα.

This is translated in the Loeb edition as follows:

> When I weighed the water from the Corinthian spring I found it to be lighter than any other in Greece.

This an odd statement, but one that is difficult to dismiss completely as Athenaeus is at pains to point out that this is something he has checked personally. The only way water could be 'lighter' is with dissolved gas, that is carbonated, but the waters from the spring, which still exists are not aerated now. Could Athenaeus have been referring to waters that became aerated by dissociation of the dissolved bicarbonate when heated in the process of *baptein*? However the whole section concerns the potability of water generally and the verb could also mean easy, relieving, easy to digest.

The significance of the ancient comments on the fountains of Peirene have been greatly increased by the excavations in Corinth by the American School of Classical Studies, which have not only located the base of the fountain but traced the channels running from it to the metalworkers' quarter where the remains of large basins were located (Stillwell *et al.* 1941 pp. 22–27) (Figure 9.3), as well as some other metalworking areas (Mattusch 1977). The use of large basins set in the ground would suggest that the water or solutions made from it were not heated, confirming as far as possible Pausanias' description that it was the metal which was heated. Caley sampled and analysed the waters finding them to contain:

Figure 9.3. *Paved basin in the metalworking quarter at Corinth, the basin connected to the Peirene drain. (Photo from Stillwell* et al. *1941)*

	ppm
Calcium	62
Magnesium	57
Potassium	48
Sodium	94
Iron	0.2
Bicarbonate	550
Chloride	106
Sulphate	48
Dissolved or colloidal silica	63
Nitrogen compounds and organic matter	Present but not estimated
Total dissolved matter	1030

The water, as Caley noted, has an unusually high concentration of a number of salts. Once again he could make no significance out of this in terms of a tempering process, but as the medium for a patination process, it stands some comparison with the solutions used in the *nikomi chakushoku* process, although it is rather weak and the copper salts are missing.

These then are some of the more important references to Corinthian bronze and black bronzes in the mainstream of classical literature, but there remains one other source of written information to consider – the alchemists. Surface treatments of base silver and copper alloys were an integral part of the transformation processes carried out by the Alexandrian alchemists. Thus one might expect their surviving works to be a rich source for this study. However, most of the texts are vague and even deliberately misleading, and in practice only one author, Zosimos, who probably lived in Egypt in the 3rd century AD, specifically mentions a process in any way analogous to the Japanese *nikomi-chakushoku* process. The majority of the other descriptions of processes give little real technical information (although a later Syriac translation of a text ascribed to Zosimus gives a much fuller description and specifically identifies Corinthian bronze as black copper or bronze. Because much of the detail of the text is almost certainly medieval the translation has been included in the Islamic section, see below p. 120). One of the fundamental processes of the alchemists was the transformation of copper alloys, the so-called process of *iosis* (Shepard 1964). In this the metal was usually heated and then chemically treated passing through several colour changes, from black through white, yellow, red and finally achieving the *iosis*. Now this term can mean either patinating or, more specifically, purpling, and in the context of the sequence of colour changes outlined above, purple seems more appropriate. Practical details of the processes are predictably virtually non-existent, but one place in the Zosimos texts states that the process of *iosis* is to be carried out with pimpernel and rhubarb (Zosimus III 16, Details of the process, Berthelot 1887–8). Rhubarb is rich in malic, acetic and oxalic acids, the occidental equivalent of the Japanese bitter plum, and presumably it performed the same function in the patination process.

We must now consider the evidence for metalwork that has survived from classical antiquity fitting our postulated description of Corinthian bronze.

Surviving examples of black patinated bronze from classical antiquity

Amongst the most renowned and splendid of the Mycenaean artefacts excavated by Schliemann from the shaft graves of Mycenae itself were the inlaid bronze daggers. Matt black panels were set into the dagger sides and into these were inlaid scenes depicted in gold and silver. In the absence of scientific investigation, speculation has continued for years over the nature of the black panels. Many have suggested that they must be of niello although without any supporting analyses (Maryon 1956 p. 481, followed by Laffineur 1974, and more recently by Xenaki-Sakellariou and Chatziliou 1989). In antiquity niello was a silver or copper sulphide applied hot as a pasty mass, although its use at all in Mycenaean times is problematic (La Niece 1983). Cooney (1968) refuted the suggestion that niello was used on the inlaid daggers, observing that 'the inlays are set on a background of black bronze' and presciently linked them with the Egyptian *ḥsmn-km*. Less than 20 of the inlaid daggers are known but very recently a previously unknown example appeared on the market from an old collection and was analysed by Ogden (1990 and this volume). This analysis amply confirmed Cooney's observation, showing the black panels to be of a black patinated bronze containing 93% of copper, 5% of tin, 1.7% of gold, 0.53% of silver and 0.5% of arsenic. The black layer was principally made up of cuprite. Other inlaid daggers in the National Museum, Athens have at last been analysed and are reported to be of a similar composition (Photos-Jones, personal communication).

Cooney was also able to identify many other potential candidates for the Egyptian *ḥśmn-km* by their colour and inlay, pointing out several later classical Greek bronzes that were inlaid and seemed to have the correct black patina. These included a mirror support of the 5th century BC in the form of a siren now in the Cleveland Museum of Art (John L. Severance Fund 67.204) and an infant Hercules of the 1st century BC now in the City Art Museum of Saint Louis (36:26) (Cooney 1968). Although black patination by itself is not necessarily an indication of deliberate ancient treatment (see below p. 117), given the accuracy of his observations generally these examples should be seriously considered, but until very recently no pieces had been tested analytically. The results of these first investigations reported here and by Ogden (this volume p. 41) show that Cooney was quite right in maintaining that these bronzes were the deliberate product of a specific treatment, even if he assumed, in common with many others through the ages, that the process must have something to do with the formation of copper sulphides.

Few other pieces have yet been recognized. One example is the extraordinary drapery fragments from a lifesize bronze statue found at the Roman city of Volubilis (Boube-Piccot 1969 pp. 54–64). The fragment is richly decorated with complex polychromy employing gilding, inlays of silver, brass and other copper alloys and a variety of patination, described in the publication as *damasquinure*. The patinations included shades of olive-brown, orange-yellow and *'une belle patine noire a reflets violet'* on an alloy containing 90% of copper, 5% of lead, 4% of silver and 0.5% of calcium, and 0.2% of antimony (note 187). Unfortunately it does not state whether gold or arsenic were sought. It is perhaps significant that this silver-rich patinated copper was let into a silver inlay. Very few other examples of this degree of polychromy survive and the implication must be that it was very rare.

The first classical piece to be consciously examined as an alloy similar to *shakudo* and being potentially of Corinthian bronze was the small Roman plaque now in the British Museum, GR 1979.12-13.1 (Craddock 1982 and 1985). This has the now familar dense black patina on one side only, heavily inlaid with gold, although some inlays are now missing (Plate 9.2).

Summary of previously published evidence

Thus to summarize the evidence from the contemporary Greek and Roman sources, and from previous investigations: Homer refers to black copper, in association with both gold-copper alloys and gold and silver inlays. From the 4th century BC onwards the term Corinthian bronze is used in contexts that suggest that it was very distinctive and highly prized by collectors, although fakes and imitations abounded amongst the genuine Corinthian pieces. It was valuable both in terms of the materials used and the workmanship displayed. Descriptions given in the *Deipnosophistae* suggest the Corinthian wares were inlaid with precious metals and from Seneca we get the first hint of patination. This is reinforced by Plutarch, from whom we learn that it was an alloy predominantly of copper with some gold and silver, and this was then treated to produce Corinthian bronze. The much longer and more detailed accounts given by Pliny in the *Natural History* include a reference in a section on dyeing (Book 9) with specific reference to purple. In Book 34 the material is described in much greater detail, and clearly Pliny was referring to the combination of copper, gold and silver in two quite different ways, as an alloy of the metals but also as additions of gold and silver to the bronze alloy where the metals remained distinct, that is as inlays. He also described several different types of Corinthian bronzes, the text suggesting something more than just inlay, and the choice of precious metals used in the inlays may have been linked to the body composition. Once again there is an association with purple materials, *hepatizon*. Pausanias actually locates the site of production near the fountains of Peirene in Corinth, and mentions the use of the waters in the colouring or patinating process. The Alexandrian alchemists, notably Zosimos, make frequent reference to the process of *iosis*, which seems to show many parallels with the later Japanese *irogane* processes.

Thus the various descriptions suggest copper alloyed with varying but small amounts of gold and silver patinated to produce a black or purple surface which in turn was inlaid with precious metals. This would of course serve as an exact description of the Japanese *irogane* alloys.

Cooney suggested several examples of inlaid classical bronzes with black patinas similar to *ḥśmn-km*, but until very recently few have been analysed or recognized for what they are.

Scientific examination of pieces from the British Museum

For this project some Egyptian and Roman items were selected for scientific examination. Six Egyptian statuettes were chosen on a purely visual basis (Figure 9.2, Plate 9.5). They are all Late Dynastic or Ptolemaic, that is ranging from about 600 to 100 BC. On most of them the whole body is blackened, but on the Osiris figure, EA 64477 the black bronze is inlaid into the bronze to show off a multi-strand gold necklace to better advantage (Figure 9.1).

Three Roman items were selected, one of a pair of small inkpots, GR 1853.2-18.7 (Walters 1907 Cat no. 90) (Plate 9.1), one of a pair of inlaid bronze couch ends, GR 1784.1-31.4 (Plate 9.3), and bronze forceps inlaid with bands of black bronze, GR 1814.7-4.969 (Plate 9.4). The inkpot is in two parts, a cylindrical body and base and a separate top. The body is inlaid with silver and the top with gold, furthermore the black patina on the top is much denser and finer than that on the body, which in fact is a very dark green, a difference reflected in the composition (Table 9.2).

Methods of examination

The surfaces of the pieces were all analysed by X-ray fluorescence which showed that some did indeed include several per cent of gold and silver in the black patinas.

Samples of the patina were taken for X-ray diffraction analysis to determine the nature of the minerals present (see below). Where possible samples of the body metal underlying the patina were drilled for analysis by atomic absorption spectrometry (AAS, details of the methodology are given in Hughes *et al.* 1976). The figures have a precision of ±2% for the copper, ±5-10% for the alloying elements and ±30% for the minor and trace metals. The detection limits for the individual elements are given in the notes to Table 9.2.

It was not possible to obtain drilled samples from the black bands around the neck of the Osiris figure (EA 64477), or from those around the Roman forceps (G R 1814.7-4.969), or from the top of the Roman inkwell (Cat 90, G R 1853.2-18.7), without doing unacceptable damage. These pieces were examined and analysed by micro-analyser (EDX) in the scanning electron microscope. The results are the average of three separate determinations; where the figures in the tables are in brackets this indicates that the element was not detected in all three determinations. This is because the concentration of the element concerned is at its detection limit of about 0.2%. The couch end, G R 1784.1-31.4, was analysed by X-ray fluorescence. These analyses are necessarily confined to the patinated surface, but concentrating on small areas such that contamination from either inlays or surrounding body metal could be avoided. The results of these analyses are given in Tables 9.1 and 9.2.

Discussion of the results

Composition

The pieces show a wide range of composition; some do indeed have levels of gold or silver that are way above those normally encountered in copper alloys of any period, but which accord very well with the levels of precious metals found in the Japanese *shakudo* alloys. The arsenic is more problematic, some of the pieces, such as the Egyptian figure of Ptah (Plate 9.5) or the Aegis of Ra and the Roman plaque do have arsenic contents that are high relative to the contemporary bronzes, but arsenic is a common minor constituent of ancient bronzes, and it is possible that these high arsenic levels are fortuitous. However, as noted above (p. 102), analyses of the Japanese *shakudo* show similar enhanced levels of arsenic (La Niece 1990), although neither arsenic or the natural arsenical copper alloy *shirome* do not usually appear in recipes for *shakudo*.

Five out the total of eleven pieces analysed here by AAS and EDX have gold, silver and arsenic contents that are typical of the ordinary contemporary bronzes and three of these have completely spurious modern organic patinas suggesting that they were never originally intended to appear black. However the other two pieces, the body of the inkpot and the Aegis of Ra, are inlaid and have a black patination suggestive that this was their original appearance (although see below, p. 117, for a discussion of recent restoration and conservation). Clearly it was possible to obtain a black patination of sorts without the addition of precious metal to the alloy, and this could explain something of the trouble the ancients had in distinguishing 'real' Corinthian bronze from inferior copies (*Satyricon* 50, *Natural History* 34 7, etc.). From these and other comments we are led to believe that only the true connoisseur could distinguish and appreciate the patina developed on the real Corinthian alloy. Naturally a similar situation pertains today with genuine *shakudo* and its imitators.

Patina

X-ray diffraction identified cuprite as the principal crystalline mineral present in the ancient black patinas, although other minerals were present, including paratacamite, a copper chloride mineral commonly found on corroding Egyptian bronzes (Table 9.3). Cuprite was the principal mineral detected by La Niece (1990) in her recent study of *shakudo*, in common with Notis (1988) and Murakami (p. 90 and 1988) in their more detailed studies on the nature of the *shakudo* patina. However, there is a practical problem here in studying the surface phenomena of ancient buried material. The Egyptian bronzes for example were quite corroded and have been extensively conserved, often followed by a 'restoration' of the patina. Three of the bronzes (EA 63581, Amun-Re, 60342, Montu-Re and 24718, Osiris) appear to have been electrolytically stripped. These treatments are unrecorded, but since records have been kept bronzes EA 24718 and 64477 have been treated for minor outbreaks of active corrosion with silver oxide (the analyses were performed on deep drillings thus avoiding the possibility of contamination). The surfaces of the three stripped Egyptian bronzes and the Osiris figure which has a real inorganic black patina band around the neck (this of course was originally intended to contrast with the bronze of the rest of the figure) have been darkened with a coloured lacquer in recent times. Presumably the lacquer was applied after electrolytic reduction, although Plenderleith (1956 p. 251) recommends an inorganic treatment 'If the bronze (embellished with silver) has been reduced it should be boiled in distilled water for some hours at the conclusion of the treatment to darken it if necessary, so that the dark metal will act as a foil to the silver.' It would seem that Plenderleith was unconsciously recommending a treatment quite close to the original for producing the black surface, but using distilled water instead of that from the fountain of Peirene, etc.

This matter of corrosion and of conservation and repatination of stripped and reduced surfaces does raise important questions concerning the reality of the surfaces we are describing here. Clearly there is now little evidence that the Egyptian figures of ordinary bronze with only a black lacquered surface were orginally black, and almost certainly the body metal of the black-inlaid Osiris would have been kept bronze coloured. Conversely the fine inlaid figure examined by Fishman and Fleming (p. 105) with 4.7% of gold in the alloy may well originally have had a black patina, even though that which it now has is apparently modern.

One of the perennial and most difficult problems in art history is to discover the original surface appearance of ancient metalwork. In most cases corrosion over the ages, especially of buried metals, will ensure that the present colouring and texture bear little resemblance to the original. The patination on the pieces we identify as being Corinthian bronze is certainly very different from the usual green or brown corrosion, but could it be no more than a natural corrosion developed during burial, or in some cases of wilful repatination?

The idea that some classical metalwork was deliberately patinated is by no means new. Early collectors, including some of the most famous such Caylus and Payne Knight firmly believed that most classical bronzes originally had a black patina and even went so far as to replace the natural patina they had acquired over the centuries with a quite spurious black patina. Elsewhere in this volume (Chapter 2), Born argues persuasively that the patina we admire on many ancient bronzes is the original chemically treated surface, but there seems little supportive evidence, as we discuss in our article on the same subject (Chapter 3). Similarly Vickers (1985) argues that much Greek silver was given or allowed to develop a fine black patina; that is, the patina that much ancient silver possessed on discovery before it was removed to reveal the silver beneath to make it acceptable to modern tastes and perceptions. Vickers' hypotheses have not found general acceptance (Boardman 1987), but once again they do highlight this problem of ascertaining the original appearance of a metallic surface especially where there is the possibility of an original deliberate patination in addition to the usual problems of corrosion, conservation and restoration.

Possession of a fine patina, which can now be one of the most attractive and prized features of an ancient bronze, can suggest that it was an original feature. However, in itself it is no proof that this was the case, as for example with the Chinese mirrors with superb black patination, which is almost certainly the result of a completely natural corrosion phenomena (see Meeks Chapter 6 this volume). Given this scepticism about other claims for early patination, we must state the reasons why we believe the pieces we are publishing here have their original synthetic patination.

There are two separate approaches: first, many of the pieces have highly unusual, but precise, compositions, for which there is no explanation or parallel except with the modern *shakudo* alloys; in other words these are quite specific alloys for patination. Second, the black patina appears as part of a more complex polychrome scheme of surface decoration. Either the black patinated bronze is itself inlaid into the body metal of ordinary bronze, as for example on the Roman forceps, or more usually the black bronze itself is heavily inlaid with gold and silver in a way which would only make sense if the background was intended to be dark and matt. Thus both the alloy and decorative schemes of the pieces we identify as being *ḥsmn-km* or Corinthian bronze independently suggest original and deliberate patination.

However, we do have to consider the possibility that an alloy produced, for example, by the chance melting down of scrap gilded bronze could have developed a patination similar to that produced by treatment during burial. The small sub-Roman bracelet from Cannington, England examined by Bayley and McDonnell (Bayley 1992) is a case in point. It has no inlay or decorative elements that require a patinated surface, and the possibility does exist that the black patina is the result of natural corrosion.

Origins and cultural transmission

This paper has so far concentrated on the identification of *ḥśmn-km* and Corinthian bronze and their similarity to the Japanese *irogane* alloys. It is now appropriate to consider the possible links between the ancient Occident and the apparently more recent Orient. First the evidence for the origins of these black patinated alloys will be discussed, followed by examples of similarly treated metals from cultures spanning the gap between the world of Classical antiquity and Edo Japan.

The Middle East

Cooney (1966) was of the opinion that black patinated bronzes originated in Mesopotamia and cited an inlaid bronze bull with a black patina now in the Louvre (A 02151), dating to about 2500 BC, and thus about a thousand years earlier than the earliest Egyptian reference to *ḥsmn-km*. Unfortunately it has not been possible to recognize any potential Sumerian or other Mesopotamian examples in the collections of the British Museum. One problem is that due to the saline soil conditions prevalent in Mesopotamia most of the bronzes are heavily mineralized, a problem compounded by the often drastic conservation, including electrolytic stripping, they may have received in the past.

If we discount the unidentified copper chisel with 4% of gold analysed by Bannister (pp. 104–5) then the earliest surviving Egyptian examples also seem to commence around the mid-second millennium BC both in Egypt and Mycenae. Cooney (1966 p. 47) noted a superb bronze axe socket from Ugarit dating to about the 15th century BC. The socket takes the form of an animal and the pelt is inlaid with gold strips (Schaeffer 1939 pp. 107–113). The bronze apparently has a dark blue patina, but analysis of the body metal by M. Leon Brun (reported in Schaeffer 1939) shows it to contain 98.3% of copper, 1.4% of iron, 0.4% of nickel, 0.22% of tin, 0.1% of lead and traces of arsenic and sulphur. Silver, gold, zinc and antimony were sought but not detected. Thus if it was a deliberate patination there is nothing distinctive about the metal composition. Partington (1935 p. 393) noted amongst some Luristan bronzes 'A large plain goblet, figured cups of Assyrian style and a fragment of of a mirror have a highly polished curious black finish produced by some unknown technique.'

When we move beyond the Middle East and the classical world forward in time and further to the East there are a number of related materials, and some actual contenders to be the same material.

India

Little direct evidence for *shakudo*-type alloys has yet been found in India although there are some early literary references. A Greek text by Philostratos of the 3rd century AD describes bronze tablets at Taxila, which illustrated the life of Alexander and Poros (Conybeare 1960 2, 20). The scenes depicted on them were of brass, silver, gold, iron and black copper and were compared to the work on the Shield of Achilles (2, 22). The same text mentions that the Indians had coins of brass and black copper or bronze (2, 7). There is an important reference in the *Ramayana*. This concerns the Aryan horse sacrifice ceremony, for which the officiating Brahmin priests were rewarded with *jambunada*, a special 'native gold, of a peculiarly dark and brilliant hue, which was compared to the fruit jamba (not unlike a damson)' (Tod 1829, 1914 edition p. 66). This was a native gold that came from the Himalayas and is echoed in some of the Tibetan, and later Chinese alloys of dark natural gold discussed below. Although vedic in origin this ceremony is described in a sanskrit text which should be dated to around the 2nd century AD, by which time India had been in contact with the Hellenistic world for some centuries. However, as yet no other literary references have been found nor have examples of the black patinated alloys, until the much later *bidri* alloys.

Indian *bidri* ware has a dense black patinated surface, invariably heavily inlaid with silver and more rarely with gold or brass (Stronge this volume Chapter 11 and 1985). The alloy, of zinc with a small amount of copper, is treated in hot aqueous ammoniacal solutions to develop the black patina, which provides a superb setting for the silver inlays, very similar in concept and appearance to Corinthian bronze. The technique has always been practised by Moslem craftsmen and its manufacture confined to the Moslem areas of India, the material taking its name from the city of Bidar in the Deccan from where the process traditionally spread. The earliest firm evidence for the production of *bidri* is no earlier than the 17th century, although popular tradition takes it back to the 14th century, still a very long way from classical antiquity. The present *bidri* makers all assert that the technique began much earlier in Arabia or Persia and came to India with the Moslem invaders. Now it would seem inherently unlikely that true *bidri* could have originated elsewhere than in India as zinc metal was not readily available in the Middle East. It is, however, possible that black patinated bronzes, based on the classical Corinthian bronze were made in the Middle East and that these were the inspiration for the *bidri* wares. Other suggestions have included the inlaid and patinated steels, either the indigenous Indian *kuftgari* or the Islamic 'false damascene' Toledo wares (Untracht 1982 p. 313).

The Islamic world

The 13th century Persian technical author, Al Kashani gives recipes for coloured gold, including black gold, to be made by adding copper to molten gold (Allan 1979 p. 9). The term Corinthian bronze continued to be used well into the medieval period, thus the Syriac lexicographer Bar Bahlul, writing in the 10th century describes it as being part silver, part gold and part copper (Duval 1888–1896).

A far more important source is included in the Islamic texts of Zosimos (Berthelot 1893). This text, known from a 15th century Syriac version, was clearly translated from a Greek original in the 10th or 11th century. However, the length and clearness and, above all, detail of the description are totally unlike that of any antique alchemist and this in itself suggests that the medieval transcriber was very familiar with the process. Indeed much of the practical detail may well be an addition, although the use of the term Corinthian bronze and the association with Zosimos does suggest there was an original, probably more abbreviated text. It states:

> Production of black metal sheets or Corinthian alloy. It is used for working images or statues you want to make black. It works the same on statues, or trees, or birds, or fishes, or animals, or on the objects you want. Cyprian copper one mina, silver eight drachmas, gold eight drachmas. Melt and strew with 12 drachmas of sulphur, 12 drachmas of untreated ammonical salts. Take it and put into a cleaned vessel, by placing ammonical salt underneath. Then strew on ammonical salt that has been projected (purified). Let it cool down. Then take it heat it up and throw this preparation into half a bushel of vinegar, some live black vitriol (acidic iron sulphate ?): all this for a mina of copper.
>
> If you want to work on more or less take the preparation in proportion and let it cool in the ingredient.
>
> Take it, laminate the metal, but do not roll for a length of more than two fingers. Then heat it and each time you heat it, throw it into the ingredient and remove the dirt in order to make it shine, This copper will retain its blackening when it will be filed and powdered; if melted it will remain saturated of its black colour. (Zosimos VI 2)

This is the most explicit statement that Corinthian bronze was indeed a black patinated alloy, similar in composition and treatment to the Japanese *irogane* alloys. Cyprian copper is normally taken to mean pure copper, following Pliny's remarks in the *Natural History* (34.94). The quantities equate to one pound of copper to an ounce each of gold and silver, approximating to an alloy containing about 6.5% each of gold and silver in copper. This figure is a little high compared to the composition of the ancient pieces, but still well within the range of the modern *shakudo* alloys. The patination process is to some extent similar to the later Japanese process, with the exception of sulphur, which was unnecessary and for which no evidence has been found on the surviving metalwork; we wonder if this could be a later interpolation from someone, who, like Professor Middleton (page 104), thought that sulphur must be essential to the process. Heating in the ammonical solution would be an effective degreasing treatment, similar to the Japanese first stage with wood ashes. The metal was then to be heated (cf. Pausanias 2,3,3 Jones 1918 p. 261) and put into a solution of vinegar and iron sulphate. Once again these are somewhat different chemicals from those in the Japanese recipes, copper salts being notably absent, but the process is basically similar. Even the comment that the metal can be worked and melted after treatment, which must be wrong, can be paralleled in Gowland's comments on *shakudo* where after describing the patination process continues that 'moreover it has excellent casting and deformation properties'. Presumably both of them meant that the metal could be easily shaped before treatment.

None of these putative Islamic Corinthian bronzes have yet been identified. The recipes for the special Persian *haft josh* alloys specify seven metals including gold, but analyses of surviving examples of these alloys contain no gold but instead are high tin bronzes, gaining their special properties from the alloy and the methods by which they had been fabricated (Craddock 1979). Although they were invariably inlaid there is no mention of patination or of black surfaces in the medieval descriptions of *haft josh* and it seems that the presence of gold and indeed of most of the other seven metals was apocryphal.

Special alloys made from mystically significant numbers of metals are of course well known in other cultures, notably India and Tibet where eight seems to have been an especially propitious number.

Tibet and China

There are frequent references in Tibetan records from the 10th century AD onwards to special alloys that seem to have been made to emulate the appearance of prized, almost mythical, native coppers which had once been found in the ground (Lo Bue 1981 pp. 41–44). Metals such as *zi-khyim* were described as being especially favoured materials from which to cast statues. It was apparently of a red hue and when treated with 'poisonous water' (Lo Bue suggests acid for this) became iridescent. It came in two varieties, the true metal which was dug as a metal from the earth like gold, and the artificial variety, which was an eight-fold alloy containing copper with gold, silver, white iron, rock crystal, black and white lead (lead and tin?) and mercury. Clearly this description must be treated with great caution, but there do seem to be the familar elements of a copper alloy with gold and silver and of surface treatment. Most Tibetan words for metals and alloys are of Sanskrit origin, but *zi-khyim* is not. Tucci (1959 p. 180 n. 2) suggested that it derived instead from the Chinese *ch'ih chin*, a term in use from the 6th century AD which is translated as 'deep coloured gold, copper' in the standard Chinese–English Dictionary (Mathew 1969 145 p. 1048), or as scarlet gold by Needham (1974 p. 261). This is especially significant in view of Needham's association of the related material *tzu mo chin*, purple sheen gold, with *shakudo* (although the usual meaning of *shakudo* is red copper, which may suggest that *ch'ih chin* is a better candidate). Here it is only necessary to note that the Tibetans understood *zi-khyim* to be a copper alloy containing amongst other things gold and silver. About one hundred Tibetan bronzes from the British Museum have been analysed and the unusual prevalence of detectable traces of gold noted (Craddock 1981 p. 25), as indeed it was by Werner (1972 pp. 146–9) who reported that small amounts of gold were sometimes added to the molten metal purely as an offering with no metallurgical significance at all. In most of the analysed pieces gold was present in small traces of the order of 0.01%–0.1%, but in one group of a *Yama* and *Yami* (OA 1880–4072), made in the 18th or 19th century, the alloy was of copper with 1% of silver and 1.5% of gold. The piece is now extensively covered by leaf gilding, and it does not seem that surface patination was really intended for this or any of the other pieces examined.

However, in the late 18th century the British traveller Turner (1800) noted in his descriptions of the metal statuettes at the Tashilhunpo monastery that although the majority were copper gilt or of brass, 'some of the images were composed of that metallic mixture which in appearance resembles Wedgwood's black ware'. One might be tempted to suggest this was *bidri* ware, but *bidri* was never used for statuettes, and anyway an experienced traveller such as Turner must have been familiar with *bidri*, and clearly thought it was something else. It is possible that instead they were of *zi-khyim*, the Tibetan equivalent of the Chinese *ch'ih chin*, or *chhih-chin* to which we must now turn.

In Needham's profound study of Chinese spagyrical science and its interaction with the West (1974 pp. 257–266), he produced a wide-ranging and as it turns out now, prescient study comparing the process of *iosis* as described by the Alexandrian alchemists with the Chinese process used to produce their 'purple sheen gold' and identifying the latter with *shakudo*. From the first millennium AD there are frequent references in the Chinese Buddhist sutras to coloured golds, notably *chhih chin*, scarlet gold, *tzu chin*, purple gold and *tzu mo chin*, purple sheen gold. An alchemic text written in 712 AD describes a seven-fold transformation of gold, the last stages being successively yellow, *huang chin*, red, *hung chin*, and then scarlet, and finally purple. This is of course very similar to the sequence of colour transformations described by the Alexandrian alchemists on various copper alloys in the process of *iosis* in the production of 'our gold'. We have already noted the identification of the term *chhih chin* with the Tibetan *zi-khyim* which was a copper alloy containing amongst other materials, gold and silver. The Chinese sources also state that both *tzu chin* and *tzu mo chin* once occurred as native metals in the ground, but these supplies having been exhausted, a synthetic alloy was now prepared. This exactly parallels the Tibetan traditions for *zi-khyim*, and both echo the legendary origin of Corinthian bronze. What then are the origins of *tzu mo chin*? It could of course be indigenous, although Needham notes the Chinese tradition that it had originated in the West and was once imported into China, and that the barbarian name for it was *yang mai*, according to the *Shui Ching Chu*, which dates from around 500 AD. If this is so then in the early first millennium AD knowledge of it could have travelled east along the Silk Route. Although India may well have also had a similar material and process by this time and there was certainly extensive contact between India and China with the strong Buddhist influence, there is no evidence that *chhih-chin* came from the Subcontinent. Indeed the Chinese origin of the term used by the Tibetans rather suggests the reverse, that *zi-khyim* was introduced into the Himalayas **from** China.

There is, however, much more direct evidence for black patinated inlaid copper alloys containing small amounts of gold and silver in China. The poet Wang Shizen recorded how in the Xuande period of the 15th century a fire burnt down a Buddha hall and melted the bronze, silver and gold figures. The resultant alloy was used to cast vessels. As in the classical world, the story was treated with scepticism (Kerr 1990 p. 28), but does suggest that such alloys were at least current.

Such alloys do indeed exist, at least from the more recent past, known as *wu tong*. This was brought to the notice of the occidental world by Collier (1940) who has formed a collection of the small ink and tobacco boxes which are the principal products of this ware (Plate 10.1). He drew attention even then to the possibility of a connection with the black bronzes of ancient Egypt. Examination of these by Collier (1977) and latterly at the British Museum (see Wayman *et al.*, Chapter 10) showed that the black patinated areas were of copper with small percentages of gold, iron, silver and arsenic. These thin sheets were braised to thicker sheets of brass and the unpatinated bases were of copper. The silver inlay was highly unusual, a silver–brass alloy, probably a silver solder, which had been melted into place. *Wu tong* was made in the western province of Yunnan in the recent past, and indeed is still being made. The material is also attracting the attention of Chinese scientists (Mang Zidan and Han Rubin 1989), Han Rubin (personal communication) states that *wu tong* first became popular in the Ming period, and that about 2% of gold was necessary before a satisfactory black colour developed. She also informed us that the patination developed quite naturally just by handling with sweaty hands, which of course is exactly paralleled by *bidri*, and that *wu tong* was still made in Yunnan and could be obtained from certain art dealers in Beijing. Yunnan is adjacent to Tibet and the *zi khyim* alloys, but the origins of *wu tong* and the connections, if any, with *chhih chin* or *tzu mo chin* are very uncertain. The Chinese Encyclopaedic Dictionary (*Ci Yuan* published in Shanghai 1932) has the following entry, (brought to our attention and translated by Dr Collier):

> *WU JIN* Black metal
> Take copper 100 parts and add gold one part. Melt to produce a purple black colour. Old reputation. Can be used for ornamental (decorative) articles. [Jin is the old European transcription of chin.]

The pieces from the Collier collection examined so far are likely to have been made within the last 100 years, but the British Museum has recently acquired a small box with a fine black patina with a gilded sunk decoration (Plate 10.2), which should be dated to around 1700 AD. The patinated areas are of copper with small amounts of gold, silver and arsenic. This is the classic *shakudo* alloy, but should we regard this piece as a late example of the putative traditional *tzu mo chin* alloys or one of the first pieces of foreign art to be influenced by the Japanese *shakudo* metalwork?

Conclusions

This study has traced the evidence for the nature and occurrence of black patinated, inlaid copper alloys of specific composition, typically containing a few per cent of gold, silver and often arsenic. The technique would seem to have developed first within the Middle East-eastern Mediterranean region at some time in the Bronze Age. The earliest certain examples are found in both Egypt and Greece dating from the mid-second millennium BC, and from that time on there are sporadic surviving examples and documentary evidence for the distinctive material. It would seem that this material, the *ḥśmn km* of the Egyptians, became possibly the *kyanos* of the early Greeks before becoming the celebrated Corinthian bronze of the classical world, and it thrived and continued in circulation at least until the 6th century AD. In the East the earliest references to a dark or purple copper occur in China from the mid-first millennium AD, and this alloy would seem to equate with the Tibetan copper alloy *zi-khyim* first mentioned in the 10th century AD and which contained silver and gold. The first sure reference to Japanese *irogane* alloys are no earlier than the 14th century AD, although after that the technique was brought to new peaks of technical and artistic excellence. It is necessary now to examine whether the material sprang from a single origin and spread across the world through the centuries or if the many diverse examples we have examined are the products of independent development.

The question of the prevalence of diffusion or independent discovery continues to be one of the central problems of archaeology and to a lesser extent art history. At one time mankind's most basic technologies and synthetic materials, farming, architecture, ceramics as well as metallurgy were believed to have originated in the Middle East from whence they were spread by conquest, trade or imitation to less-favoured parts of the world. This was always little more than an assumption, almost an axiom, grounded as much on the prestige of the ancient historical record, much of it biblical, as on the magnificence of the surviving monuments and artefacts. Scholars such as Smith (1923) could claim for Egypt the direct origin of all the civilizing elements of European culture, and what is more gain a fair measure of acceptance for their ideas. However, continuing research over the last half-century in Europe and elsewhere has shown that broadly the idea of *ex oriente lux* is no longer tenable. Physical dating methods have pushed the dates for European prehistory way back beyond the first civilizations of the Nile and Euphrates, and detailed study of the spread of basic technologies such as agriculture and metallurgy have shown that the old model was certainly simplistic if not just plain wrong in many areas. Indeed, research on the inception of European metallurgy would suggest independent discovery at a number of places, although possibly under the stimulus of imported metal artefacts (Craddock 1992).

In the realm of alchemy and the history of science and technology, Needham was the first and certainly the most powerful voice against the heliocentric view of the development of scientific thought. However, Needham was not prepared to ditch diffusion, merely to shift its centre eastwards by several thousand miles to China. In his discussion of the rival processes of diffusion and independent discovery he has stated that although 'It will be readily admitted that the simpler the idea, invention, technique, or machine the more likely it may be that it originated independently in different parts of the world' yet 'the most probable supposition is that they originated in one place, radiating out in all directions from there'. He makes the important point that the likelihood of diffusion is a function of the sophistication of the process, etc. relative to the rest of the culture 'the degrees of complexity of inventions likely to diffuse, should follow up the rising curve of general cultural attainment' (Needham 1965 pp. 228-9). That is, the more complex and sophisticated the process the more likely it is to have had a single origin. Could this argument apply to the undoubtedly sophisticated and technically complex inlaid and patinated bronzes?

Just at the time when the tide is running very strongly in the archaeological world against technical diffusion, in art history cross-cultural links of styles and motifs are being increasingly stressed. Paradoxically this has come about through the self-same process of detailed examination and stylistic analysis of material over broad areas — in archaeology the technical differences and chronological discrepancies became ever more apparent, whereas in art the stylistic links began to emerge. The examples are many but perhaps most pertinent to this work is the recent study made by Rawson (1984) of the origins of those familar elements of Chinese art the lotus, peony and the dragon. In this she shows that these have been derived from elements of Hellenistic art such as the acanthus motif, which itself has a long history in Middle Eastern art.

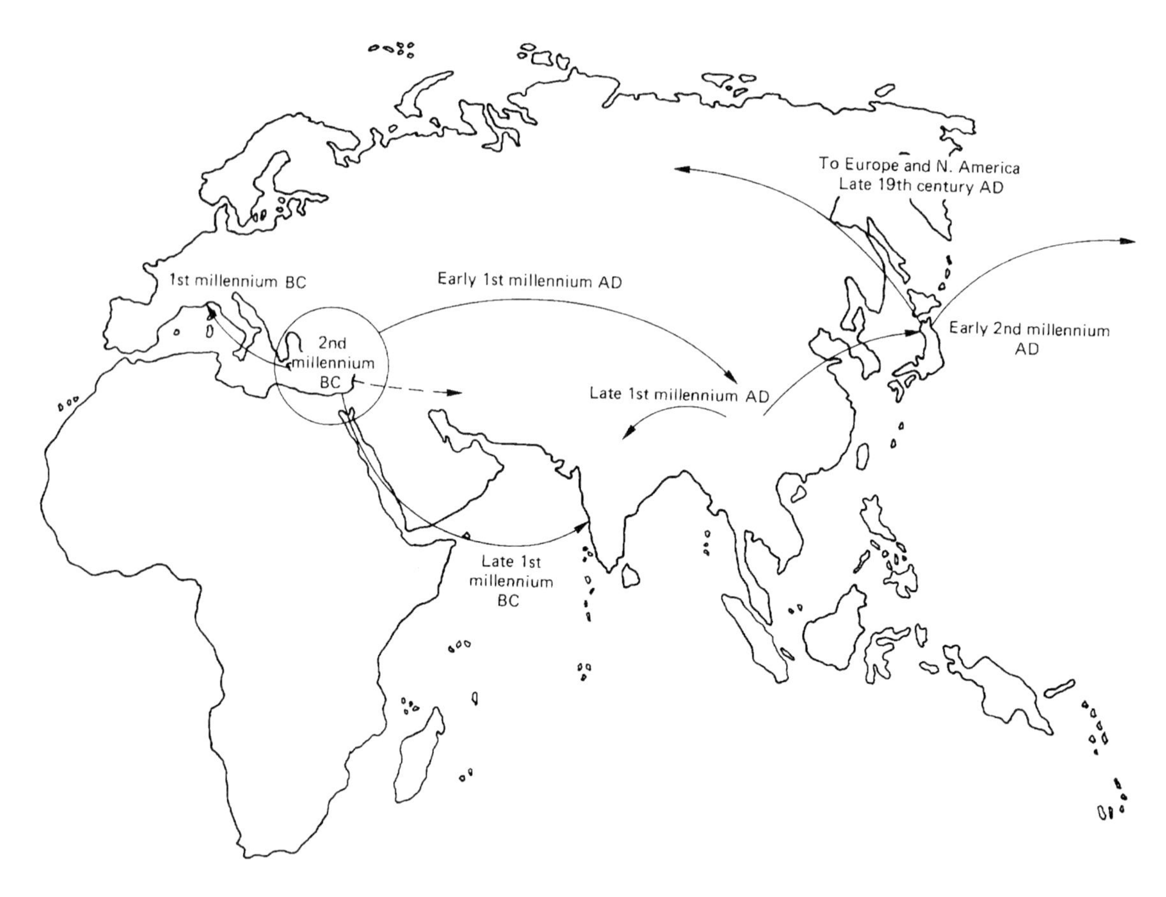

Figure 9.4. *Map showing the putative spread of the black-patinated copper alloys around the Old World over a period of almost four thousand years.*

Mindful of this background of conflicting paradigm and ideology, what, if anything, can we say of the possible cultural affiliations of this international family of patinated and inlaid bronzes? Are they a family and if so what is their family tree? They fall into both the main categories discussed above: clearly they are metals and technically complex, but equally clearly they are luxury items, heavily decorated and very much part of the art world as a distinctive material and finish, even if not a style. To further complicate the matter their method of production had strong links with the world of alchemy.

We will consider the material itself first. It is characterized by a highly unusual and often specific composition. Moreover it is significant that in modern practice only the gold seems to be strictly necessary, yet in examples from Mycenaean Greece across the Roman world to Japan, silver and arsenic are also present almost as if following an inherited recipe.

In the classical world at least, and one suspects elsewhere, it was an esoteric luxury material that catered to a highly sophisticated taste, a taste that was perhaps educated to appreciate it. The patinated surface does not have immediate appeal, for all the incorporation of precious metals, and in fact an acceptable black surface could be produced on most copper alloys that was barely distinguishable from the real thing. It was a true connoisseurs' material, its prestige resting on limited availability, and qualities that only the elite could detect and appreciate. Thus, initially at least, it was likely to win adherents as much by its prestige as by any immediate appeal. It was something one had to appreciate and at least appear to be able to pick out the genuine from the imitators if one was to claim any pretentions to taste at all. Many of the classical references are slightly mocking, there is a hint of emperor's clothes about the whole subject. All this could suggest a material that spread as much by established reputation as by immediate attraction and thus a candidate for diffusion rather than independent discovery. If this is the correct model, then it cannot just have been examples of the material itself that inspired production elsewhere because the composition is so clearly replicated, and the treatments as described by Gowland and others of the Japanese processes seem to echo those of Zosimos and Pausanias. Thus not only the product but the process itself seems to have spread, the exception being the Indian *bidri*, where, if there is a connection, the black patinated copper alloy inspired a similar product but with a very different material and technology.

It may be argued that the absence of any stylistic links in any of the various manifestations of the material militate against conscious imitation. The main gulf is in linking the classical Corinthian bronze with the Japanese *shakudo* pieces. There the main links are the Tibetan *zi-khyim* and the Chinese purple sheen gold for which the evidence is literary, no early examples having yet been identified to establish any vestigial stylistic connection such as Rawson has established on other Chinese material. There is anyway no intrinsic reason why the style has to follow the technique, and the only well-recorded transmission of the technique, that from Japan back to Europe in the late 19th century, provides an excellent illustration of this. The Victorian sculptor Alfred Gilbert worked closely with Roberts-Austen, after the latter's return from Japan, on patination experiments directly inspired by the Japanese *irogane* processes (Roberts-Austen 1888). Gilbert began to incorporate *shakudo*-type patinated alloys into his work in the 1890s. However, despite the direct technical links there is not the slightest hint of the orient in Gilbert's work, as exemplified by pieces such as the polychrome bronzes he made for the Graham memorial in Glasgow cathedral, or the Versey memorial at the Abbey Leix in Ireland (Dorment 1985).

Thus on balance the evidence would suggest a single origin for the material, tentatively represented by Figure 9.4, which became appreciated and adapted around the world to a wide variety of objects and styles, although apparently preserving the essentials of composition and inlay very precisely from Mycenaean Greece through to present day Japan. As the whole basis of the material is patination, many pieces must lie unrecognized in collections around the world, their carefully patinated surfaces mistaken for natural corrosion. Hopefully, as the full history and extent of the process becomes better known it will lead to the identification of many more examples of this most sophisticated and beautiful material.

References

Allan, J. W. (1979) *Persian Metal Technology*, Ithaca Press, London

Arevalo, F. (Trans.) (1878) Isidorus: *Etymologiae* **16**. In *Patrologiae Cursus Completus.* Garnier, Paris

Babbitt, F.C. (trans.) (1936) *Plutarch: Moralia*, 5, (Loeb edn), Heinemann, London

Bannister, C.O. (1929) Report of the Sumerian Copper Committee. *Proceedings of the British Association for the Advancement of Science for 1928*, p. 438

Bassore, J.W. (trans.) (1932) Seneca: *Moral Essays, De Brevitae Vitae* (Loeb edn), Heinemann, London

Bayley, J.C. (1992) *Non-ferrous metalworking in England: Late Iron Age to Early Medieval.* Thesis submitted for degree of PhD, University of London.

Bernal, M. (1987) *Black Athena*, Vol. 1, Free Association Press, London

Berthelot, M. (1887/8) *Collection des anciens Alchimistes Grecs*, Georges Steinheil, Paris

Berthelot, M. (1893) *La chimie au Moyen Age* I-III, Georges Stenheil, Paris.

Blümner, H. (1922) *Realencyclopedie d. class Altertumwissenschaft*, **22** A, Druckenmüller Verlag, Stuttgart

Boardman, J. (1987) Silver is white *Revue Archeologique*, **2**, 279–285

Boube-Piccot, C. (1969) *Les bronzes antiques du Maroc, Études et Travaux d'Archéologie Marocaine* IV, Direction des Monuments Historiques et des Antiquites, Rabat

Butterworth, G.W. (trans.) (1960) *Clement of Alexandria: The Exhortation to the Greeks*, (Loeb edn), Heinemann, London

Caley, E.R. (1941) The corroded bronze of Corinth. *Proceedings of the American Philosophical Society*, **84**(5), 689–761

Collier, H.B. (1940) Black copper of Yunnan. *Journal of Chemical Education*, **17**. 19–21

Collier, H.B. (1977) X-ray fluorescence analysis of black copper of Yunnan. *Naturwissenschaften*, **64**. 484

Conybeare, F.C. (1912) *Philostratus: The Life of Apollonius of Tyana*, (Loeb edn), Heinemann, London

Cooney, J.D. (1966) On the meaning of . *Zeitschrift für Ägyptische Sprache und Altertumskunde*, **93**, 43–47

Cooney, J.D. (1968) Siren and Ba, birds of a feather. *Bulletin of the Cleveland Museum of Art*, **55**, 262–271

Cowell, M.R. (1986) Report on the analyses of some Egyptian material. In *Catalogue of Egyptian antiquities in the British Museum 8 (Axes)* (ed. V. Davies) BMP, London

Craddock, P.T. (1979) Copper alloys of the mediaeval Islamic World. *World Archaeology*, **11**(1) 78–89

Craddock, P.T. (1981) The copper alloys of Tibet and their background. In *Aspects of Tibetan Metallurgy* (eds W.A. Oddy and W. Zwalf) British Museum Occasional Paper No. 15, London, pp. 1–32

Craddock, P.T. (1982) Gold in antique copper alloys. *Gold Bulletin*, **15**(2), 69–72

Craddock, P.T. (1985) Three thousand years of copper alloys. In *Application of Science in Examination of Works of Art* (eds P.A. England and L. van Zelst) Museum of Fine Arts, Boston, pp. 59–67

Craddock, P.T. (1992) Copper production in the Bronze Age of the British Isles. *Bulletin of the Metals Museum*, **18**(2), 3–28

Dorment, R. (1985) *Alfred Gilbert*, Yale University Press, London

Duval, R. (1888–1896) *Lexicon Syriacum, auctore Hassano Bar Bahule*, Paris, p. 1238

El Gayar, El Sayed and Jones, M.P. (1989) Old Kingdom copper smelting artifacts from Buhen. *Journal of the Historical Metallurgy Society*, **23**(1), 16–24

Engels, D. (1990) *Roman Corinth*, University of Chicago Press, Chicago

Emery, W.B. (1963) Excavations at Buhen. *Kush*, **11**, 116–120

Fishman, B. and Fleming, S.J. (1980) A bronze figure of Tutankhamun. *Archaeometry*, **22**(1), 81–86

Garland, H. and Bannister, C.O. (1927) *Ancient Egyptian Metallurgy*, Charles Griffin, London

Gowland, W. (1894) A Japanese pseudo-speise (Shirome). *Journal of the Society of Chemical Industry*, **13**(5), 1–26

Gowland, W. (1915) Metals and Metal-working of old Japan. *Transactions and Proceedings of the Japan Society of London*, **13**, 19–100

Giumlia-Mair, A.R. and Craddock, P.T. (1993) Corinthium Aes -Das Schwarze gold der Alchimisten, *Antike Welt*, Sonderheft

Gulick, C.B. (trans.) (1927–1941) Athenaeus: *The Deipnosophistae*, (Loeb edn), Heinemann, London

Halleux, R. (1981) *Les Alchemistes Grecs I* Papyrus de Leyde. Soc. d'édition *'Les Belles Lettres'*, Paris

Heseltine, M., (trans.) (1913) Petronius: *The Satyricon*, (Loeb edn), Heinemann, London

Higgins, R.A. (1974) *Minoan and Mycenaean Art*, Thames and Hudson, London

Hughes, M.J., Cowell, M.R. and Craddock, P.T. (1976) Atomic adsorption techniques in archaeology. *Archaeometry*, **18**, 19–36

Jacobson, D.M. and Weitzman, M.P. (1992) What was Corinthian bronze? *American Journal of Archaeology*, **96**, 237–247

Jones, H.L. (trans) (1918) *Pausanias: Description of Greece*, (Loeb edn), Heinemann, London

Kent Hill, D. (1969) Bronze working. In *The Muses at Work* (ed. C. Roebuck) MIT, Cambridge, Mass, 60–95

Kerr, R. (1990) *Later Chinese Bronzes*, Victoria and Albert Museum, London

Laffineur, R. (1974) L'incrustation à l'époque mycénienne. *L'antiquité classique*, **43**, 5–37

La Niece, S. (1983) Niello: An historical and technical survey. *The Antiquaries Journal*, **58**(2), 279–297

La Niece, S. (1990) Japanese polychrome metalwork. In *Archaeometry '90* (ed. A. Hauptmann, E. Pernicka, and G. Wagner) Birkhauser Verlag, Basel, pp. 87–94

Lo Bue, E. (1981) Statuary metals in Tibet and the Himalayas. In *Aspects of Tibetan Metallurgy* (ed. W.A. Oddy and W. Zwalf) British Museum Occasional Paper no. 15, London, pp. 3–67

Lucas, A. (1962) *Ancient Egyptian Materials and Industries*, Edward Arnold, London

Lyman, B.S. (1890–1) Japanese swords. *Numismatic and Antiquarian Society Proceedings*, 23–60

Mang Zidan and Han Rubin (in chinese) (1989) Studies of the blackening on copper–silver–gold alloy. *Studies in the History of Natural Sciences*, **8**, 4

Maryon, H. (1956) Fine metalwork. In *A History of Technology*, 2nd edition (eds C. Singer, E.J. Holmyard, A.R. Hall and T.I. Williams) OUP, Oxford

Mathew, R.H. (1969) *Chinese – English Dictionary*, Harvard University Press, Cambridge, Mass

Mattusch, C.C. (1977) Corinthian metalworking : the forum area. *Hesperia*, **46**, 380–389

Middleton, J.H. (1887) Comments on a figure exhibited March 10, 1887. *Proceedings of the Society of Antiquaries of London*, **11**, 1–3

Murakami, R., Niiyama, S. and Kitada, M. (1988) Characterization of the black surface on a copper alloy coloured by traditional Japanese surface treatment. *Proceedings of the IIC Congress, Kyoto* (ed. N.R. Bromelle) IIC, pp. 133–136

Murphy-O'Connor, J. (1983) Corinthian bronze. *Revue Biblique*, **80**(1), 80–93

Murray, A.T. (trans.) (1963) *Homer: The Iliad*, (Loeb edn), Heinemann, London

Needham, J. (1965) *Science and Civilisation in China I*, CUP, Cambridge

Needham, J. (1974) *Science and Civilisation in China V Pt 2*, CUP, Cambridge

Notis, M.R. (1988) The Japanese alloy *shakudo*: its history and its patination. In *The Beginning of the Use of Metals and Alloys* (ed. R. Maddin) MIT Press, Cambridge, Mass, pp. 315–327

Ogden, J. (1990) Report on Lot 259. In *Gold*, Habsburg and Feldman's sale catalogue 14 May 1990, Geneva, pp. 262–4

Partington, J.R. (1935) *Origins and Development of Applied Chemistry*, Longmans, Green and Co., London

Plenderleith, H.J. (1956) *The Conservation of Antiquities and Works of Art*, OUP, London

Rackham, H., (trans.) (1940 and 1952) *Pliny: The Natural History*, III and V. (Loeb edn), Heinemann, London

Rawson, J. (1984) *Chinese Ornament*, British Museum Publications, London

Riederer, J. (1978) Die naturwissenschaftliche Untersuchung der Bronzen des Ägyptischen Museums Stiftung Preussischer Kulturbesitz, Berlin. *Berliner Beiträge zur Archäometrie*, **3**, 5–42

Riederer, J. (1982) Die naturwissenschaftliche Untersuchung der Bronzen der Staatlichen Sammlung Ägyptischer Kunst in München. *Berliner Beiträge zur Archäometrie*, **7**, 5–34

Riederer, J. (1983) Metallanalysen der ägyptischen Statuetten des Kestner-Museums, Hannover. *Berliner Beiträge zur Archäometrie*, **8**, 5–17

Roberts-Austen, W.C. (1888) Cantor Lectures on Alloys: Colours of Metals and Alloys Considered in Relation to their Application to Art. *Journal of the Society of Arts*, **36**, 1137–1146

Roberts-Austen, W.C. (1892) *Report on the Analysis of various examples of Oriental Metalwork in the South Kensington Museum*, HMSO, London

Roberts-Austen, W.C. (1898) *An Introduction to the Study of Metallurgy*. Griffin and Co., London.

Schaeffer, C. (1939) *Ugaritica III* Mission de Ras Shamra, Paris

Shearman, F. (1988) An original decorated surface on an Egyptian bronze statuette. In *Conservation of Ancient Egyptian Materials* (eds S.C. Watkins and C.E. Brown) IAP, London, pp. 29–34

Shepard, H.J. (1964) Colour symbolism in the Alchemic Opus. *Scientia*, **58**, 1–24

Smith, G.E. (1923) *Ancient Egypt and the Origins of Civilisation*, Harper, London

Stillwell, R., Scranton, R.L. and Freeman, S.E. (1941) *Corinth*, **1**(2) Architecture, Cambridge

Stronge, S. (1985) *Bidri Ware*, Victoria and Albert Museum, London.

Tod, J. (1829, reissued 1914) *Annals and Antiquities of Rajast'han*, Routledge, London

Tucci, G. (1959) A Tibetan classification of Buddhist images according to their style. *Artibus Asiae*, **22**, 179–187

Turner, S. (1800) *An Account of an Embassy to the Court of Teshoo Lama, in Tibet,* London

Untracht, O. (1982) *Jewelry Concepts and Technology*, Robert Hale, London

Vickers, M. (1985) Artful crafts: the influence of metalwork on Athenian painted pottery. *Journal of Hellenic Studies*, **105**, 108–128

Walters, H.B. (1907) *Catalogue of Greek and Roman Silver in the British Museum*, Trustees of the British Museum, London

Werner, O. (1972) *Spektralanalytische und Metallurgische Untersuchungen an Indischen Bronzen*, Brill, Leiden

Whitehouse, H. (1987) A statuette of Ptah the Creator. *The Ashmolean*, **12**, 6–8

Xenaki-Sakellariou, A. and Chatziliou, C. (1989) *'Peinture en Métal' a L'Époque Mycénienne*, Ekdotike Athenon, Athens

10

Wu tong, a neglected Chinese decorative technology

Michael Wayman and Paul Craddock

Abstract

Wu tong is a Chinese ornamental material in which a rich black surface layer, with silver alloy inlay, is formed on the surface of a copper alloy sheet. This chapter describes the methods used to create *wu tong*, based on metallographic analysis. The analysis shows that the sheet on which the surface decoration is created is actually a composite, in which a thin surface layer of copper alloyed with small amounts of gold and silver is bonded, by silver soldering and subsequent mechanical working, to a thicker backing layer of cheaper low-zinc copper alloy. This composite material is then suitable for inlaying, assembly into objects such as small decorative boxes, and finally patinating to create the black surface background. Also discussed is the similarity between *wu tong* and other black surfaced copper alloys including the Japanese *shakudo* material.

Introduction

While living in China during the 1930s, Dr H. B. Collier was able to assemble a collection of small boxes in a material known as *wu tong* ware which, according to his sources, was made in Yunnan province, southwest China. These objects, typically ink or tobacco boxes, but including small containers for jewellery, etc. (Plate 10.1), are made of copper whose surfaces have been inlaid with silver in a background which has been patinated to a deep rich black, providing an attractive contrast with the inlay. In at least some cases, the boxes were clearly produced to order for presentation in honour of such events as school graduations (Collier 1940).

Despite the fact that this is a distinctive and attractive material, it seems to have attracted very little attention, with no previous publications in the west beyond those of Collier (1940, 1977). This is all the more surprising given its similarity to the very familiar Japanese *shakudo* alloys, and the speculation by Needham, discussed below, that the Chinese *Tzu Mo Chin* could be ancestral to *shakudo*.

Some literary references to *wu tong* do exist, thus the Chinese Encyclopaedic Dictionary Ci Yuan, published in 1932 at Shanghai, has the following entry, which we are grateful to Dr Collier for bringing to our attention:

> *Wu Jin*, black metal (Jin is an old form of chin, gold):
> Take copper 100 parts and add gold 1 part. Melt to produce a purple black colour. Old reputation. Can be used for ornamental (decorative) articles.

From the dealers who supplied the *wu tong*, Collier also obtained a printed information sheet, apparently provided by one of the manufacturers, which gives a limited amount of information on how it was made and should be looked after. His translation and summary is as follows:

> Yoh Fu-hsing's *'Wu Tong'* Shop
> 1. Any design can be produced — flowers and grasses, human figures, birds or animals, feathers and hair; landscapes, curios; characters or quotations — all exactly according to pattern.
> 2. The method of manufacture is as follows. First melt ordinary copper, then add gold, silver, iron, tin and so forth, and the various chemicals. Afterward hammer into a plate and then engrave grooves and inlay the silver or gold wire. This 'black copper' is not produced by applying a chemical after the completion of the article.
> 3. If the base is black copper, silver or gold wire is inlaid. If the base be gold or silver, a black copper pattern may be set in, and then gold or silver wire inlaid on this, if desired.
> 4. Upon completion the surface is red in colour, and the pattern dull. Grasp the article firmly in the hand, so that the perspiration may soak into the surface. Within a day or two the copper will turn black. The colour is quite permanent, the pattern clear and distinct, in all, a thing of beauty.
> 5. Acid perspiration will prevent the formation of the black colour. In order to obtain a good colour and bright pattern, acid, greasy or dirty substances must not be rubbed upon the surface. But if an article has been thus damaged, it may be restored by cleaning the surface with an abrasive, when the perspiration of the hand will bring back the black colour.
> 6. If the pattern or colour is not very good, it may be improved by abrading the surface and re-forming the black film as described above. Only genuine *wu tong* will turn black on contact with the perspiration of the hand. Imitation articles cannot be turned black in this manner, if they have once been rubbed red. (Collier 1940: 20-21)

The results of our examination of some pieces are not in exact accord with these descriptions, which are nevertheless interesting. More reliable information on the production of *wu tong* is given in the *Records of the Yunnan Province* for 1939 (**142**, pp. 7-8, reproduced in Mang Zidan and Han Rubin 1989). There it states that the best black ware was made in Shiping (a city in Yunnan). Typical products were vases, boxes and inkwells, decorated with floral patterns or ideograms. The decoration was cut into the surface, silver wires or pieces were put into the channels, and the piece heated so that the silver melted and filled the grooves. Then it was filed flat and polished. Finally the piece was patinated, this being the last operation, as with other patinated copper alloys.

The *Record of Shiping Town* for 1938 (**16** p. 9, Mang Zidan and Han Rubin 1989) gives an interesting insight into the organization of the production of *wu tong*. There was a small village near Shiping, called Yuejiawan, in which particularly fine examples of *wu tong* and other unusual polychrome copper alloys were made, and apprentices came to learn the craft in the workshops there. The name of the village was said to be derived from the Yue family, who were the only people who possessed the secret of how *wu tong* was made. This recalls the similar situation with the Japanese irogane alloys and the Goto family (see p. 98).

Wu tong is now undergoing active study by the Archaeometallurgy Group of Beijing University of Science and Technology, as well as by the Department of Scientific Research in the British Museum; some preliminary results are reported here.

Technology of *wu tong*

Although *wu tong* is firmly placed within the growing family of black patinated copper alloys containing precious metal both in the alloy and in the inlay (Craddock and Giumlia-Mair, this volume Chapter 9), our studies show that it is technically very different. In *wu tong* the patinated metal turns out to be a composite material consisting of a thin layer of the special alloy, intimately bonded to a thicker sheet of a low zinc brass; a further difference is that here the inlay was melted into place.

This information was obtained from the metallographic study of a segment of *wu tong*

Figure 10.1. *Microstructure through the cross-section of the composite sheet with the black patinated surface uppermost. A shallow inlaid groove is visible, penetrating into the sheet surface. About a quarter of the way down from the upper surface can be seen a band of lighter contrast (corresponding to a higher average atomic number); this is the silver-rich zone which remains from the original silver soldering of the upper copper–silver–gold alloy sheet to the lower copper–zinc alloy sheet. The high content of non-metallic inclusions in this zone is another consequence of the soldering process. Scanning electron micrograph, backscattered electron image. Sheet thickness 0.4 mm.*

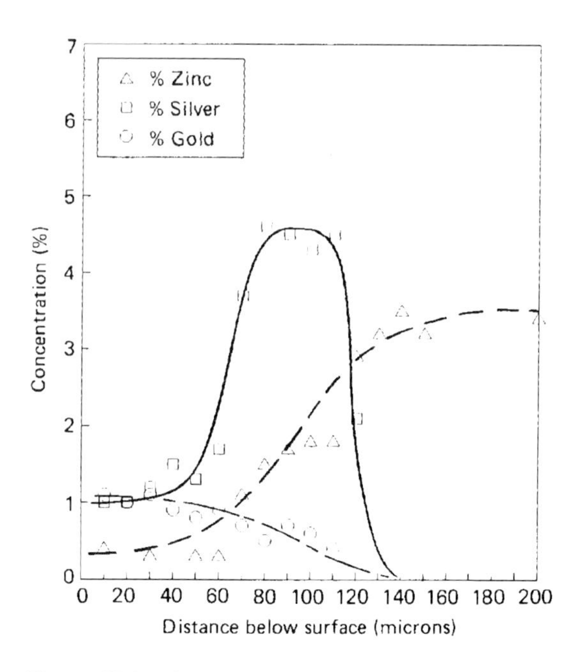

Figure 10.2. *Elemental distribution of silver, gold and zinc across the composite sheet. The left-hand edge of the plot represents the black patinated surface of the sheet; the right-hand edge of the composite sheet is well off the plot to the right at a distance below surface of about 400 microns (0.4 mm). The silver-rich zone at a depth of 70–110 microns is the remnant of the silver soldered bond which joined the copper–silver–gold alloy on the left to the copper–zinc alloy on the right.*

which was kindly provided by Dr Collier. The segment was mounted in epoxy resin in an orientation chosen to permit investigation of a plane passing through the cross-sections of several areas of inlay. The mounted section was then subjected to standard metallographic sample preparation procedures, including grinding to 1200 grit, polishing with 6 and 1 micron diamond paste abrasive, and finally with 0.05 micron alumina slurry. Examination was carried out by optical microscopy, and by means of a scanning electron microscope equipped with an energy dispersive X-ray analyser (the SEM-EDX technique).

This sample as well as the boxes shown in plate 1 were also subjected to quantitative X-ray fluorescence analysis.

As mentioned above, the sheet metal into which inlays were melted is complex, being formed of two layers soldered together (Figure 10.1). The thicker support layer is a brass consisting of copper with about 4% of zinc while the outer patinated layer is of copper with about 1% each of silver and gold. The two layers are joined by a region enriched in silver (Figure 10.2). It is believed that this microstructure has arisen from the two layers having originally been soldered together using a silver solder containing copper, silver and zinc. The diffuse nature of the soldered region suggests that after soldering the composite was reduced in thickness by rolling or hammering, performed either hot or cold with periodic annealing; the combination of mechanical work and high temperature has allowed the silver solder filler metal to diffuse outwards into the two bonded layers leaving a maximum of about 5% silver in the soldered region. After this thickness reduction, the composite sheet was left about 0.4 mm thick of which the copper-zinc layer is about 0.3 mm. In this state, fully softened by the final heating, the composite sheet was ready to be inlaid.

Inlay

It is likely that the designs would have been worked onto the flat prepared composite sheets and inlaid before they were made into boxes, a procedure which involves bending to shape. It is not clear how the designs were made. They are quite fine and in the soft copper they could simply have been scribed with a steel point; alternatively they could have been engraved or chased. Examination of the inlaid lines on the boxes revealed that some of them appear to be made up of discontinuous short segments suggestive of chasing, while others have a tapering gouged shape at one end and an abrupt termination at the other end, typical of engraving. Thus it appears that more than one technique may well have been used.

Examination of the cross-section through an inlaid area showed clearly that the inlay metal has been molten and has solidified in the prepared groove (Figure 10.3). The composition of the inlay metal is about 55% silver, 40% copper and 4% zinc with about 1% of lead; the inlay metal also contains about 0.4% gold which is likely to have come with the silver rather than representing a deliberate addition. These figures are approximate and furthermore, as the metal was molten in the copper groove, some dilution by the copper is likely. However, despite this, the composition corresponds well with that of silver solders (CDA 1952) such as that used to join the two layers of alloy during the preparation of the composite sheet. The inlaying could have been accomplished by placing thin strands of silver solder in the grooves as stated in the manufacturer's information quoted above with the additional step of melting *in situ*, or alternatively by flushing the whole surface with molten silver solder and then removing as much of the excess as possible while liquid, leaving the solder mainly restricted to the grooves. In either case silver solder would have been left proud of the sheet surface, to be filed or abraded flat. Indeed long straight scratches can be seen to run continuously over the surfaces of both body metal and inlay.

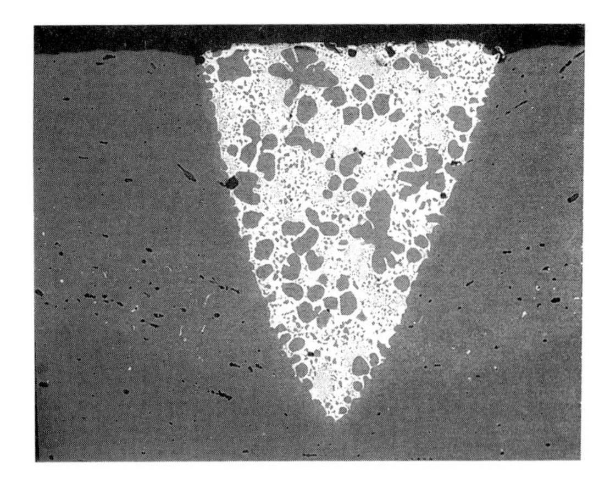

Figure 10.3. *Microstructure of the inlaid sheet showing a deeply inlaid groove. The large dark regions within the inlay are dendrites of copper-rich copper–silver alloy (83% copper, 10% silver, 6% zinc), the first constituent to solidify during the freezing of the inlay. The brighter speckled regions are silver-rich copper–silver eutectic (72% silver, 24% copper, 3% zinc). Based on its composition, the inlay would have a melting temperature of about 820°C. The silver-rich zone in the sheet can be seen crossing near the tip of the inlay; the deflection of this zone away from the inlaid surface is a consequence of the deformation of the sheet under the pressure of the engraving or chasing tool. Some of the inlaid grooves, including that shown in Figure 10.1, are much shallower than this example. Scanning electron micrograph, backscattered electron image. Inlay depth 0.18 mm.*

The *wu tong* sheets were then ready to be cut, bent and assembled to form the box and then, finally, patinated.

Construction

The boxes such as those shown in Plate 10.1 are constructed of conventional copper sheet, to which the *wu tong* composite sheet metal is attached to form the decorated outer surfaces as desired. In a typical case, as illustrated in Figure 10.4, the body of the box consists of a base and an inner body with rim stiffener; the inner body is wrapped in a thin layer of *wu tong* composite sheet. Similarly the lid consists of a basic top, onto which is attached a layer of *wu tong*, and side walls which are also made of *wu tong*. Decorative silver moulding strips

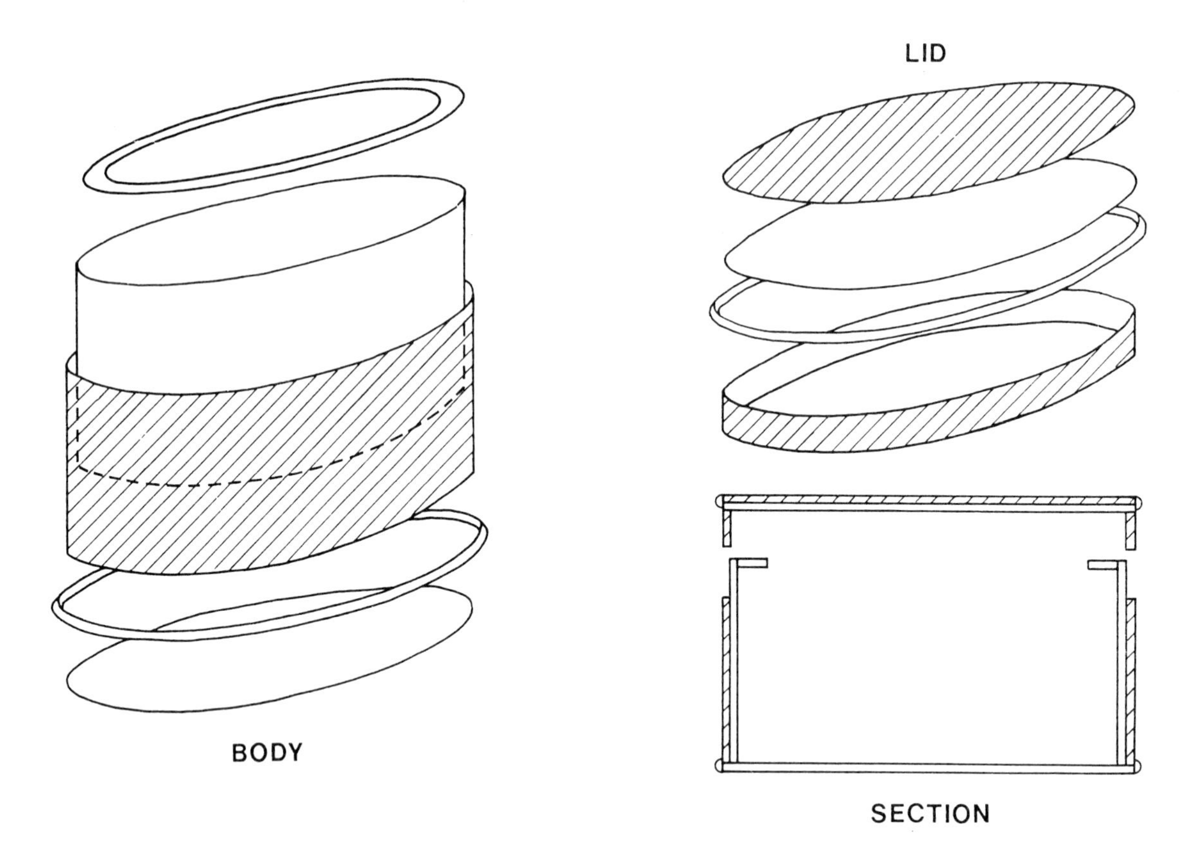

Figure 10.4. *Construction of a* wu tong *box, using the central box in Plate 10.1 as a model. Left: exploded view of body. Right: (above) exploded view of lid; (below) cross-section through body and lid. The hatched components are the* wu tong *composite sheets, inlaid and patinated, which are soldered to the underlying supporting sheets. (Drawn by Brenda Craddock.)*

are applied around the peripheries of the lid and the body bottom.

The analytical study of one box showed that in order to join these components several different solders were used with a wide range of melting temperatures; these include brass brazing alloy, silver solder and soft lead–tin solder. This is common practice, utilizing the high melting solder first and then lower melting solders for subsequent joints, thereby reducing the risk of the earlier joins melting. For example, the attachment of the *wu tong* composite sheet to the underlying body and lid structures was accomplished late in the assembly process by soft soldering.

Patination

There is little direct information available on this stage, but it is currently being studied by the Archaeometallurgy Group of the Beijing University of Science and Technology. These workers have tried treating the alloy with tea water, and with weak aqueous solutions of acetic, tannic and lactic acids, individually and in combination (Mang Zidan and Han Rubin 1989). They found that the hotter and more concentrated the solution, the quicker the black patina forms, but if it forms too quickly it is likely to be loose. After some days under ideal conditions, near neutral pH at about 37–42 °C, a continuous firmly adherent patina has formed. It is noteworthy, however, that the films reported by these workers contain silver oxide, as determined by X-ray diffraction, whereas only cuprite (copper oxide) was found on the boxes analysed here.

The Japanese treat similar alloys by boiling the carefully polished and degreased metal in weak aqueous solutions of copper sulphate, copper acetate and alum. Murakami (this volume, Chapter 7) gives figures of 1.9 g of natural verdigris, 1.2 g of copper sulphate and 0.2 g of alum per litre. The clean material is boiled in this solution for about 30 minutes to develop the patina.

The Chinese documentary sources noted by Mang Zidan and Han Rubin and by Collier mention perspiration as another patination agent. This is applied by handling the object with sweaty hands over a number of days. It seems likely that this is really only a remedial operation used to repatinate areas where the surface has been damaged, and recalls the similar instructions which are given with the Indian zinc *bidri* alloys at present.

Origins and cultural affinities

In his profound work on Chinese alchemy, Needham (1974) dwelt in detail on the identification of the metal described in the early literature as *Tzu Mo Chin*, purple sheen gold. He speculated that this could perhaps be identified with *shakudo*, the well-known Japanese *irogane* alloy. Collier, in his 1977 paper, drew attention to the composition of the *wu tong* boxes and suggested them as a candidate for Needham's purple sheen gold. The present study, coupled with that of Craddock and Giumlia-Mair (this volume, Chapter 9), has strengthened this connection but there is still a long historical gap between the literary reference of the early centuries of the first millennium AD and the materials studied herein, bridged only by some medieval references to copper alloys containing small amounts of gold and silver (Kerr 1990). What is believed at present to be the earliest Chinese piece is a small box (Plate 10.2) now in the British Museum (OA 1983.6-16.1) which is tentatively dated to about 1700 AD. The designs on this suggest that it may have been produced for the European market (J. Rawson, personal communication). The *wu tong* from the Collier collection was probably all made within the past 100 years, although the literary references do suggest that *wu tong* is a traditional material with some history. Han Rubin (personal communication) believes that the material originated in the Ming period. The location of the industry in Yunnan may well be significant for its earlier history, as the name of a comparable alloy (containing gold and silver) from adjacent Tibet, *zi-khyim*, is derived from the Chinese *tzu chin*, scarlet gold, which in turn is clearly related to the *tzu mo chin*, purple sheen gold (Craddock and Giumlia-Mair, Chapter 9). However, there is at present a chronological gap of over 1000 years, during which time *shakudo* developed in Japan, almost certainly from a Chinese progenitor, and it is conceivable that the *wu tong* studied here is a re-introduction from Japan.

Further study of this neglected but fascinating material should serve to establish its history both within China and beyond.

Acknowledgements

This study has its origins in the collections and work of Dr H.B. Collier. His continued support, not least in loaning this material for study and display at the British Museum and providing additional information and comments has proved invaluable. We are also grateful to B. Craddock for drawing Figure 10.4, and to Dr A. Giumlia-Mair for the information which she obtained in discussions with the Archaeometallurgy Group at the Beijing University of Science and Technology.

Note

Since this paper was written Dr Collier has died, and the bulk of his collection has been donated to the University of Alberta,

Edmonton, except for the items illustrated here which have been purchased for the British Museum (reg. nos. OA 1992.11-9.1-6).

References

CDA (1952) *The Welding, Brazing and Soldering of Copper and its Alloys*, CDA Publication 47, Copper Development Association, London

Collier, H.B. (1940) Black Copper of Yunnan. *Journal of Chemical Education*, **17**, 19–21

Collier, H.B. (1977) X-ray fluorescence analysis of black copper of Yunnan. *Naturwissenschaften*, **64**, 484

Kerr, R (1990) *Later Chinese Bronzes*, Victoria and Albert Museum, London

Mang Zidan and Han Rubin (1989) Studies of the blackening on copper–silver–gold alloy. *Studies in the History of Natural Sciences*, **6**(4)

Needham, J. (1974) *Science and Civilization in China* **5**(2), 257–271, Cambridge University Press, Cambridge

11

Bidri ware of India

Susan Stronge

Abstract

Bidri ware is inlaid Indian metalwork, made from a high-zinc alloy and characterized by its black surface colouring. The technique takes its name from the Deccan city of Bidar, though its origins are not known. The earliest pieces date from the late 16th or early 17th century and their style incorporates Deccani decorative elements. Mughal influence is also apparent, however, making it difficult to assign a provenance to pieces which are not usually inscribed. The technique reveals intriguing information: the zinc content is so high that it means metallic zinc was used in its manufacture. The early Indian source for zinc was Zawar in Rajasthan but the Deccani craftsmen may have found it more convenient to use Chinese zinc, bought from the much nearer Coromandel Coast ports. Analysis has been carried out on the surface coloration and, although replication experiments have been successfully carried out, it has not been possible to explain exactly how the metal changes colour.

Bidri is the name conventionally given to a class of Indian metalwork. It describes objects made from a high-zinc alloy and inlaid with silver or brass, or a combination of both, or overlaid with patterns of silver wire hammered on to a cross-hatched ground. The brass used for inlaying has a golden appearance and is sometimes mistaken for real gold. The most characteristic quality of *bidri*, and the one which seems to make the technique unique to the Indian subcontinent, derives from the final stage of the manufacturing process. The finished surface of the object is covered with a mud or clay paste containing one part of ammonium chloride, mixed with one quarter of unrefined potassium nitrate and the same amount of salt (Untracht 1968 p. 147). When the paste is removed, the zinc alloy has changed colour to a deep black which enhances, without altering, the brightness of the inlay.

How and when the technique originated is not known. Its name links it with the city of Bidar and it is said, traditionally, to have developed under the Bahmani rulers of the Deccan. The city was indeed under Bahmani control and, under Shihabuddin Ahmad I (reign: 1422 to 1436), became the sultanate's capital. The Bahmanis were from Iran; Persian was spoken at court and the culture and artistic products of the court were strongly Iranian. The monuments, for example, were built after Iranian models and were covered with glazed tiles which owe nothing to indigenous Indian traditions (Crowe 1986 p. 91).

Iran may also have provided a model for base metal wares inlaid with silver but there is, as yet, no evidence for the manufacture of *bidri* at such an early date. The word does not appear in Persian language dictionaries of the 16th and 17th centuries but it cannot be discounted that the technique may have been known by another name which has not so far been identified.

The earliest surviving pieces date from the late 16th or early 17th century and are heavily influenced by Iranian forms and design. A bowl in the collection of the Victoria & Albert Museum (V & A), for example, (Figure 11.1) has a shape which Iranian craftsmen copied from Chinese porcelain in metal, jade and pottery and has a marked Safavid character in the decoration of flower heads within cusped, foliate cartouches. Similar motifs are found in Iranian pile-carpets of the same period which must have been used in the Deccan; the sultans who commissioned miniatures painted with the costliest materials (including gold and imported lapis lazuli ground to provide an intense blue), would certainly have been able to afford the finest carpets of the Islamic world to which they belonged (Zebrowski, 1983). However, carpets were also prized by the Mughal emperors and there is nothing in the form and decoration of the bowl which would preclude it being a product of Mughal India, where artistic influence from Safavid Iran was marked and a great many Iranian artists and craftsmen were employed.

An extremely fine ewer in the same collection (Figure 11.2) dates to the mid-17th century and includes some specifically Deccani motifs, though the shape is also found to the north. Around the neck, single flowering plants are contained within an arcade. The tall, narrow four-centred arches with straight sides were introduced into the Deccan from Iran but the particular proportions of the arches on the ewer, as well as the ogee at the apex, are features of those commonly found in the Deccan and well represented at Bidar (Michell 1986 Figures 13, 14). Mughal arches of the same period differ in being broader and cusped and, although this form is also found on the ewer, on the attachment plate for the handle (Stronge 1985 figure on p. 38), this may be explained by the expansion of the Mughal empire which was now seriously threatening the independent existence of the sultanates and which, in 1656, absorbed Bidar. Aesthetic influence seeped across the borders and, as the

Figure 11.1. *Bowl:* bidri *inlaid with silver and brass; Deccan; late 16th or early 17th century. Victoria and Albert Museum: IS 10-1973. H 6.2 cm Diam. 14 cm. Given by Mr Simon Digby. (Photo courtesy of the Trustees of the Victoria and Albert Museum.)*

Figure 11.2. *Ewer:* bidri *inlaid with silver and brass; Deccan; mid-17th century. Victoria and Albert Museum: 1479-1904. H 28.5 cm W 18.4 cm. (Photo courtesy of the Trustees of the Victoria and Albert Museum.)*

Mughal style was by now well established, confident and dynamic, it must have been eagerly copied by Deccani artisans working in a rather decadent Persianizing manner. The decoration on the ewer appears somewhat old-fashioned when compared to artistic trends in Mughal India, where the floral ornamentation of the emperor Shah Jahan's new monuments of the 1630s and 1640s overwhelmed the arts.

Apart from illustrating the quality of early *bidri*, the ewer is interesting in another respect: round its rim, inlaid in silver wire, is a Persian inscription (Stronge 1985 figure on page 40) claiming it belonged to no less a personage than Timur-i Leng, the great ruler of Samarqand known in the west as Tamberlaine, and that it was made by Husayn Isfahani. The inscription also includes a date, 809 of the Islamic era, which corresponds to 1406 or 1407 of the Christian era. As the shape and decoration are anachronistic this can be discounted as entirely fraudulent. The inscription is probably an addition of the 19th century, when there seems to have been a market for objects and paintings of high quality which purported to be linked in some way with the Mughal emperors, who were descended from Timur (Jones 1990, pp. 227–29).

Pieces contemporary with the ewer demonstrate, however, that Mughal influence could completely displace any obvious Deccani features. A bowl from a *huqqa* or water-pipe (Figure 11.3), also in the V & A, is inlaid with flowering plants within cusped cartouches in a style typical of Mughal India in the mid-17th century and deriving from the decoration of the monuments commissioned by Shah Jahan (reign: 1628–56). His most famous architectural legacy, the Taj Mahal at Agra, was begun in 1632 and bears dates of 1635, 1636 and 1637 (Begley and Desai 1989). It was built from white marble and has surfaces carved or inlaid with coloured hardstones to provide decorative details where flowering plants predominate. The plants are rendered according to a peculiar convention which gives an illusion of realism because certain blossoms are identifiable and there is a feeling of gentle movement in the drooping stems and the curled-over petals and leaves. Yet the identifiable blossoms are usually combined with alien

Figure 11.3. Huqqa *base:* bidri *inlaid with silver and brass; Deccan or Northern India; second half of the 17th century. Victoria and Albert Museum: IS 27-1980. H 18.6 cm Diam 16.8 cm. (Photo courtesy of the Trustees of the Victoria and Albert Museum.)*

leaves, and the elements of the plant are carefully and unnaturally arranged to show, simultaneously, the different stages of its evolution and to allow it to fill the space allotted to it with consummate elegance rather than to imitate the random, often untidy forms of nature. These floral motifs dominated the arts of the Mughals from this period onwards; infinitely adaptable, they could be used with equal success on flat or curved surfaces, woven or embroidered, carved on to jade or inlaid into wood or metal.

The proliferation of the style and its encroachment into Deccani territory make it extremely difficult to establish where this piece, and others with different versions of the floral style, might have been made. Tobacco first came to India in the 16th century, when the Portuguese introduced it into the Deccan, and was smoked in *huqqas*. In 1604, Akbar sent an

Figure 11.4. *Detail from a page of an Atlas prepared for Col. Gentil Faizabad, 1770. India Office Library: Add.Or. 4039 folio 13. (By permission of the British Library.)*

envoy to the ruler of the Qutb Shahi sultanate at Bijapur to arrange a matrimonial alliance between the two royal houses as a preliminary step in his plan to conquer the entire Deccan. The envoy returned with the required promise that the ruler, Ibrahim Qutb Shah, would allow his daughter to marry Akbar's son, and with tobacco (Joshi, 1950). Certain members of the court quickly became addicted and, by the mid-17th century, the *huqqa* was used all over India. Although the *bidri huqqa* has decoration clearly related to that on the Deccani ewer, its motifs are predominantly Mughal. The question therefore arises as to whether it was a Deccani piece made under Mughal aesthetic influence (Bidar had fallen to the armies of Shah Jahan in 1656) or whether Deccani craftsmen took the technique to the Mughal court. The former may seem more likely but it is known that Deccani craftsmen were taken into imperial Mughal service. The emperor Aurangzeb, Shah Jahan's son and successor (reign: 1658–1707) completed the conquest of the Deccan by defeating the remaining independent sultanates of Bijapur and Golconda, in 1686 and 1687 respectively. A history of the reign notes:

> It would require another volume to describe in detail the coming of the Haidarabadis to the imperial Court and the admission into the imperial service of professional men, men of skill, and artisans of every kind. (Sarkar 1947 p. 184)

This states clearly that men from the capital of Golconda sultanate, Haidarabad, moved to the Mughal court; Haidarabad is about 60 miles from Bidar.

So far, although there is no evidence to link *bidri* production with Bidar, certain aspects of the decoration are exclusively associated with the Deccan. It is not until the early 18th century that documentary proof becomes available. Sir Jadunath Sarkar, the great historian of Mughal India, noted a reference to *bidri* in a history of India, the *Chahar Gulshan*, written in Persian in 1759 and translated by him into English. The *Chahar Gulshan* includes a statistical account taken by Sarkar, on internal evidence, to be from a compilation of about 1720 (Sarkar 1901, p.xxi). A copy of the manuscript in the British Library contains the following passage, kindly translated by Dr Melikian-Chirvani, concerning Bidar in the section on the regions, or *subahs*, of the Deccan:

> In this *subah* the fine and rare *bidri* vessels are made...the craftsmen of this place make them with such delicacy that even a painter could not imagine them.

The author mentions specifically the manufacture of bowls, ewers, boxes and spittoons (Stronge 1985 p. 33). This passage is corroborated by an illustrated atlas produced in India in 1770. Commissioned by a French officer, Col. Jean-Baptiste Gentil who served as military adviser to the Nawab of Oudh from 1763–75, it includes a map of the *subah* of Bidar and includes a tiny drawing of a man inlaying a *huqqa* (Figure 11.4). A second drawing of the wares produced in Bidar depicts *huqqas*, boxes, a ewer and basin, and mouthpieces for *huqqas* (Stronge 1985 p. 18).

From Bidar, the craft spread during the 18th century to Purnea (in present-day Bihar), Murshidabad (in Bengal) and to Lucknow (Uttar Pradesh). In the 19th century, *bidri* ware became widely known outside India. The Great Exhibition held at London's Crystal Palace in 1851 gave *bidri* international acclaim and illustrations of motifs on the pieces displayed in the Indian Court were included in Owen Jones' influential *Grammar of Ornament*. This exhibition, and the series of similar exhibitions which followed, stimulated a demand for *bidri* which led to new centres being established, though these did not survive long. The London store, Liberty & Co, established in 1875, also included *bidri* from Bidar and Purneah in its catalogues (Stronge 1985 p. 28). The craftsmen expanded their repertoire to include items for their new western clientele and continued to produce the traditional *huqqas*, spittoons and boxes, but were forced to restrict their range of patterns which became limited to a narrow range of stock motifs. The focus of the industry was always the Deccan and the products of the second half of the 19th century are dominated by the poppy pattern which is also ubiquitious in the paintings of the region. In recent years, the industry has been confined to Bidar, now in Karnataka, and to Hyderabad in Andhra Pradesh. When the States were reorganized in 1956 Bidar became part of Karnataka, and was thus separated from its main market of Hyderabad. As a result, most of the Bidar craftsmen moved to Hyderabad where there is now a considerable industry (Figure 11.5) (Gowd 1964), though a small number remain in Bidar (personal communication, Dr Barbara Brend, April 1991 and p. 146). The craftsmen continue to adapt their wares to local demand, particularly the tourist industry, and are also now producing high quality reproductions of items such as *huqqas*, in the forms and patterns of the 18th and 19th centuries (Figure 11.6).

Figure 11.5. *Craftsmen at the Yaqoob Bros.* Bidri *Works, Hyderabad, 1991. All stages of manufacture are carried on here, with a number of specialist craftsmen working together. Many establishments are much smaller with only two or three craftsmen and apprentices. (Photo by Paul Craddock.)*

The technique

Despite the lack of information concerning the early development of the industry, it is certain that what was known as *bidri* by the early 18th century was being produced to a very high standard by at least the beginning of the 17th century. An examination of the technique reveals some intriguing information although, frustratingly, leaves many crucial questions unanswered.

The alloy from which it is made has a very high zinc content: analysis carried out by Graham Martin on 27 pieces in the V & A's collection (see Table 11.1) showed a minimum content of 76% and a maximum of 98%, with

Figure 11.6. *A modern spittoon in mid-19th century style, Gulistan/Mumtaz* Bidri *Works, Hyderabad, 1986. (Photo by Paul Craddock.)*

a margin of error estimated at ±3.5% (La Niece and Martin 1987 pp. 97–98). The other significant ingredients found were copper and lead, but the lead may not all have been intentionally part of the alloy as it occurs naturally with zinc ore, although undoubtedly some lead was added as a cheap adulterant. Some of the pieces analysed were of the 17th century; the bowl in Figure 11.1, for instance, yielded a result of 93.1% zinc. This means that the craftsmen had access to supplies of metallic zinc considerably earlier than in the west.

The production of metallic zinc from its ore presents significant problems. Unlike copper or iron ore, for example, where a traditional simple shaft furnace will allow the metal to settle as an ingot at the bottom, zinc boils at a very low temperature and vaporizes. It would therefore re-oxidize in the flue and be lost unless special arrangements were made to condense it. In the west, William Champion of Bristol patented a zinc distillation process in 1738 which allowed the metal to be produced on an industrial scale (Day 1973 pp. 73–4). In India zinc was being produced on an industrial scale centuries before this, as proved by an Anglo-Indian team formed to investigate disused zinc mines at Zawar in Rajasthan, (Craddock *et al.* 1990, pp. 29–72). The existence of these mines had been known for some time: in 1908, for instance, the *Imperial Gazetteer of India* commented that 'the lead and zinc mines are said to have yielded up to 1766 a net revenue of 2 lakhs of rupees (i.e. 200 000 rupees) and were worked until the famine of 1812' (*Imperial Gazetteer* 1908 p.

Table 11.1 Atomic absorption analysis of *Bidri* metal

Victoria & Albert Museum acquisition nos.	*Description*			%		
		Zinc	*Copper*	*Lead*	*Tin*	*Iron*
2539-1883 I.S.	*Huqqa*	91.2	3.2	2.9	1.2	0.2
1578-1904	*Huqqa*	98.0	3.3	1.4	0.1	0.2
45-1905	*Huqqa*	88.2	3.0	2.9	0.7	0.5
I.S.46-1977	Weight	92.6	4.7	1.4	5.7	0.1
1479-1904	Ewer	89.3	4.8	0.8	<0.1	0.8
I.S.10-1973	Bowl	92.1	3.6	1.4	0.3	0.1
02942 (I.S.)	*Huqqa*	84.6	3.3	1.0	0.1	0.5
I.S.131-1958	Pan box	95.0	3.1	1.5	0.1	0.1
I.S.181-1965	*Huqqa*	76.1	10.1	8.2	11.4	0.6
I.S.31-1976	Bottle	98.6	2.6	1.0	0.4	1.2
857-1874	*Huqqa*	92.1	3.6	0.8	0.6	0.4
I.S.17-1970	Pan box	81.3	2.0	2.0	<0.1	0.5
02949	Bottle	83.8	2.5	6.6	<0.1	0.2
02949	Bottle	93.2	2.4	1.1	0.1	0.2
855-1874	*Huqqa*	99.2	2.6	0.4	0.2	0.6
02941 (I.S.)	Basin	91.2	2.8	2.1	0.1	0.2
I.S.4-1977	*Huqqa*	83.6	3.7	0.8	<0.1	0.6
I.S.19-1978	*Huqqa*	85.3	4.6	3.0	1.3	<0.1
120-1886	Bottle	86.7	5.3	1.9	0.2	0.1
I.M.224-1921	Bottle	91.7	3.6	0.7	0.7	0.7
2066-1883	Box	79.6	2.7	19.9	0.1	0.5
1402-1903	*Huqqa*	95.3	2.7	1.1	3.0	0.5
I.S.11-1973	Vessel	97.6	3.4	1.8	<0.1	0.1
I.S.39-1976	*Huqqa*	89.2	3.4	0.5	0.7	0.7
I.S.19-1980	*Huqqa*	80.6	3.6	7.3	6.4	0.9
856-1874	*Huqqa*	77.4	4.4	0.9	0.4	0.1

(Analysis by Graham Martin)

96). More recently, the metal-producing territory of Jalor, mentioned in a history of India completed in 1596 for the Mughal emperor Akbar, was identified as modern Zawar (Habib 1982 p. 20). In this book (the *A'in-i Akbari* or Institutes of Akbar), the author Abu'l Fazl lists the metals and various alloys known in Hindustan (Blochmann and Jarrett 1977 I, pp. 41–2). Most of these are so far unidentified due to the lack of technical analysis on the one hand and a dearth of dated, or closely datable objects on the other. Abu'l Fazl includes a metal called *jast*, which in later periods specifically meant zinc: '*jast*, which resembles lead, is nowhere mentioned in philosophical books, but there is a mine of it in Hindustan, in the territory of Jalor'. The team discovered that smelting on a vast scale had taken place in this area, indicating a huge industry, and from the discovery of several intact furnaces they were able to identify the exact process and to link it with that described in early Indian alchemical texts. They concluded that the main period of zinc production on an industrial scale began some time in the 14th century and continued, with interruptions, for the next four centuries (Craddock *et al.* 1990 pp. 49–50).

It would seem that there were no other large-scale centres of zinc production in the Indian subcontinent and that, if using indigenous supplies, the *bidri* craftsmen must have used Zawar zinc. So far isotope studies have not revealed any zinc from this source in *bidri* artefects (Craddock *et al.* 1990 p. 52) but the number of samples is extremely small and cannot be said to be statistically conclusive. It has, however, been a perpetual problem when considering the reasons why *bidri* production should have developed in the Deccan that there is no local supply of the basic raw material. A source far away in the north, in the Hindu strongholds of Rajasthan, is not the most obvious for a Muslim sultanate in the Deccan, but seemed possible given the level of production. The negative evidence of the isotope testing may suggest that the metal was, after all, imported from abroad; imports of Chinese zinc into southern India are known from the mid-17th century (Craddock *et al.* 1990 p. 53) and Iranian supplies could also have been used. Either could have been shipped to the Coromandel Coast ports of Nizampatan and Machhlipatan, both of which supplied Bidar (Habib 1982 Figure 15B). This still does not, however, explain the choice of Bidar as a centre of *bidri* production and, unless historical texts eventually provide the answer, it may never be satisfactorily explained.

Analysis has also been carried out on the surface coloration of *bidri* ware (La Niece and Martin 1987). The clay paste which is smeared over the finished object in order to blacken the zinc has been known to contain ammonium chloride and potassium nitrate since at least 1809, when the first western commentator listed the clay's active ingredients (Buchanan 1928 p. 534). However, the process by which the change was effected had never been explained. Replication experiments were carried out to test the hypothesis that the colour of the patina depended on the presence of copper in the alloy: a mixture of potassium nitrate and ammonium chloride applied to pure zinc produced only a pale grey patina. When copper sulphate was added a superficial black patina formed almost instantly. When zinc was alloyed with about 3% of copper and the first two ingredients were applied, it acquired a deep black surface immediately which adhered well (La Niece and Martin 1987 p. 99). It was concluded that the use of clay did not seem to be a necessary part of the process, yet the Deccani craftsmen of Bidar and Hyderabad went to great lengths to obtain exactly the right kind of clay. In 1989 I interviewed a craftsman who had moved from Hyderabad in the Deccan to Hyderabad, Sind, at Partition in 1947; he had taken a supply of local clay with him to guarantee his results. At Bidar, soil from the fort walls was insisted upon and, at the Hyderabad Gulistan Mumtaz works, 'soil from old places' was used (La Niece and Martin 1987 p. 99). This would seem to suggest that soil rich in nitrates, obtained from animal or human waste, is necessary to provide the required chemicals in the most economical way; analysis of earth taken from the walls of Bidar fort in 1956 concluded that the active ingredient in the earth used to coat *bidri* was an alkali nitrate (Gairola 1956 pp. 116–18). It may be the case, though this hypothesis has not been tested, that the longevity of the coloration depends on it having been applied by means of the mud paste. The authors of the 1987 study of *bidri* patination suggest that the clay may also serve as a poultice to absorb the unwanted zinc chloride formed during the process and that it may have to be of fine texture: a coarse clay produced an uneven patina. This all suggests the kind of advanced, almost instinctive, metallurgical understanding for which the subcontinent, and the Deccan in particular with its renowned crucible steel manufacture, has been famous for centuries.

Acknowledgements

I am greatly indebted to Dr A.S. Melikian-Chirvani for translating the passage from the *Chahar Gulshan*, and to Dr Barbara Brend for information about the modern *Bidri* industry in Bidar.

Appendix: Contemporary production of *bidri* ware in Hyderabad and Bidar

Paul Craddock

This description of the process as it is practised today is largely based on visits to the Gulistan/Mumtaz *bidri* works in 1985 and to the Yaqoob Bros. *bidri* works at Hyderabad in 1991. The author is extremely grateful to the proprietors and craftsmen in these and other workshops for their patience in answering detailed questions, and allowing full access to the workshops and their work to be interrupted as they were posed for photographs, etc.

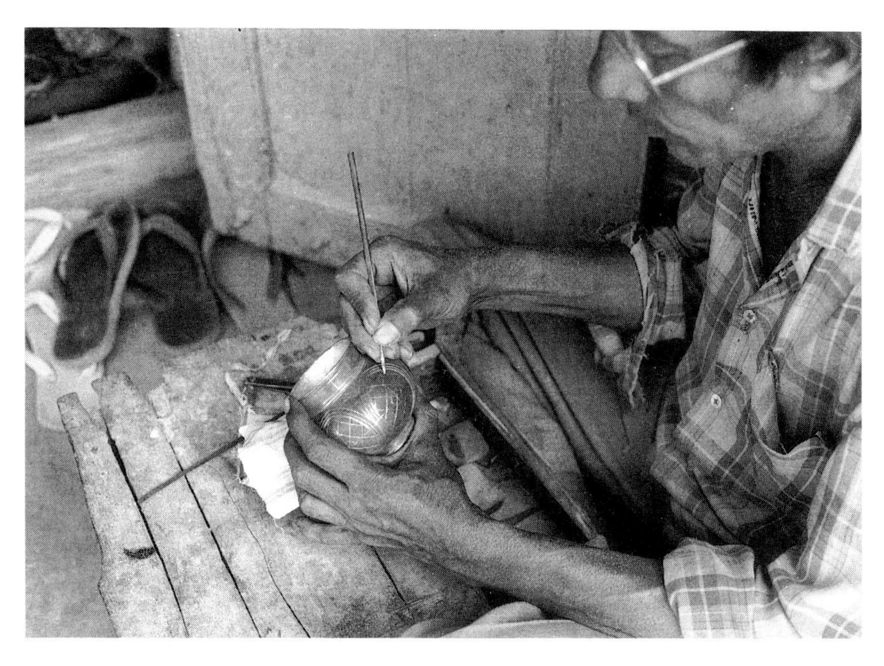

Figure 11.7. *Scribing the design with a steel point onto the surface (Yaqoob Bros.) (Photo by Paul Craddock.)*

An excellent account of *bidri* making, also based on interviews with Hyderabad craftsmen is given in Untracht (1968 pp. 138–149), although with the primary object of instructing western craftsmen rather than producing a precise record of the Indian process. Other good descriptions are to be found in Yazdani (1947 p. 20) and Lal (1991 pp. 1–8).

The Yaqoob Bros. consume about 200 kg. of zinc a month, obtained from the market and direct from Hindustan Zinc Ltd., the Gulistan works also use some scrap zinc obtained from lithography plates, etc., and both workshops use old electrical cable for the copper. The alloy used at the Gulishtan works contains 5% copper whereas the Yaqoob Bros use 6.25% of copper with the remainder being of zinc in both cases. The alloy given by Untracht contained rather less copper (4%), but also had about 3% of lead, and that quoted by Yazdani contained about 12% of copper and 3% of lead. Variable quantities of lead are found in old *bidri* wares (see Table 11.1), but none of the many workshops visited by the author mentioned lead, and none of the pieces purchased were found to contain any lead on analysis. Powdered steel is recorded as an ingredient of the *bidri* alloy in recipes from Lucknow in quantities up to 20% by weight, and 1.27% was found in a piece from Purnia in Bihar (Lal 1991). However, iron has never been reported in *bidri* from the Deccan and metallurgically it seems inherently unlikely as a deliberate constituent, being very difficult to dissolve in the zinc as well as creating serious embrittlement in any alloy so formed. If the reports are accurate, possibly the iron powder was intended to stay in suspension in the *bidri* alloy and patinate separately at the surface creating a speckled brown polycromy on the black *bidri*.

All of the *bidri* pieces are sand castings. Nowadays items such as the sides and bases of boxes are cast separately and are then joined together with soft solder. The moulding medium is ordinary fine sand to which is added a little castor oil containing some tree resin in the ratio of 200 g resin to a litre of oil, to give cohesion to the sand. Standard procedures of moulding and casting are followed and are described and illustrated in detail by Untracht (1968). Small items are cast in batches of up to

Figure 11.8. *The scribed pieces are set in wax and the design chiselled out using tracers (Yaqoob Bros.) (Photo by Paul Craddock.)*

Figure 11.9. *Inlaying silver wire using a punch into the undercut channels cut to hold it (Yaqoob Bros.) (Photo by Paul Craddock.)*

a dozen but medium and large items are cast individually. To produce a piece from template to finished casting can be done in under 10 minutes by an experienced craftsman.

After the casting has been trimmed, filed and assembled it is ready for the inlay work to commence. First the surface is temporarily blackened by rubbing with a damp cloth containing crystals of copper sulphate. The design is drawn on using compasses and scribers (Figure 11.7). The piece is then set in wax and the design traced and chiselled out with edges undercut to hold the inlay (Figure 11.8). Where there is to be sheet inlay the surface of the depressions are roughened and the inlay is prepared by placing a piece of sheet over the depression and tapping it lightly with a mallet to mark the shape on the sheet so that it can be cut to fit the depression precisely. Very soft fine silver is used which can be hammered quite securely in place. A great deal of the designs on modern *bidri* are

Figure 11.10. *The piece is warmed up and the slurry of mud and ammonium chloride smeared over the surface. The patination develops almost immediately (Gulistan/Mumtaz Works) (Photo by Paul Craddock.)*

made up exclusively from wire inlays (Figure 11.9). Although these can often be very complex, this is regarded as less skilled work and is consequently much cheaper.

After inlaying, the article is rigorously scraped and filed and in one of the few concessions to the 20th century, mechanically buffed.

The final stage is the formation of the black patina. This is done with a hot slurry of mud and ammonium chloride; the mud is not just any mud but that from 'under the walls of old places', which as Sue Stronge has already made clear in India are likely to be pretty rich in salt, nitre and urea. The Gulistan workshops used mud that tasted salty. This practice is not peculiar to the *bidri* trade, nitre for gunpowder was also collected during the dry season after the monsoons from around mud walls from where it had crystallized. Indeed India, especially Uttar Pradesh, was one of the world's main suppliers up until the discovery of the deposits in Chile in the 19th century. The Yaqoob workshops actually bring in about 50 kg of soil per month from under the old fort walls of Bidar itself. On the face of it it seems inherently unlikely that it is necessary to have to bring soil over 100 km. However, there may be firm scientific reasons for this beyond sentiment. Hyderabad, in common with most of the state of Andhra Pradesh, is on granite whereas the distinctive latteritic soils of the Deccan, used at Bidar, only start just before that town.

The patina develops within a few seconds if hot. Thus either the slurry is boiled and the pieces dipped in the pan, or the piece itself is placed in the hearth for a few seconds to warm up and then the cold slurry is smeared over it (Figure 11.10). The latter method is exactly paralleled in the description given by Pausanius of the Corinthian bronze patinating process (see p. 112). The mud is then washed from the piece and a very little coconut oil rubbed in to complete the process. *Bidri* is one of the few materials that benefits from being handled with sweaty greasy hands, and the silver can be cleaned by gently rubbing in a little Silvo or any other non-abrasive ammonia-based silver polish.

Present state of the industry (based on conversation with Yaqoob Bros.)

The craft of *bidri* making almost died out in central India in the late 19th–early 20th century, but was revived by the Nizam's government who instigated a training course. The craft has always been carried out by Moslems, but the only person who could be found to teach the craft was Sri Ramana Maslei, who was a Hindu. However the craftsmen are now almost

Figure 11.11. Bidri *plate with typical modern floral design of sheet inlay (Yaqoob Bros.); now in the British Museum, OA 1992.7-31.1 (Photo by Tony Milton.)*

exclusively Moslem. The craft continued to be supported by the state both before and after Independence with training programmes, loans to establish workshops and sales outlets in the the various State Craft Emporia as detailed in Stronge (1985). The industry is now in a healthy state in the Deccan employing about 125 workers in Hyderabad, 35 in Bidar and a few in Aurangabad.

Much of the production goes to the various State Emporia or in privately run craft shops although specialist *bidri* shops such as *Bidri Crafts* in the Gunfoundry area of Hyderabad are rare. There is some sale from the works themselves as well as from direct commissions and many of the smaller establishments double as a shop.

The range of artefacts produced now has probably never been wider, and can be broken down into four main areas.

1. Small, relatively easily made (but never mass produced) items, such as letter openers, key-rings, ashtrays etc. that sell for a few rupees to tourists.
2. Straight copies of traditional designs, although often with wire rather than sheet inlay. These cater for the more discerning tourist and typically sell for a few hundred rupees.
3. Items designed in a contemporary idiom to suit modern Indian middle class taste, and these can sell for a few hundred up to several thousand rupees (Figure 11.11).
4. Special commissions, both private but particularly from the state for pieces for government offices, embassies, and presentation and commemorative works, where the continuing skill in both design and execution are given full expression.

References

Begley, W. and Desai, Z.A. (1989) *Taj Mahal. The Illumined Tomb*, The Aga Khan Program for Islamic Architecture. Harvard University and Massachusetts Institute of Technology, Cambridge, Mass.

Blochmann, H. and Jarrett, H.S. (trans. and ed.) (1977) *The A'in-i Akbari*, by Abu'l Fazl, 3rd edition. Oriental Books Reprint Corporation, New Delhi

Buchanan, F. (1928) *An Account of the District of Purnea in 1809–10*, Patna

Craddock, P.T., Freestone, I.C., Gurjar, L.K., Middleton, A.P. and Willies L. (1990) Zinc in India. In *2000 Years of Zinc and Brass* (ed. P.T. Craddock) British Museum Occasional Paper No 50, London, 29–71.

Crowe, Y. (1986) Coloured tilework. In *Islamic Heritage of the Deccan* (ed. G. Michell) Marg Publications, Bombay

Day, J.(1973) *Bristol Brass: A History of the Industry*, David & Charles, Newton Abbott

Gairola, T.R. (1956) Bidri ware. (Ancient India *Bulletin of the Archaeological Survey of India*, **12**, 116–18

Gowd, K.V.N. (1964) Selected crafts of Andhra Pradesh: Bidri ware'. *Census of India 1961*, vol. II, part VII-A (3), New Delhi, pp. 3–21

Habib, Irfan (1982) *An Atlas of the Mughal Empire*, Delhi, *Imperial Gazetteer of India*, vol. XXIV, new edition. OUP, Oxford, 1908, p. 96

Jones, M. ed. (1990) *Fake? The Art of Deception*, British Museum Publications, London

Jones, O. (1856) *The Grammar of Ornament*, Day & Son, London

Joshi, P.M. (1950) Asad Beg's mission to Bijapur, 1603–1604. In *Prof. D. V. Potdar 61st birthday*

commemoration volume (ed. S. Sen), Poona, pp. 184–96
La Niece, S. and Martin, G. (1987) The technical examination of bidri ware. *Studies in Conservation*, **32**, 97–101
Lal, K. (1991) *National Museum Collection Bidri Ware*, National Museum, Delhi.
Michell, G. (1986) Bidar. In *Islamic Heritage of the Deccan* (ed. G. Michell) Marg Publications, Bombay, pp. 42–57
Sarkar, Sir J. (1901) *The India of Aurangzib*, Bose Brothers, Calcutta
Sarkar, Sir J. (trans. and ed.) (1981) *Maasir-i Alamgiri*, by Saqi M. Khan, Calcutta, 1947; reprint Lahore, Ever Green Press, 1981
Stronge, S. (1985) *Bidri Ware. Inlaid Metalwork from India*, Victoria and Albert Museum, London
Untracht, O. (1968) *Metal Techniques for Craftsmen*, Doubleday, New York
Yazdani, G. (1947) *Bidar*, Published under the authority of the Nizam
Zebrowski, M. (1983) *Deccani Painting*, Philip Wilson Publications, London

12

Special finishes on non-ferrous metals at the National Maritime Museum

Laurence Birnie

Abstract

Scientific instruments are a central area of study at the National Maritime Museum and research into surface finishes and their preservation has been continuing for some years. This paper discusses why many metal surfaces of scientific instruments and related objects were finished either by deposition of another metal or by chemical colouring. The chemical colouring of scientific instruments is of great interest to conservators of these objects. What can appear to be tarnished surfaces are in fact, intentionally coloured and must be preserved. Metal finishing methods and recipes are given in this paper, including bronzing, gilding, silvering, and lacquering.

Introduction

The main aim of this paper is to explain how and why metal surfaces of many scientific instruments were finished either by deposition of another metal or by chemical colouring. The latter case is of special interest because what appear to be tarnished surfaces can, in fact, have been intentionally finished. Irrevocable damage can be done if such a surface is cleaned without careful prior inspection. The techniques and recipes discussed and explained in this paper represent a cross-section of the main types of metal finishing processes but many others may be encountered. It must be stressed that some of the substances used in the processes described are highly toxic and dangerous to health and therefore appropriate safety precautions must be observed at all times.

There are three main reasons for colouring the metals used in scientific instruments; aesthetic appearance, corrosion resistance and to prevent them from being too reflective. The colours used on scientific instruments include brown, blue, black, grey and green. These colours may be applied by dipping, brushing or spraying. The colouring can be detected when the instrument is dismantled; parts that have been pressed together will show unexposed areas with the original finish.

The explanation of methods of metal finishing included in this paper should be useful to the conservator who is concerned, when a surface is to be conserved, that the metal finishing present will remain intact. In addition, awareness of the recipes and formulations of metal finishing solutions may be helpful in determining the types of corrosion encountered on instruments which have deteriorated. If a decision is made to restore surfaces to their original condition, then formulation of recipes may also be helpful in their restoration. The matching of new parts may also be required and therefore a knowledge of the original type of metal finish and how it was obtained is essential.

Mechanical finishing

Most instruments do not have a mirror bright finish, but rather they are given a slight satin or a machined finish. This is achieved with decreasing grades of abrasive, the final grade of abrasive depending usually on the quality of the instrument and its purpose. The lines produced usually follow the lines of the instruments and skill is required to ensure that the sharp lines of the instruments are retained. On large areas patterns were sometimes produced using fine abrasives.

The finishing of scientific instruments requires precision work. The surface as it leaves the superfine file is finished by using Water of Ayr stone to matt the surface and finally by rubbing with soft grey slate stone. Washemery and fesh polishing paper are used for polishing where brightness is desirable, particularly for steel. Heads of screws and small turned parts were finished by a clean cut or with a burnisher on a lathe.

Bronzing

For the protection of finished metalwork on instruments to be used in bright sunlight, a dark grey bronze finish is pleasant to the eye, leaving bright only the parts which are required to be easily seen, such as milled heads, heads of screws etc. The bronzing is effected by the application of a liquid which corrodes the metal and at the same time, leaves a dark satin deposit upon it. A great number of 'bronzes' are available, ranging through all colours of the spectrum, but for scientific instruments, only dark greys and occasionally dark browns were applied. The bronzing solutions used to produce these shades are largely based on either platinum, arsenic or mercury solutions.

Platinum chloride solutions are permanent and do not cause corrosion of the substrate metal after their application. A typical recipe is given by Hiorns (1929 p. 138):

platinum chloride	1 part
water	5 parts

It produces a black/grey colour by depositing a thin film of platinum metal on the surface. This bronzing solution was well known, but due to its cost it was not used as frequently as modern conservators might have wished.

Bronzing solutions based on arsenic oxides deposit a thin, adherent film of arsenic. A variety of solutions and techniques are available and the shade and depth of finish can be varied. Moreover, the process is versatile in that most copper alloys are amenable to treatment. A typical solution is the following given by Fishlock (1962 p. 221):

arsenic trioxide	13oz (370 g)
hydrochloric acid	10 fl.oz (284 ml)
sulphuric acid	2½ fl.oz (71 ml)
water to 1 gallon	(4550 ml)

Arsenic solutions are relatively inexpensive and act rapidly. The resultant finish is sufficiently wear resistant to withstand mopping with leather (polished work) or dry scratch brushing, but should be protected for service by lacquer or wax. They can be applied to most copper alloys including nickel silver, although the most satisfactory results are probably obtained on yellow brass.

The bronzing solutions containing mercuric chloride cause the most problems. These were very cheap and gave a fine dark surface but they were certain to rot the brass and produce a pitted or spotted appearance. If the blisters are removed then a bright red compound can be observed under a microscope. This is mercuric oxide. A simple spot test with dimethylamino-benzylidine rhodanine can confirm the presence of mercury. The following mercuric chloride based solution was commonly used to bronze instruments (Hiorns, 1929 p. 159):

mercuric chloride	150 grains (9.7 g)
ammonia nitrate	150 grains (9.7 g)
water	3½ fluid ozs (99 ml)
ammonia	until the precipitate formed dissolves

This solution instantly coats both copper and brass with mercury when immersed in it. As can be seen from the constituents of this bronzing solution, it has many highly corrosive compounds which will attack copper and copper alloys.

Gilding

The old adage that all that glitters is not gold should be taken as a warning; the conservator cannot assume that if a metal is bright and gold in colour it is a gilded copper alloy. Only on very fine ornamental scientific instruments would gold have been applied. A lacquer can create the appearance of gold due to added tints and dyes, but lacquer wears off more easily than gold from high spots.

The presence of gold can be detected by X-ray fluorescence analysis or by chemical tests. Unlike silvering, gilding would not be used to give a contrast of colour to the copper alloys, as both would appear bright and yellow when polished, although a different shade. In addition to the physical attachment of a layer of gold to the surface of an object, gilding can be accomplished either by amalgamation or by immersion.

Gilding by amalgamation is highly dangerous due to the emitted mercury fumes. As a result its use is now prohibited. This process is described by Spon (1873 p. 307).

> A quantity of mercury is put into a crucible or iron ladle, which is lined with clay and exposed to heat till it begins to smoke. The gold to be mixed should be previously granulated, and heated red hot when it should be added to the mercury and stirred about with an iron rod till it is perfectly dissolved. If there should be any superfluous mercury it may be separated by passing it through clean soft leather; and the remaining amalgam will have the consistence of butter, and contain about three parts of mercury to one of gold.

The object to be gilded was cleaned in boiling dilute nitric acid, then dipped in a solution of mercury in nitric acid. This process is called 'quicking' and makes the surface white. The amalgam could then be applied with a soft brush until all the surface was covered. The object was exposed to a gentle degree of heat and brushed to prevent irregular dissipation of the mercury. When the mercury had been driven off by the heat, the gold was left on the surface of the metal.

In the immersion gilding process, the gold is deposited onto the surfaces of objects which are simply immersed in particular solutions. In all cases, the deposited layer is very thin. The following recipes are taken from Hiorns (1929 pp. 295-6):

potassium hydrogen carbonate	60 parts
gold chloride	1½ parts
water	200 parts

The mixture is boiled for 2 hours during which period the solution, at first yellow, assumes a green colour. Brass and copper articles are gilded by dipping them for about half a minute in the solution after careful cleaning.

The early solutions for immersion gilding did not contain cyanides, but, later potassium cyanide crystals were used which gave better results:

crystallized sodium phosphate	2¾ drachms (9.8 ml)
pure caustic potash	1½ drachms (5.3 ml)
gold chloride	½ drachm (1.8 ml)
potassium cyanide	9 drachms (32 ml)
water	1 quart (1300 ml)

Dissolve the sodium phosphate and the potash in part of the water and the gold chloride and the potassium cyanide in the other part. Mix the two solutions together and boil for use as before. It is best to dip articles in a partly used solution and finish in a freshly prepared one.

After the discovery of electrolysis in the early 1830s most gilding was applied by electro-deposition which gives superior results (Child, this volume, Chapter 24).

Silvering

Silvering is applied to parts of an instrument where contrast is needed, for example on compass dials and scales. A silvery coloured plating of nickel was sometimes plated onto instruments for corrosion resistance (see Chapter 24). A plating of true silver can be applied either from a paste or by electro-deposition.

A typical silvering paste is composed of equal parts of fine ground silver chloride, sodium chloride and cream of tartar (potassium hydrogen tartrate) and can be applied with a pad of damp cotton wool. The following method is used by the author:

> The object is first rubbed over with sodium chloride on a pad of damp cotton wool (a new piece of wool is used for each chemical). Equal amounts of silver chloride, sodium and cream of tartar are mixed, then applied to the brass or copper surface with a movement of the pad that will follow the mechanical surface finish of the instrument. The surface is constantly washed during this process. The silver will gradually appear on the surface as a silver white deposit. If the suface begins to darken, a fresh damp pad of cream of tartar is applied which will whiten it and also remove all traces of sodium chloride which causes green spots to form. Finally, the surface must be thoroughly washed and lacquered for protection.

Bright silver is electro-deposited from a cyanide solution based on silver cyanide and sodium or potassium cyanide together with sodium or potassium carbonate and hydroxide.

Lacquering

The first questions conservators must ask with regard to lacquer are whether the original lacquer is an integral part of the instrument and whether it must be conserved along with the rest of the instrument. In some cases, of course, these questions may be irrelevant as there may be little original lacquer remaining. Lacquer on the surface of an object is its first line of defence against corrosion. If it is lost then the instrument is not protected and this may jeopardize other surface finishing such as coloured finishes, graining marks and patterns which were originally under the lacquer.

It must also be noted that the lacquer present at the time the instrument arrives at the conservator's bench may not be the original lacquer, as some scientific instruments have been re-lacquered many times during their lifetime. However, the lacquer should not be removed simply because the conservator *suspects* that it is not original. A decision must be made with regard to the degree of protection which remains on the instrument. If two-thirds of the original lacquer remains in good condition then an attempt should be made to conserve it. In this case if the area that has been left unprotected is corroded, this corrosion should be removed or made stable. The lacquer can be patched in after the tarnish has been removed, but this takes a great deal of skill. Identification of the original lacquer is essential for matching the colour and body with the original and the way in which the lacquer is applied should be noted.

The main argument for leaving a lacquer on a scientific instrument is for research purposes. This criteria can be satisfied by leaving one part of the instrument in an untouched, supposedly original condition. This can be a part that, in comparison to the rest of the instrument, is in good condition. The part that has been left untouched must be noted on the conservation and curatorial recording system and, if possible, a written record should be kept with the instrument. A decision to remove the lacquer must be made only after full consultation with the curatorial department concerned.

Conservators must also decide whether to replace the old decayed lacquer with a lacquer of similar constituents or with a modern synthetic lacquer. The original lacquers have proved to be long lasting and relatively stable. Their only drawbacks are that they were prepared only for brushing and they contain many dyes and additives. Modern lacquers, although mixed in a way which makes them ready to spray, may have hidden dangers for the future. For example, the cellulose nitrate based lacquers are easily applied by brush or spray but they are very unstable.

Modern acrylics should be investigated as an alternative to cellulose nitrate, and modern refined sprayable shellac which is water white may also be used. The dyeing of these modern lacquers to match original dyed lacquers also need to be examined and their stabilities tested.

In the past, the lacquers used on scientific instruments were made up of three or four types of constituents. They were based on gums and resins, with a solvent. The solvents used include methanol, ethanol, oil of turpentine and white spirit (a turpentine substitute). Dyes and other additives in various proportions altered colour, hardness and thickness. Rouveyre (1911) gives recipes for lacquers coloured by such ingredients as saffron, turmeric and extract of red sandalwood.

Reversible synthetic cellulose nitrate lacquers have gradually replaced resin and gum based lacquers during this century. The main lacquers now used to protect metals are cellulose nitrate and acrylic based lacquers. Cellulose nitrate is an ester of nitric acid. It was first produced by Schoenbein in 1845, who treated cellulose with a mixture of nitric and sulphuric acids. The cellulose from cotton or wood pulp is now treated with a mixture of nitric acid and sulphuric acid. The acid residue is then washed out of the crude cellulose nitrate to ensure stability of the material. The cellulose nitrate fibres are then dehydrated in alcohol and dissolved in a solvent, normally amyl acetate or acetone. The viscosity of the cellulose nitrate solution can be varied to suit the method of application. The disadvantage of cellulose nitrate is that because it is an ester of nitric acid it is particularly susceptible to photochemical degradation into acids such as nitric and formic. Cellulose nitrate may undergo hydrolysis by acids or by alkaline reagents, resulting in denitration and reduction of molecular chain length; this may also be attributed partly to traces of acid impurities remaining from the nitration process. It should also be noted that metallic oxides cause cellulose nitrate in amyl acetate solution to undergo irreversible gelation.

The technique of application of all lacquers is important. Traditionally, scientific instruments were lacquered by brushing with a flat camel-hair brush. The metal was raised to 76°C then the varnish applied in single, long strokes. This was highly skilled work requiring many years of practice to perfect, and hence it is not reasonable to expect the conservator to accomplish this. If not properly done, lacquering does not give a coherent coating, leaving unprotected metal which will be exposed to future corrosion.

Application of lacquer by spraying on a cold metal is also a process requiring skill. The lacquer must be of the right viscosity and the air pressure of the spray-gun at the correct level. The author uses a small 'Devilbiss MP' at 30–40 pounds per square inch. All spray lacquering is carried out in a 'Devilbiss' spray booth with a 'Devilbiss' compressor. The spray booth should contain a turntable to rotate the work as it is being sprayed so that a coherent coating is applied.

All solvents and lacquers should be used in a fume cupboard or extractor system. Contact with the skin must be avoided especially in the case of toluene and xylene solvents. Government regulations for storage and use of cellulose nitrate must be followed at all times because of its highly flammable nature.

Surface cleaning

The main reason for cleaning is to remove dirt and corrosion products from the instrument's surface. Cleaning is a process which must be given very careful thought before execution. Extensive, usually irreparable, damage can be done to the surface of an instrument and evidence of methods of manufacture lost by overzealous cleaning and polishing. Before cleaning, the surface of the instrument must be inspected using at least ×10 magnification. This will give the conservator much visual information on the state and finish of the instrument.

The degree to which the dirt and corrosion products are removed depends on several factors. For example, it may be that the instrument is an example of its type or is a relic of specific historic significance by association with a person or event. In the former case the dirt and corrosion products may be removed, while in the latter the dirt and corrosion products may themselves be of historical significance. In any case, this is a subject that should be fully discussed by the conservator and the curator. In all cases the instrument must be treated in a way which will render any remaining dirt and corrosion products stable so that no further damage is done to the surface of the instrument.

Finger prints can do great harm to both lacquered and bare substrate metal surfaces of the instruments. Analysis indicates that perspiration contains 2 to 17 g per litre of lactic acid and 2.5 g per litre of sodium chloride, plus fatty and amino acids; a very sticky and corrosive mixture. Therefore it must be stressed that when handling scientific instruments gloves of some kind must be worn. Cotton gloves are acceptable for a short period but they do accumulate perspiration which eventually seeps through to the outside of the glove. Cotton gloves must therefore be frequently washed in non-ionic detergent or disposed of. For long periods of wear, rubber surgical gloves are best as they give complete protection to the object. Ethanol will remove much of the perspiration from newly placed fingerprints.

The danger of damage caused by fingerprints as well as by the use of the wrong tools for dismantling should be considered not only by the conservator but also by the curator who handles the instruments and may be tempted to remove parts for research purposes.

Cleaning processes used on scientific instruments may be purely mechanical, purely chemical or a combination of the two. Commercial cleaning pastes, wadding and solutions can consist of many different abrasives and chemicals mixed together and so can be mechanical and chemical in their action upon metallic and organic surfaces. Some can be very efficient cleaners but others are harmful to scientific instruments.

The following is a list of available products used to clean metal surfaces. Their commercial name is given in italics and the maker given in brackets, followed by the composition.

Brasso (Reckit and Colman) Abrasives, white spirit, ammonia, fatty acid.
Bluebell (Reckit and Colman) As 'Brasso' but coarser abrasives.
Duraglit Metal polish (Reckit and Colman) Wadding cotton, abrasive, white spirit, fatty acid, perfume, ammonium soap complex.
Duraglit Silver polish (Reckit and Colman) Wadding cotton and 'Silvo'.
Silvo (Reckit and Colman) Siliceous earth, water, alcohol, fatty acid, perfume and anti-foaming agent.
Solvol Autosol (Solvolene) Petroleum, fatty acids, solution of ammonia, fullers earth, sulphurized oil derivatives.
No.1 Watch Cleaner (Horological solvents) Ammonia, methylated spirits, acetone, oleic acid, butyl acetate.
Horolene Clock Cleaner (Horological solvents) Ammonia, methylated spirits, distilled water, oleic acid.
Silver Dip (Goddards) 0.9% sulphuric acid, thiourea, water, perfume.

The products that contain ammonia must be avoided because of the risk of stress-corrosion cracking (Shreir 1963) of copper alloys.

Conclusions

The subject of metal finishing on instruments is a complex one because there can be several types of finish on one instrument. It is a subject that the conservator must be very aware of in attempting to conserve the surface of any instrument.

Bibliography

Birnie, L.A.A. (1980) Colouring of copper and its alloys. Dissertation for Licentiate of Metal Finishing examination. L.I.M.F. A.M.I. (corr) T. Polytechnic of the South Bank, London. Unpublished

Blackshaw, S. and Daniels, V. (1979) The testing of materials for use in storage and display in museums. *The Conservator*, **3**, 16

Boxall, J. and von-Fraunhofer, J.A. (1977) *Concise Paint Technology*, Elek Science, London

Couzens, E.G. and Yarsley, V.E. (1956) *Plastics in the Services of Man*, Penguin Books, London

Fishlock, D. (1962) (reprinted 1970) *Metal Colouring*, Robert Draper Ltd., Teddington

Forbes, E.G. (1974) *The Birth of Navigational Science*, Maritime Monographs and Reports No.10, National Maritime Museum, Greenwich

Higgins, R.A. (1977) *Properties of Engineering Materials*, Hodder and Stoughton, Sevenoaks

Hiorns, A.H. (1929) *Metal-Colouring and Bronzing*, Macmillan and Co., London

Hudson, M. (1973) *Structure of Metals*, Hutchinson Educational, London

Koob, S.P. (1982) The instability of cellulose nitrate adhesives. *The Conservator*, **6**, 31

Mills, J.S. (1977) Natural resins of art and archaeology, their sources, chemistry and identification. *Studies in Conservation*, **22**, 12

Morgans, W.M. (1969) *Outlines of Paint Technology*, Griffin and Co., London

Newey, H. (1976) Aspects of the Conservation of Early Scientific Instruments and Apparatus. Institute of Archaeology, University of London. Unpublished Dissertation

Pearsall, R. (1974) *Collecting and Restoring Scientific Instruments*, David and Charles, Newton Abbot

Rollason, E.C. (1973) *Metallurgy for Engineers*, Edward Arnold Ltd, London

Rouveyre, E. (1911) *Cinq Cent Soixante Recettes et Procedes (560 Recipies and Trials)*, Flammarion, Paris

Shreir L.L. (ed.) (1963) *Corrosion*, Volumes 1 and 2, Butterworth, London

Spon, E.F.N. (1873) *Workshops Receipts*, E. and F.N. Spon, London

Stanley, W.F. (1914) *Surveying and Levelling Instruments*, E. and F.N. Spon, London

Turner, G.P.A. (1967) *Introduction to Paint Chemistry*, Chapman and Hall, London

Von Fraunhofer, I.A. (1976) *Basic Metal Finishing*, Elek Science, London

13

The patination of iron by bower-barffing

Martha Goodway

Abstract

The bower-barff process was a patented method of developing a protective, aesthetically pleasing velvety black surface on iron. The process had two essential steps. The first was the intentional production of a light layer of flash rust by exposing the iron to steam. This was followed by conversion of the rust to magnetite, which is black, by hydrogen reduction using ordinary producer gas. The method had many different applications, from utilitarian water pipes to highly decorated 'fire proof' cast iron libraries. Hydrogen reduction as a conservation treatment was also a 19th century development that had similar features but made no reference to bower-barffing.

Architectural cast iron

Iron smelting with coke was developed in 1709 by Abraham Darby, thereby solving the crisis in fuel that limited the production of cast iron in England. An immediate application for the resulting cheap and abundant supplies of cast iron was in structures. One famous example is the picturesque Ironbridge of 1779 whose elements were cast in Darby's shop.

Compared with its widespread use in England from about 1720 (Hamilton 1949), the architectural use of cast iron in the United States was considerably delayed. Part of the problem lay in longer distances and higher transportation costs, another was difficulty in smelting with the readily available anthracite fuel. This was overcome by the adoption of the hot blast in 1840 (Lee 1983). (The hot stove had been patented in England in 1828 (Tylecote 1976) but in the States where fuel was more abundant the need for it had not been felt.)

Cast iron soon became a prominent feature of mid- and late 19th century architecture in the United States (Gayle *et al.* 1980). Certainly the most widely known of these is the dome of the United States Capitol Building (Plate 13.1) which, with its supporting structure, was constructed entirely in cast iron.

Cast iron architectural elements from this period are still very much appreciated, some decorative elements such as capitals having been mounted on stone plinths as free-standing sculpture. Unfortunately these elements have not been painted. Acquaintance with weathering steels such as CorTen and their use by artists as renowned as Picasso seem to have resulted in an aesthetic appreciation of rust and the firmly entrenched notion that the surface of cast iron needs no protection.

Fortunately this was not the attitude at the time these elements were made. They most certainly had been painted to protect them from atmospheric rusting; cast iron lighthouses such as the one at Cape Hatteras and Gibbs Lighthouse in Bermuda still are. Often the colour was chosen and the paint skilfully applied to give the appearance of some other material, in a tradition similar to the sand-loaded paint applied to wooden buildings of the Federal period in imitation of masonry. The Capitol dome, for example, is painted to match the sandstone of the building, but a nearby cast iron fountain by Bartholdi of The Four Oceans is painted to resemble bronze.

Interiors, where weather proofing requirements were not so stringent, could make use of plating. This approach was adopted by Louis Sullivan in his design for staircases in the old Chicago Stock Exchange. This building, like so many other examples of 19th century architectural cast iron, no longer stands but various parts of the building have been preserved. One of the staircases (Plate 13.2) is now installed in New York as a functional exhibit in the American wing of the Metropolitan Museum of Art. In the course of conservation treatment it was discovered that the staircase had been cast entirely in iron, plated with copper, and the copper plate patinated to look like cast bronze.

A large amount of cast iron was used in the Old Executive Office Building, which is located in Washington adjacent to the White House. Completed in 1888, it is considered one of the best examples of Second Empire architecture in the United States (Wright 1984). Among its features are three particularly fine late Victorian cast iron libraries. One of these, the War Department Library (Plate 13.3), was recently put back into service as the White House Law Library.

The design of the War Library took advantage of cast iron's fluidity in rendering detail, as well as its structural strength. Many of the decorative elements in the War Library were plated in either brass or 'bronze', i.e. copper. The colours of the plating are set off against a velvety-black matte surface of sand cast iron, which had been treated by a method known commercially as 'bower-barffing' (Plate 13.4).

According to the language of the specifications the intent was 'to avoid painting and the consequent dulling of the lines', so 'heavy electroplating' and 'Bower-Barffing' were specified. The express object was 'to have a large amount of brightwork in the combination to secure a permanently brilliant and lively effect'. After years of benign neglect the Library presented a rare example of genuinely low maintenance construction, remaining in good condition despite roof leaks, floor scrubbing, and some overpainting.

The design that was intended to produce good ventilation of the books seems also to have operated to keep dampness from injuring the ironwork as well. In fact unusual attention was paid to ventilation in the overall design of the building. For example pilasters of classical design in the hallways, also of painted cast iron, are hollow. They conceal ductwork that is part of an elaborate ventilation system which, in the days before air conditioning, allowed air to move freely through this massive building, providing some measure of comfort during the notoriously hot and humid Washington summers.

Bower-barffing

The first patent which applies to the bower-barffing process was awarded to an Englishman, Frederick Settle Barff, in 1876. Barff (1823–1886) was a graduate of Cambridge who at the time of his invention was professor of chemistry at the Royal Academy and at University College in London, and author of a school text on chemistry that went into numerous editions (Barff 1869). Barff's process used superheated steam to generate an aesthetic, durable and protective black layer on iron (Barff 1876).

It was while working with his friend Hugh Smith on the problem of corroding steam boilers that Barff developed the method of using steam at various temperatures to produce magnetite on the surface of iron that would, in the words of the patent, prevent the 'incrustation and corroding of steam-boilers' (Barff 1877). In his experiments, Barff found that iron objects at 500°F (260°C) treated with superheated steam for 5–7 hours withstood indoor atmospheres without rusting. If treated at 1200°F (650°C) for 6 or 7 hours they would safely endure outdoor exposure.

Professor Barff's method was similar to a standard laboratory exercise to produce hydrogen, but in Barff's method its aim was, for the first time, the conversion of the iron surface to magnetite. Magnetite (Fe_3O_4) is harder than iron, having a Moh's hardness of

6 (Hurlbut and Klein 1977), and is not so easily scratched or abraded. Best of all it is unaffected by a moist atmosphere, as attested by the ocean beaches in many parts of the world whose black sand is chiefly magnetite.

In the laboratory exercise hydrogen is produced by passing steam over red-hot iron filings in an iron tube. The filings are converted into magnetite, but apparently their fine size led to an erroneous assumption that magnetite produced on larger areas would also be in a finely divided form rather than a coherent layer. In fact the magnetite layer is usually so coherent that corrosion occurs only where it has been broached, by drilling for example. The corrosion can be accelerated because the exposed iron acts as a small anodic area in an otherwise electropositive surface, and corrosion pits occur in the exposed area. However, even then the rust remains localized and does not continue under the magnetite surface layer.

There had already been considerable interest both in the effects of superheated steam on iron pipes and in the black conversion finish on Russian sheet iron. Around 1861 the Admiralty had been concerned with the safety of apparatus for superheating steam because of the likelihood of producing potentially explosive hydrogen gas. Michael Faraday among others (Nursey 1877) investigated the action of superheated steam on iron tubes and found that the surface was covered with a closely adherent layer of oxide that was protective. In investigations on Russian sheet iron Percy (1877) had analysed the black surface by heating 'in a current of dry hydrogen, when steam ... was evolved'. From this he concluded that 'the iron was more or less superficially oxidized'. The sightly and protective coating was also magnetite, though as the patentees later pointed out (Bower 1881), on Russian sheet iron it had been produced simply as an artifact of manufacture.

Barff's method, which used superheated steam and required the iron parts to be brought to red heat and treated for 5–8 hours, was widely publicized, as for example in *The Times* of 6 March 1877. Professor Barff's paper, read before the Society of Arts, was immediately reprinted in the United States (Barff 1877). It also moved Dr Percy (1877) to review the contemporary information on protective oxides.

The report in *The Times* also attracted the attention of George Bower (Bower 1881), a successful gas works contractor. He specialized in gas illumination, supplying all the necessary equipment from his factory in St. Neots, Cambridgeshire (Bower 1865, 1880). The factory building is still standing (Smith 1983). The equipment manufactured there ranged from the retorts used for producing gas from coal, wood or peat, steam boilers, condensers and holding tanks, to stoves and street lamps. George Bower and his son Anthony set to work with this equipment and very soon were able to commercialize the process that Barff himself (1879) admitted was beset with difficulties.

George Bower began with the idea that if an oxide was to be produced, air, cheaper than superheated steam, could be used as the oxidizing agent and beginning in 1877 received a series of patents for his process (Bower 1877, Bower and Bower 1878). Barffing required three sources of heat, one to heat the steam boiler, the second to heat the coils in which the steam was superheated, and the third the furnace containing the iron to be treated. But Bower's method, which need only the furnace with suitable control for the entry of air, had difficulties of its own.

In extensive experiments mainly conducted by his 18-year-old son Anthony it was found that the amount of air allowed to enter the furnace had to be closely regulated according to the surface area of the objects being treated. Too much air produced ordinary red rust on the objects, which were then referred to as 'lobsters' from their bright red colour (Tweedie 1881). The procedure could give good results with cast iron and examples of Bower's method received medals at industrial exhibitions (Bower 1880). However, it proved to be hardly less expensive than Barff's, though faster, and was difficult to apply to wrought iron and polished steel. A third approach was suggested to his father by Anthony Bower, to use carbon monoxide as well as air for these materials. Eventually these methods were applied in combination (Bower 1883).

One aspect which is little discussed is the general acceptance of black as the appropriate colour of iron, so much so that the appearance of statuary bronze when given a commercial 'antique black' finish is disturbing. Perhaps this explains why Bower's suggestion that statuary cast in iron be given his treatment (Barff 1877) seems not to have been taken up. Instead such statues were painted, often to simulate patinated or weathered bronze.

By 1882 Bower had bought out Barff's patents and had begun licensing the bower-barff method on the continent and also in the United States. An example is the exclusive licence obtained by Towne in 1887 for the Yale and Towne Manufacturing Company for the application of bower-barffing to builder's hardware (Towne 1904).

Where the War Library was treated is not known but by 1886 there were four furnaces for bower-barffing in the United States, one in Philadelphia, one in New Jersey at Little Ferry, and two in Brooklyn (Wells 1886); one of these was at the Hecla iron works at North Third and Eleventh Street (AIME 1883).

The method used for the War Library may not have been as Bower described. Most likely it differed in many details. Every reference to bower-barffing by those who actually operated it mention difficulties in controlling the process to get the intended result. Towne (1904) wrote that 'many unexpected difficulties were encountered and much time and expense involved in overcoming them'. Wells (1886) went into a highly detailed account of American practice, reporting the use of petroleum vapour to create the reducing atmosphere which converted the red oxide to the black. Twenty-minute periods of reduction were alternated with 40 minutes of oxidation so as to build up a reasonable depth of converted surface. This took about 8–10 hours. A more uniform colour was obtained by finishing with barffing for 1 hour with steam.

After the turn of the century very little mention of the method is made. Handbook entries (e.g. Lock 1896, Spon 1932, Lyman 1948, Osborne 1956) tended to be retained without alteration over many editions (cf. McGannon 1957 vs. US Steel 1985). There are also historical references to its use in cast iron architecture (Gloag and Bridgwater 1948, Gayle *et al.* 1980), and there is even one unlikely reference to bower-barffing as a possible historical treatment of the famous iron pillar of Delhi (Bardgett and Stanners 1963).

Bower-barffing and conservation

Barff's patent of 1877 ('Protecting and Cleaning Iron Surfaces') suggests that 'to clean the surfaces to be treated from rust...place the articles to be cleaned in a hot chamber into which...pass the hydrogen obtained from the water decomposed in the...treatment of the articles whilst being coated with black oxide'. In his claim to the novelty and originality of his invention he states that this is in part 'The employment of hydrogen for the purpose of reducing the oxide or oxides of iron or rust....' George Bower (1881) also claimed that 'if the iron were rusted, the rust itself could be converted into a rustproof covering of magnetic oxide' and exhibited a specimen so treated as proof of his assertion.

Conservators of the day did not immediately take up these suggestions. There may have been some sort of connection between the introduction of bower-barffing and the development of the hydrogen reduction of rusty iron artifacts but there is no evidence of a direct one. The concept, however, was in the air.

In 1887 Hartwich reported that 'Sehr gute Resultate habe ich auch erhalten, wenn ich die Sachen, im Wasserstoffstrome glühte....' Krause had already identified chlorine as the active agent in destroying archaeological iron (Krause 1882, Rosenberg 1917, Jakobsen 1988). Hartwich's method, in which the object was heated to redness in a stream of hydrogen, proved to be an excellent method of stabilization for iron that contained chlorine, yet most subsequent treatments (Rathgen 1905) involved steeping instead. Krefting's method added electromechanical reduction by zinc in caustic soda during steeping. The choice of treatment perhaps was dictated more by the safety of the operator than of the object.

For reasons of safety the first known use of Hartwich's method in the United States (Bright 1946) used producer gas instead of hydrogen. In this it more closely resembled Bower's procedure. Willard M. Bright was a graduate student in chemistry at Harvard University when he treated a number of excavated iron artifacts from Yugoslavia in the collections of the Peabody Museum. After treatment they had what he described as a 'dark matallic lustre'. Dr Bright, whom I

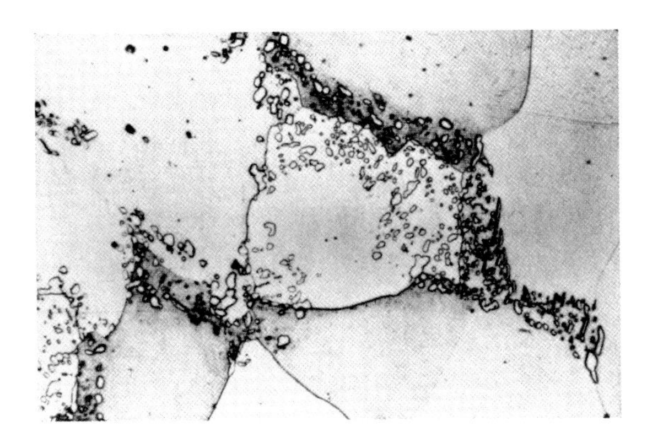

Figure 13.1. *Microstructure of a fibula treated by Bright (1946) and later studied by Cyril Stanley Smith (Brown* et al. *1977). The structure shows grain boundary carbides speroidized by lengthy high temperature hydrogen reduction treatment (cf. structures illustrated by Archer and Barker 1987).*

talked with recently, is still of the opinion that his treatment produced iron carbide rather than magnetite on the surface although he did no analyses. One of the objects that he treated was a fibula, later studied metallographically (Figure 13.1) by Cyril Stanley Smith (Brown 1977). In ignorance of its conservation treatment, Professor Smith correctly inferred from its microstructure that the artifact had been heat treated subsequent to its manufacture.

There have been no formal studies, to my knowledge, of the effect of bower-barffing on the microstructure of the treated iron. Sir Joseph Whitworth is reported to have tested the strength of bower-barffed iron, finding that there had been 'no alteration whatever' (Bower 1883), but no data are given. There have been, however, several studies on the effects of hydrogen reduction on the microstructure of treated artifacts. Tylecote and Black (1980) discussed the possible effects of heat treatment and different atmospheres on various types of artifacts, especially laminated blades. They concluded that the loss of microstructural detail in carburized wrought iron was too great to warrant the use of hydrogen reduction for objects likely to have been hardened in this way. A series of experiments by Archer and Barker (1987) on carburized iron samples quenched and tempered at several temperatures, then subjected to isothermal treatment at temperatures used in hydrogen reduction, gave microstructures more or less similar to that in Figure 13.1. Some treatment regimes resulted in identical microstructures, regardless of initial differences among the samples. The most conspicuous change was the development of cementite (iron carbide, Fe_3C) at the grain boundaries, first as a film which then began to spheroidize, the size of the globules increasing with temperature up to 700°C. No experiments on cast iron have been reported.

The present use of hydrogen reduction came about through the post World War II development of underwater archaeology and the very large number of iron artifacts being recovered from the sea, as for example from the Wasa (Barkman 1977) and the Mary Rose (Barker 1982). The hydrogen reduction practice developed for these finds also aimed at the production of magnetite rather than full reduction to metallic iron. This is for very much the same reason given for bower-barffing: the development of a protective and lasting patina with an appearance appropriate to iron.

References

American Institute of Mining Engineers (1883) *Transactions*, **11**, 606

Archer, P.J. and Barker B.D. (1987) Phase changes associated with the hydrogen reduction conservation process for ferrous artifacts. *Journal of the Historical Metallurgy Society*, **21**, 86–91

Bardgett, W.E. and Stanners, J.F. (1963) The Delhi pillar: a study of the corrosion aspects. *Journal of the Iron and Steel Institute*, **201**, 3–10

Barff, F.S. (1869) *An Introduction to Scientific Chemistry Designed for the Use of Schools and Candidates for University Matriculation Examinations*, third edition 1871, reprinted 1881, Groombridge, London

Barff, F.S. (1876) Protecting and cleaning iron surfaces. British Patent 1876 no. 862

Barff, Professor (1877) Zinc white as paint, and the treatment of iron for the prevention of corrosion. *Journal of the Society of Arts*, **25**, 254–261; reprinted as The treatment of iron for the prevention of corrosion. *Van Nostrand's Engineering Magazine*, **16** (April 1877), 300–302

Barff, F.S. (1879) Rust, and Can science conquer rust? In *Science for all*, volume 2 (ed. R. Brown) Cassel, Petter and Galpin, London, pp. 41–47 and 241–249

Barker B.D. (1982) Conservation of ferrous archaeological artefacts. *Industrial Corrosion*, **3**, 9–13

Barkman, L. (1977) Conservation of rusty iron objects by hydrogen reduction. In Brown *et. al.* (eds) 1977, 155–166

Bower, G. (1865) *The Gas and Water Engineer's Book of Reference*, illustrated with numerous engravings and containing description and prices of all articles required for, and in connection with the system of gas illumination, as manufactured and supplied by George Bower, engineer and gas-work contractor, office 222, Great Portland Street, London; works and business address: St. Neots, Huntingdonshire; together with information useful for gas engineers, managers of gasworks, and others. George Bower, St. Neots

Bower, G. (1877) Treatment of metallic surfaces. British patent 1877 no. 2051

Bower, G. and Bower, A. (1878) Protecting iron and steel from oxidation. British patent 1878 no. 1280

Bower, G. (1880) *The Gas and Water Engineer's Book of Reference* (2nd edition), George Bower, St Neots. Section Z, the preservation of iron and steel from rust, 171–172

Bower, G. (1881) On the preservation and ornamentation of iron and steel surfaces. *Journal of the Iron and Steel Institute*, **1**, 166–182

Bower, A. (1883) The Bower-Barff process. *Transactions of the American Institute of Mining Engineers*, **11,** 329–339

Bright, W.M. (1946) The treatment of iron antiquities. *The Museums Journal*, **46**, 1–5

Brown, B.F., Burnett, H.C., Chase, W.T., Goodway, M., Kruger, J. and Pourbaix, M. (eds) (1977) *Corrosion and Metal Artifacts*, NBS Special Publication 479, Washington DC

Gayle, M., Look, D.W. and Waite, J.G. (1980) *Metals in America's Historic Buildings*, US Department of the Interior, Heritage, Conservation and Recreation Service, Technical Preservation Services Division, Washington DC

Gloag, J. and Bridgwater, D. (1948) *A History of Cast Iron in Architecture*. George Allen and Unwin, London, 370

Hamilton, S.B. (1949) Old cast-iron structures. *The Structural Engineer*, **27**, 173–191

Hartwich, C. (1887) Zur Conservirung von Eisenalterthümern. *Chemiker Zeitung*, **11**, 605

Hurlbut, C.S. Jr. and Klein C. (1977) *Manual of mineralogy (after James D. Dana)*, 19th edition. John Wiley and Sons, New York, Magnetite, 278–279

Jakobsen, T. (1988) Iron corrosion theories and the conservation of archaeological iron objects in the 19th century with an emphasis on Scandinavian and German sources. In: *Early Advances in Conservation* (ed. V. Daniels), British Museum Occasional Paper no. 65, London

Krause, G. (1882) Ein neues Verfahren zur Conservirung der Eisen-Alterthümer. *Verhandlungen der Berliner Gesellschaft für Anthropologie, Ethnologie und Urgeschichte*, **11**, 553-538

Lee, A.J. (1983) Cast iron in American architecture: a synoptic view. In *The Technology of Historic American Buildings* (ed. H. Ward Jandl), Foundation for Preservation Technology, for the Association for Preservation Technology, Washington DC, 97-113

Lock, C.G.W. (1896) *Workshop Receipts*, 3rd series. E. & F.N. Spon, London, 248-249

Lyman,T. (ed.) (1948) *Metals Handbook*. American Society for Metals, Cleveland, Ohio, 731

McGannon, H.E., (ed.) (1957) *The Making, Shaping and Treating of Steel*, 7th edition, US Steel Corporation, Pittsburgh, 625

Nursey, P.F. (1877). *Journal of the Iron and Steel Institute*, **2**, 13–14

Osborne, A.K., (ed.) (1956) *An Encyclopedia of the Iron and Steel Industry,* Philosophical Library, New York

Percy, J. (1877) The protection from atmospheric action which is imparted to metals by a coating of certain of their own oxides, respectively. *Journal of the Iron and Steel Institute*. Discussion of this paper. 1878, *JISI*, 2(10–16), 456–460

Rathgen, F. (1905) *The Preservation of Antiquities*, (trans. G.A. Auden and H.A. Auden), CUP, Cambridge

Rosenberg, G.A.T. (1917) *Antiquités en fer et en bronze, leur transformation dans la terre contenant de l'acide carbonique et des chlorures et leur conservation*, Gyldenhall, Copenhagen

Smith, C.S. (1976) In Brown *et al.* 1977, 229–230

Smith, T. (1983) Early product finishing: part 1. *Product Finishing*, **36**, (6), 40–41. Early product finishing: part 2. *Product Finishing*, **36** (7), 40-41

Spon, E. and F.N. (1932) *Workshop Receipts for Manufacturers and Scientific Amateurs*, volume IV, E & F.N. Spon, London, pp. 38-42

Towne, H.R. (1904) *Locks and Builders Hardware*, John Wiley and Sons, New York

Tweedie, G.R. (1881) Discussion on Bower 1881. *Journal of the Iron and Steel Institute*, **1**, 178–179

Tylecote, R.F. (1976) *A History of Metallurgy*, Metals Society, London

Tylecote, R.F., and Black J.W.B. (1980) The effect of hydrogen reduction on the properties of ferrous materials. *Studies in Conservation*, **25**, 87–96

United States Steel (1985) *The Making, Shaping and Treating of Steel*, 10th edition, US Steel Corporation, Pittsburgh

Wells, J. (1886) Rustless iron. *Popular Science Monthly*, **29**, 393–397

Wright, M. (ed.) (1984) *The Old Executive Office Building: a Victorian Masterpiece*, The Executive Office of the President of the United States of America, Washington DC.

14

Oxidized silver in the 19th century: the documentary evidence

Judy Rudoe

Abstract

The aim of this chapter is to look at the evidence provided by contemporary accounts for the widespread use of oxidized silver in Europe and America in the 19th century. Many instances are either little-known or completely forgotten, because the items have long since been cleaned or burnished so that no trace of their original surface colour remains. Oxidized silver became fashionable in France in the 1840s, for work in the neo-gothic and neo-renaissance styles, and spread to England by the 1860s. The influence of Far Eastern metalwork in the West, associated with the Aesthetic Movement of the 1870s and 1880s, led to the use of oxidized silver in imitation of the dark grounds of Japanese *shakudo* or *shibuichi* metalwork, especially in America. The taste for darkened silver surfaces continued in Europe up to the First World War.

The popular term 'oxidized' silver is generally used to describe silver that has been coloured in shades of dark-grey or blue-black, in most cases by means of a chemical dip with which the silver reacts. The commonly used solutions in the 19th century are those in which sulphur is the activating chemical, forming a metallic sulphide compound surface film that is a combination of the chemicals in the solution and the silver. This is traditionally called oxidization, although it is achieved with sulphur. Indeed, the term oxidized had become so entrenched in the vocabulary of jewellers, silversmiths and the retail trade by the end of the last century that it remains in current use, although 'coloured' would be technically more accurate.

Henry Wilson, the arts-and-crafts silversmith, thought newly whitened and polished silver unpleasantly glaring and in his manual on silverwork and jewellery of 1903, he describes the method used to darken the surface with chemical compounds of sulphur, such as potassium sulphide, ammonium sulphide and barium sulphide. Wilson recommended for general use a hot solution of ammonium sulphide, brushed over the work and left until the desired colour was reached, then washed and dried. Finally, he recommended rubbing the surface with a chamois leather, to remove the colouring from the projecting portions, giving what he described as a richer, older appearance (Wilson 1903 pp. 235–7).

Precise recipes, including the widely used liver of sulphur (potassium sulphide) are given by more recent authors (Field and Bonney 1925 pp. 185–97, Maryon 1954 pp. 159–60, Untracht 1982 pp. 717–18). Untracht notes that chemically achieved colouring is normally adherent and can only be removed with high

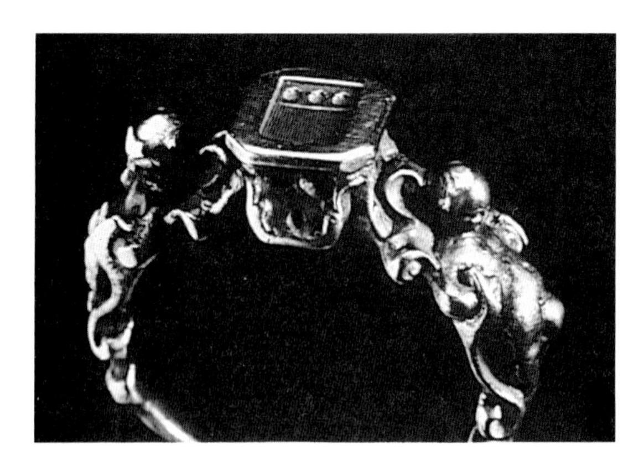

Figure 14.1. *Oxidized silver and silver-gilt finger-ring in the renaissance style, designed by F.-D Froment-Meurice in 1844. British Museum, bequeathed by A.W. Franks in 1897. D.3 cm.*

heat, acid, abrasive or by electrochemical means. Thus it can in most cases be distinguished from tarnish, which is easily removed. This is borne out by the experience of the British Museum Metals Conservation section in cleaning the 19th-century oxidized silver in the collections.

The taste for oxidized silver seems to appear first in France, in the 1840s, when it is well documented in the work of François-Désiré Froment-Meurice, the Parisian goldsmith noted for his silverwork and jewellery in neo-gothic and neo-renaissance styles. Froment-Meurice's biographer, Philippe Burty, described several oxidized items and noted that the writer, Eugène Sue, who commissioned many pieces of silversmiths' work from Froment-Meurice, discussed with the artist which parts were to be in *argent noir*. In one of his letters to Froment-Meurice, Sue even requested 'a small bottle of liquid to blacken the bright silver' (Burty 1883 p. 23). The renaissance-style silver-gilt finger-ring shown in Figure 14.1 was designed by Froment-Meurice in 1844 and bears two oxidized silver figures amongst gilded scrollwork flanking the bezel. The ring was included in the recent British Museum exhibition *Fake? The Art of Deception* (Jones 1990 cat. no. 20), not as a fake, but as an example of a revivalist piece that lost its identity and was passed off as 16th century a mere eight years later: in 1852 it was acquired in Paris by Lord Londesborough as renaissance, despite the fact that Froment-Meurice, a prominent Parisian jeweller, had exhibited it in Paris along with other *bijoux renaissances* in 1844 (Vever 1906–8 p. 180). The British Museum ring was acquired with the Franks Bequest in 1897 (AF 2578, D.3cm) as 'Modern French'. Froment-Meurice's jewellery was often highly sculptural, with three-dimensional figures in stage settings inspired by gothic and renaissance architecture. He frequently used oxidized silver in conjunction with gold as well as parcel-gilding. The gothic-style necklace shown in Plate 14.1 comprises a three-dimensional figure-group of a crusader taking leave of his lady, executed in cast and chased gold, set within a frame of oxidized silver; the darkened silver with its blue-black tones provides a dramatic contrast with the gold, but may also have been intended to imitate cast iron work or bronze sculpture. This necklace may have been assembled at a later date from elements originally designed for a chatelaine, but even if this is so, the elements appear to be of 19th-century date with contemporary oxidization (Rudoe 1991a pp. 43–52 and cat. no. 6).

A number of French firms exhibited oxidized silver at the Great Exhibition of 1851. In addition to Froment-Meurice, the firms of Frédéric Rudolphi and Alexandre Gueyton were also represented. The Victoria & Albert Museum purchased an oxidized silver casket by Rudolphi at the 1851 exhibition (Aslin 1973 no.6), but they had previously acquired a very fine piece by Rudolphi, an oxidized silver bottle of 1844 (V&A 919-1844), with superbly chased foliate ornament and sophisticated surface texturing. It was purchased by the Council of the School of Industrial Design from the Exhibition of Industrial Art held in Paris in 1844. The School of Design collections formed the nucleus of the Museum of Ornamental Art, subsequently the South Kensington Museum (Aslin 1973 no. 3). In the same Museum are two caskets exhibited by Gueyton at the 1851 Exhibition. The silver is combined with jewels or with parcel gilding, but the oxidization is barely evident, having already probably been lost in the last century (Aslin 1973 nos. 8 and 9).

The use of oxidized silver by the French and German silversmiths was much admired in 1851 by the critic, R.N. Wornum, who disliked intensely the English silversmiths' adherence to conventional trade practice in their constant use of dead and burnished silver:

> The system of *boiling out* to produce the whitest possible appearance of the silver, seems to be one essentially opposed to the display of excellence of design; and when the dead white thus produced is combined only with burnished portions, the sole effect of a work is a mere play of light without even the contrast of shadow. The result is a dazzling whiteness; pure flashiness in fact, such as precludes the very idea of modelling — for this can only be displayed by a contrast of light and shade, which, in so uniform a dazzling mass as an ordinary piece of dead and burnished silver plate, is impossible.
>
> Flashiness may be a natural refuge for vague undefined forms, to the deformities of which it is an effective cloak; and so long as our silversmiths adhere to their Rococo scrolls, and other inanities of the Louis Quinze, its aid will be indispensable. Immediately the details of design, however, are substantially reformed, frosting and burnishing, except as occasional incidental aids, must go together with the preposterous forms to which alone they owe their present popular development. If we turn from the English to the foreign silver-work, the contrast in this respect is surprising; frosting and burnishing seem to be unanimously banished from all high class design, whether French or German, and oxidising substituted in their places, and the consequence is, that in many foreign examples we have specimens of the most elaborate modelling, most effectively displayed as works of Art..... The process of oxidation, as it is termed, not only protects the silver from further tarnishing, but can convey every variety of tint from white to black, so that it is particularly well calculated to display fine modelling or chasing, which would be utterly thrown away in a dazzling white material.
>
> We hold it to be proved by the Exhibition, that all frosting and burnishing, except for occasional relief or variety among the minor details of a design, are fatal to silver-work as Art, however they may enhance its effects as specimens of a noble metal.

Wornum singled out a large vase or table ornament exhibited by Wagner & Son of Berlin, which, had it been in dead or burnished silver, might have been overlooked for the exquisite details of modelling would have been indiscernible. The works of Froment-Meurice, Rudolphi and Gueyton were contrasted with the frosted and burnished Louis Quatorze dinner service by Odiot of Paris: Wornum conceded that it was the finest of its class in the exhibition, but the value of the modelling was lost through overall matting and polishing, resulting in 'something positively vulgar in such a mere metallic blaze as a service of this kind displays'. (Wornum 1851 pp. VIII-XI).

Wornum also describes an intermediate class amongst the French work, in which the silver was left in its pure colour without the aid of frosting or burnishing and was then slightly oxidized. He includes in this class a group of Queen Elizabeth and Leicester, modelled by the French artist, Emile Jeannest, and exhibited by Messrs Elkington, apparently in the French section. Elkington commissioned other French artists such as Antoine Vechte (1799–1868) to design for them, and it is significant that among the pieces in the English section of the exhibition for which Wornum had a good word to say were a shield and vase, modelled by Vechte for the London firm of Hunt & Roskell, with 'chasing of unequalled beauty and delicacy' which 'owing to the oxidised silver surface shows with all the vigour of a fine proof engraving'.

To summarize Wornum's views, which may be taken as representative of artistic circles in England, frosting and burnishing was seen as disguising poor quality English design, while oxidization was synonymous with artistic continental silver of good design. Oxidized silver was associated with good taste, while burnished silver had a cheap and vulgar image.

This was re-affirmed at the next International Exhibition in London in 1862, at which the Parisian firm of L.G. Jarry received an honorable mention for their excellent workmanship. Their exhibits included an agate cup, a silver tazza and silver-mounted engraved glass, all of which employed oxidized silver. They are included in the three-volume record of the 1862 exhibition, compiled by J.B.Waring and lavishly illustrated with colour lithographs. The pieces by Jarry are illustrated together with an oxidized silver teapot by Fannière Frères, described as in the renaissance style, and as over-elaborate and inartistic as anything produced in England (Figure 14.2). Fannière Frères also exhibited a range of small objects such as fans, cigar-cases, brooches and pins, in finely chased oxidized silver.

Waring took up many of Wornum's sentiments, insisting on the superiority of oxidized silver 'for the production of works intended to be of an artistic character', but he had his reservations:

> Thanks to the improved taste of purchasers, and also of the manufacturers themselves, the use of oxydized silver for artistic works has become very general, although it has had to contend with the attraction that the precious metals in their polished state possess with the public in general, from their bright, cheerful, and rich look, especially when placed on the festive board, and sparkling in reflected light. It may be urged, that in silver thus treated one of its peculiar qualities is wilfully ignored or discarded; and we are fain to admit that such an objection to the use of oxydized silver is not without force, since several common metals present a perfectly similar effect; and indeed, unless the spectator were informed that the work was in silver, he might as well have believed it to be in iron, or pewter, or zinc, or a compound metal of an ordinary character. If the value of the object depends on the art displayed in its design and fabrication, and not on the material, why make use of an expensive metal, to which you give the appearance of a common one? Certainly one of the great charms of silver, as a metal, is the rich and glittering appearance of which it is capable; and wilfully to render it as dull as lead, appears an unreasonable proceeding. We are not questioning for a moment the superiority of oxydised silver over burnished silver for groups of figures and the better class of works in decorative art; but we do question the good sense, and consequently the good taste, of bringing a noble and expensive metal down to a level in appearance with a common and cheap one. (Waring 1863 vol. 3, pl. 265).

Figure 14.2. *Oxidized silver wares shown by the Paris firms of L.G. Jarry and Fannière Frères at the International Exhibition in London of 1862. (Illustration from J.B. Waring* Masterpieces of Industrial Art and Sculpture at the International Exhibition of 1862, *London 1863, Vol.3, pl.265.)*

By this time, English silversmiths had learnt from the displays of their French rivals at the 1851 exhibition, and Hancock & Company, among others, were ready with a series of oxidized silver presentation pieces in 1862. Three of these pieces were designed to illustrate 'the Poetry of Great Britain' and were included in Waring's volumes; the large central vase in Waring's illustration was dedicated to Shakespeare, the two at the rear to Milton &

Figure 14.3. *Oxidized silver presentation pieces shown by the London firm of Hancock & Company at the 1862 International Exhibition in London. (Illustration from J.B. Waring,* Masterpieces of Industrial Art and Sculpture at the International Exhibition of 1862, *London 1863, Vol. 3, pl. 232.)*

Byron (Figure 14.3) (Waring 1863 vol. 3, pl. 232). The following year, when the Danish residents of London wished to present a gift to the Danish Princess Alexandra on the occasion of her wedding to Edward, Prince of Wales, they commissioned an oxidized silver vase from the London manufactory of Jes Barkentin, a metalworker of Danish descent, providing further indication of the taste for oxidized silver in fashionable circles (J. Culme 1987 vol. I, p. 26).

The fashion spread throughout Europe. The British Museum holds examples of both Viennese and Italian oxidized silver jewellery of the 1870s and 1880s in the neo-renaissance taste. The Viennese pieces are a bracelet by Markowitsch & Scheid and a brooch by G.A. Scheid, two Viennese firms that specialized in oxidized silver jewellery in the renaissance style (Gere *et al.* 1984 cat. 1034-5). The Italian piece is a neo-renaissance brooch and matching earrings with cupids in shell niches by the firm of Marchesini of Rome and Florence (Gere *et al.* 1984, cat. 1033).

Darkened silver surfaces were almost always advertised as 'artistic', but criticism arose precisely because the colouring was not always artistically applied. The French firms, always pre-eminent, achieved a beautiful and attractive finish. But with less competent firms the finish could, indeed, be as dull as lead. The German critic, F. Pecht, criticized the over-heavy oxidization current in Germany at the time, deploring

> the tasteless mania for oxidising silver so much that it looks like lead. I have searched in vain to find one single reason for this absurdity, for making precious metal look mean and dull, especially as it would never occur to the best of the French silversmiths such as Christofle, but we are doing it nevertheless. I can only explain this through a lack of sense of colour, otherwise one could not bring such artificial qualities to silver. (Pecht 1872)

Oxidization of silver, when used on sculptural pieces in historicist neo-gothic and neo-renaissance styles, was usually, though not always, an all-over surface colouration, and continued to be practised until the end of the 19th century. But more inventive ways of using oxidized silver, in conjunction with other coloured metals, became fashionable for silver in the Oriental taste of the 1870s and 1880s. In order to distinguish sombre 'artistic' silver for aesthetes from vulgar displays of glittering metal, the surfaces were not given an all-over darkening, but were darkened in selected areas only, in imitation of a range of Islamic, Indian and Far Eastern wares, and in particular of the dark grounds of Japanese *shakudo* or *shibuichi* alloys.

Some of the most innovative pieces were made in America, where the hallmarking laws, unlike those of Great Britain, did not forbid the combination of precious and base metals. The name of Tiffany & Company of New York is the most widely known, but other American firms such as Gorham Manufacturing Company of Providence, Rhode Island, and the Whiting Manufacturing Company of North Attleboro and Newark made equally daring pieces. The Tiffany archive contains numerous working drawings indicating the different coloured alloys to be applied and this has been discussed elsewhere (Gruber Safford and Wilford Caccavale 1987 pp. 808–19). What the working instructions do not indicate, presumably because it was done as a matter of course, is that the silver grounds were oxidized. That this was so is known only from contemporary accounts, such as that of the English author and critic George Augustus Sala, who gave an evocative description of Tiffany's display at the Paris exhibition of 1878:

> Purely of American design and execution is Messrs Tiffany's tea-service in oxidized silver and variously coloured gold, adorned with an exquisite pattern in relief, embodying the apologue of 'the Spider and the Fly'. I am shown, also a teapot, in its way unique, and in which the silver has been oxidized to an inimitably delicate purple hue (Sala 1880 p. 81).

Coloured silver surfaces that might correspond to this description hardly ever survive. The British Museum, however, has recently acquired an oxidized silver tankard in the Japanese taste by Tiffany & Company, which, exceptionally, has retained its original surface colouring. It is decorated with applied silver reliefs of figures, bamboo and other plants; the background is oxidized throughout to a deep grey-black matt colour with a purplish sheen (Plate 14.2) (Rudoe 1991b cat. no. 351). X-ray fluorescence (XRF) surface analysis by Susan La Niece has shown that the alloy is 92% silver, conforming with the American standard for sterling silver, and no element other than silver and a small amount of copper was detected. The patina is thus almost certainly silver sulphide, but the coating is too thin to allow sampling for more detailed analysis. The author has seen no other example of such a well-preserved original oxidized silver surface in American Japanesque silver. It could reasonably be described as having 'an inimitably delicate purple hue' and can perhaps help us to understand what American Japanesque pieces were meant to look like, and how much more subtle the contrast of the coloured appliqués appears against a darkened ground.

The original design drawing for this tankard survives in the Tiffany archive and indicates that the shape was designed between 1870 and 1874, while the decoration was designed in December 1874. It is thus a very early example of Tiffany Japanesque silver, which was first publicly shown some two years later at the Philadelphia Centennial Exhibition of 1876. That darkened grounds were in use by 1876 can be seen from a pitcher of that year, in silver, copper and silver gilt, exhibited at the Philadelphia Centennial Exhibition and now in the Museum of Fine Arts, Boston. It is not in the Japanese taste and the floral motifs are more Indian or Islamic in inspiration, but the darkened background to the bands of relief floral ornament gives a far greater depth and vigour than if the entire surface had been polished bright (Carpenter and Zapata 1987, cat. no. 52, cover illustration).

Also in the British Museum collections is a silver tray or waiter with a frog amongst water weeds, inlaid in red copper, black-patinated copper and gold (Figure 14.4). The background is no longer coloured but surface XRF analysis by Susan La Niece has shown that the frog's black spots are inlaid in copper with a thin black patina worn in parts to reveal the copper beneath. The darker surrounding areas are inlaid with nickel silver. The design drawing in the Tiffany archive describes the spots as 'Japanese gold'. It is therefore significant that the black spots contain platinum, either in the patina or alloyed with the copper. If the platinum is in the alloy, it suggests that the blackened copper inlays are Tiffany's adaptation of Japanese *shakudo* alloys, with platinum replacing gold in the traditional Japanese recipe (Rudoe 1991b cat. no. 283). If the platinum is in the patina, Tiffany's may have been using a platinum chloride solution to produce a black-grey surface film. This method of achieving a surface colour is described in an English metal-colouring manual published in 1892 by the principal metallurgist at Birmingham Municipal Technical School (Hiorns 1892 p. 266).

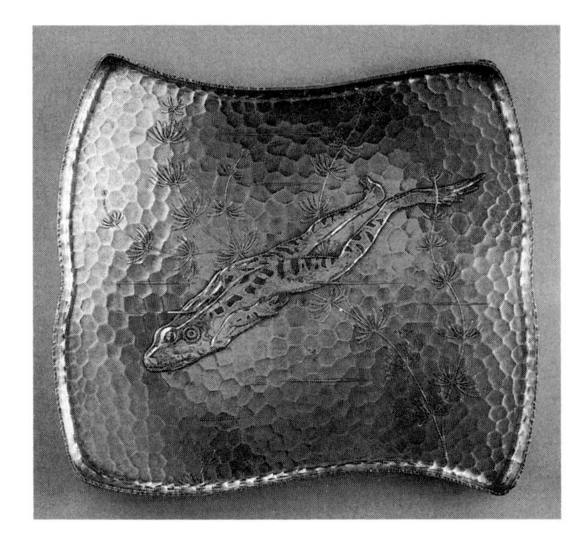

Figure 14.4. *Silver tray with inlays in gold, copper, black-patinated copper and nickel silver. This tray would originally have had an oxidized silver background. British Museum, given by Mr and Mrs J. Cox in 1981. W.23.8 cm.*

The role of Tiffany's chief designer, Edward C. Moore, is well documented. He had his own collection of both Islamic and Japanese metalwork, which provided inspiration for Tiffany silver. However, with the large-scale production of ornaments in the Japanese taste by the Birmingham jewellery trade in the 1880s, the source of inspiration lies in the small pieces of ornamental metalwork made in Japan specifically for export to the west in the late 19th century. Following the banning of the wearing of the *Samurai* sword in 1876, the skilled Japanese metalworkers who had made the decorative sword mounts turned to supplying the Western market. These small pieces were usually entirely European in taste and shape, mounted as jewellery in the West and sold as popular curiosities. The British Museum holds a group of such pieces, employing the Japanese *shibuichi* and *shakudo* alloys, both dark-coloured alloys of silver or copper, and inlaid with gold, silver and copper (Plate 14.3) (Gere *et al.* 1984, cat. 1088–93). The group illustrated includes a genuine Japanese purse-mount or pouch-fitting with a stag-beetle attacking a butterfly in high relief.

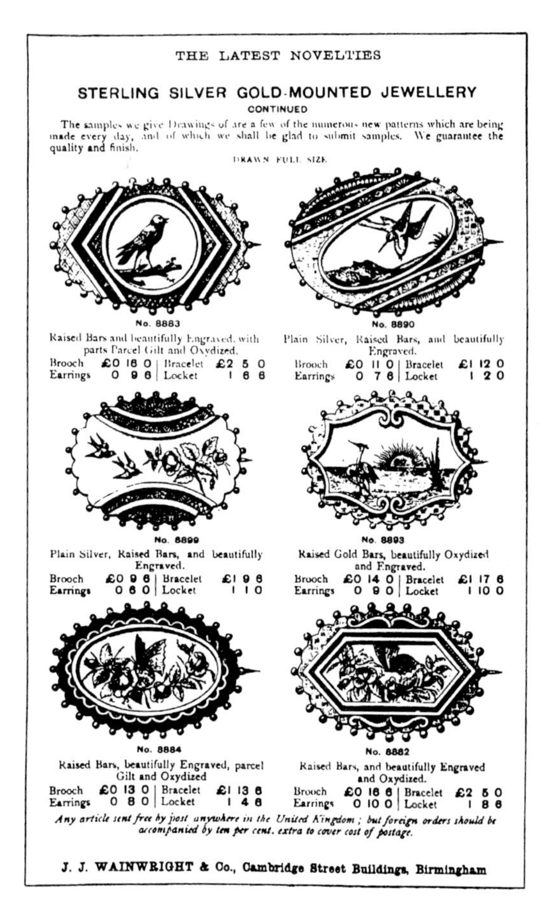

THE LATEST NOVELTIES

STERLING SILVER GOLD-MOUNTED JEWELLERY

CONTINUED

The samples we give Drawings of are a few of the numerous new patterns which are being made every day, and of which we shall be glad to submit samples. We guarantee the quality and finish.

DRAWN FULL SIZE

No. 8883

Raised Bars and beautifully Engraved, with parts Parcel Gilt and Oxydized.

Brooch	£0 16 0	Bracelet	£2 5 0
Earrings	0 9 6	Locket	1 6 6

No. 8890

Plain Silver, Raised Bars, and beautifully Engraved.

Brooch	£0 11 0	Bracelet	£1 12 0
Earrings	0 7 6	Locket	1 2 0

No. 8899

Plain Silver, Raised Bars, and beautifully Engraved.

Brooch	£0 9 6	Bracelet	£1 9 6
Earrings	0 6 0	Locket	1 1 0

No. 8893

Raised Gold Bars, beautifully Oxydized and Engraved.

Brooch	£0 14 0	Bracelet	£1 17 6
Earrings	0 9 0	Locket	1 10 0

No. 8884

Raised Bars, beautifully Engraved, parcel Gilt and Oxydized

Brooch	£0 13 0	Bracelet	£1 13 6
Earrings	0 8 0	Locket	1 4 6

No. 8882

Raised Bars, and beautifully Engraved and Oxydized.

Brooch	£0 16 6	Bracelet	£2 5 0
Earrings	0 10 0	Locket	1 8 6

Any article sent free by post anywhere in the United Kingdom; but foreign orders should be accompanied by ten per cent. extra to cover cost of postage.

J. J. WAINWRIGHT & Co., Cambridge Street Buildings, Birmingham

Figure 14.5. *Page from a trade catalogue of J.J. Wainwright & Company, Birmingham, c.1880, advertising oxidized silver and parcel gilt brooches.*

Coloured metalwork of this kind was imitated in silver, often with applied coloured gold, or parcel-gilding, engraved and, more significantly, oxidized. These popular silver brooches and bracelets were produced in huge numbers but they never survive with their original surface colouring. The illustration in Figure 14.5 is from a trade catalogue issued *c.*1880 by the Birmingham firm of J. Wainwright & Company. Another page in the same catalogue shows the latest novelty, 'Ye spider and Ye flie' jewellery, a matching suite of earrings, brooch and pendant. The accompanying text reads: 'raised gold parts, the background oxydised, with raised silver spider and gold "flie", which have a very natural and

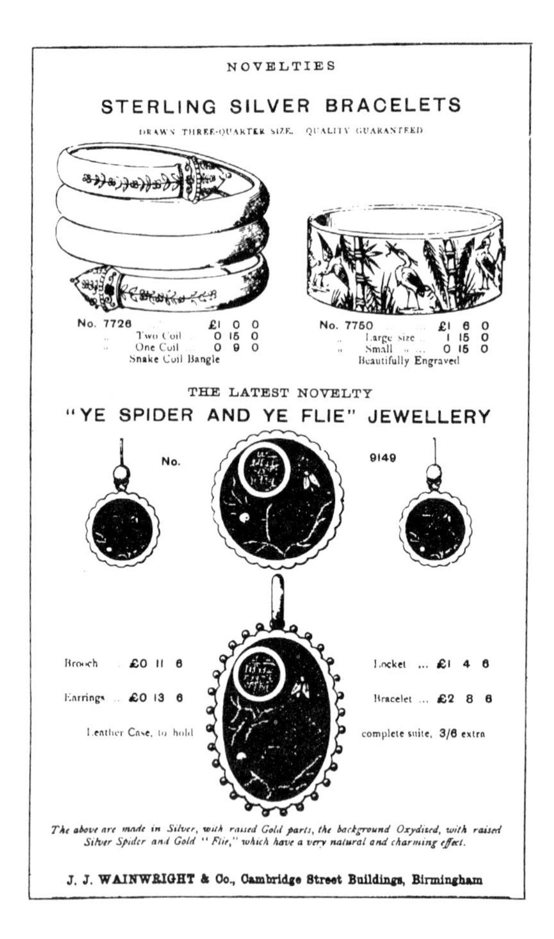

NOVELTIES

STERLING SILVER BRACELETS

DRAWN THREE-QUARTER SIZE. QUALITY GUARANTEED

No. 7726 ... £1 0 0
„ Two Coil ... 0 15 0
„ One Coil ... 0 9 0
Snake Coil Bangle

No. 7750 ... £1 6 0
„ Large size ... 1 15 0
„ Small „ ... 0 15 0
Beautifully Engraved

THE LATEST NOVELTY

"YE SPIDER AND YE FLIE" JEWELLERY

No. 9149

Brooch ... £0 11 6
Earrings ... £0 13 6
Leather Case, to hold complete suite, 3/6 extra
Locket ... £1 4 6
Bracelet ... £2 8 6

The above are made in Silver, with raised Gold parts, the background Oxydised, with raised Silver Spider and Gold "Flie," which have a very natural and charming effect.

J. J. WAINWRIGHT & Co., Cambridge Street Buildings, Birmingham

Figure 14.6. *Page from a trade catalogue of J.J. Wainwright & Company, Birmingham, c.1880, advertising 'ye spider and ye flie' jewellery, of silver, with oxidized backgrounds and raised gold parts.*

Figure 14.7. *Pewter frame with surface patina of oxidized silver, made c.1905-6 by the Württembergische Metallwarenfabrik, Geislingen, southern Germany. Designed originally as a tray and converted later for use as a mirror. British Museum, L.52 cm.*

charming effect' (Figure 14.6). A very similar 'Spider and Flie' brooch is to be found in the Hull Grundy Gift to Norwich Castle Museum: the raised rims are gold, but the oxidized background has long since disappeared.

At the end of the 19th century, many silversmiths of the Arts and Crafts movement shared Henry Wilson's preference for darkened silver. Gilbert Marks (1861–1905) did not actually colour his silver, but a contemporary critic, writing in *The Art Journal* in 1897 (p. 252) noted that it was not 'subjected to the ordinary polishing processes' and praised the 'dull yet exquisite grey of unpolished silver'. C.R. Ashbee (1863–1942), writing on tableware in the same journal in 1899 (p. 337) recommended, as a cheaper substitute for silver, silvered copper 'which, when oxidized, often looks just as beautiful'.

Even in the early 1900s, Elkington & Company of Birmingham were still advertising in their trade catalogues electroplate copies of the works of L. Morel-Ladeuil and A.A.

Willms, created three or four decades earlier in repoussé silver and damascened steel; the copies were gilded and oxidized to imitate the colours of the original. The taste for darkened silver surfaces continued up to the First World War, especially in Germany, where there is evidence for oxidized silver finishes from firms such as the Württembergische Metallwarenfabrik (WMF) at Geislingen, near Ulm. WMF household goods were exported widely to England and America; usually made of silver-plated Britannia metal (an alloy of tin, antimony and copper that was cheaper than pewter), they were produced with varying thicknesses of silver-plating and various surface finishes, either bright, or oxidized, the latter indicated by the letters 'ox' on the underside. A pewter frame in the British Museum, made by the WMF in about 1906 (originally designed as tray and later converted for use as a mirror), has a surface patina of oxidized silver, in the form of a thin silver plating, oxidized and then polished off the higher areas, so that the lower parts of the relief remain black (Figure 14.7; Rudoe 1991b, cat. 329). Darkened silver surfaces were also widely fashionable in German jewellery of the early 20th century, as shown by a silver waist-clasp designed in about 1901 by Patriz Huber (1878–1902) at the artists' colony established by Grand Duke Ernst Ludwig of Hesse in Darmstadt in 1898. The clasp was made by the Pforzheim firm of Theodor Fahrner and was imported for sale in London by Murrle Bennett & Company, an Anglo-German concern operating in Pforzheim and London. Most of their goods were sold through Liberty's in London (Figure 14.8; British Museum MLA 1991, 5-2,1).

It should also be recorded that much of the silver and pewter sold in the early 1900s by Liberty & Company in London was imported from Germany; it was oxidized, not only to show up the pattern, but also to overcome the cleaning problem (Bury 1977 pp. 14–27).

The continued popularity of oxidized wares is illustrated by a report on the British Industries Fair of 1915, at which William Hutton & Sons of Sheffield showed oxidized silver-plate, specifically to meet the demand for goods that were not being imported from Germany during the war: 'They showed some oxidized silverware of beautiful French grey finish, and lacquered to prevent tarnishing; thus the various articles only required dusting with a soft damp cloth to preserve their original freshness' (Culme 1987, vol. I, p. 249).

Figure 14.8. *Oxidized silver buckle designed* c.*1901 by Patriz Huber and made by the firm of T. Fahrner, Pforzheim, Germany. British Museum, W.7 cm.*

If only that was all that had ever been done to the many and various examples of once oxidized surfaces, most would not now be irreparably lost. These notes will have served their purpose if they draw attention to the range and popularity of oxidized silver in Europe and America over three-quarters of a century between 1840 and the First World War.

Acknowledgements

I am grateful to Janet Zapata for locating (in the Tiffany archive), the design drawings for both the tankard in Plate 14.2 and the tray in Figure 14.4; to Sue la Niece for her advice in the preparation of this paper and to the members of the Metals Conservation section of the British Museum – Hazel Newey,

Marilyn Hockey, Celestine Enderly, Ian McIntyre and Dick Ryan – with whom I have had many useful discussions over the years.

References

Aslin, E. (1973) *French Exhibition Pieces 1844–78*, Victoria & Albert Museum, London

Burty, P. (1883) *F.D. Froment-Meurice. Argentier de la Ville de Paris*, Paris

Bury, S. (1977) New light on the Liberty metalwork venture. *Bulletin of the Decorative Arts Society 1890-1940*, 1

Carpenter, C.H. Jr. and Zapata, J. (1987) *The Silver of Tiffany & Co. 1850–1987*, exhibition catalogue, Museum of Fine Arts, Boston

Culme, J. (1987) *The Directory of Gold & Silversmiths, Jewellers and Allied Traders 1838-1914*, 2 vols., Antique Collectors' Club, London

Field, S. and Bonney, S.F. (1925) *The Chemical Colouring of Metals*, Chapman and Hall, London

Gere, C., Rudoe, J., Tait, H. and Wilson, T. (1984) *The Art of the Jeweller: a Catalogue of the Hull Grundy Gift to the British Museum*, 2 vols., British Museum Publications, London

Gruber Safford, F. and Wilford Caccavale, R. (1987) Japanesque Silver by Tiffany & Co. in the Metropolitan Museum of Art. *Antiques*, October

Hiorns, A.H. (1892) *Metal-colouring and Bronzing*, Macmillan, London

Jones, M. (ed.) (1990) *Fake? The art of deception*, British Museum Publications, London

Maryon, H. (1954) *Metalwork and Enamelling*, Dover, London

Pecht, F. (1872) *Die Kunstindustrie*, Vienna, p. 272 ff

Rudoe, J. (1991a) *François-Désiré Froment-Meurice and Jules Wièse, collaborators.* In *The Belle Epoque of French Jewellery 1850–1910*, (ed M. Koch), London

Rudoe, J. (1991b) *Decorative Arts 1850–1950: A Catalogue of the British Museum Collection*, British Museum Publications, London

Sala, G.A. (1880) *Paris Herself Again in 1878–9*, 2 vols. Remington & Co., London

Untracht, O. (1982) *Jewellery Concepts and Technology*, Hale, London

Vever, H. (1906-8) *La Bijouterie Française au XIXe Siècle*, 3 vols., Paris

Waring, J.B. (1863) *Masterpieces of Industrial Art and Sculpture at the International Exhibition of 1862*, 3 vols, London

Wilson, H. (1903) *Silverwork and Jewellery*, Pitman, London

Wornum, R.N. (1851) The Exhibition as a lesson in taste. *The Art Journal Illustrated Catalogue of the Great Exhibition of 1851*, section IV, 'The precious metals', George Vertue, London (reprinted Crown Publishers, New York, 1970)

15

Gilding of metals in the Old World

Andrew Oddy

Abstract

This paper discusses the main techniques used to gild other metals in antiquity – a process which is defined as the application of a layer of gold on to the surface of a less valuable metal. The earliest gilding is found in the Middle East at about 3000 BC, when gold foil was attached to other metals by mechanical means. The use of gold foil continued into the Roman period. During the second millennium BC methods for refining gold were discovered which allowed gold foil to be hammered even thinner to convert it into gold leaf. This was used for gilding by sticking it to the base metal using an adhesive or by heating to bring about inter-diffusion of the gold and the underlying metal.

In about the 4th century BC the Chinese discovered fire-gilding, and this reached the Roman world in the 2nd century AD, from when it remained the main method of gilding silver, copper and, to a lesser extent, bronze, until the invention of electroplating in the mid-nineteenth century.

Conflicting evidence for the knowledge of fire-gilding in the West before the 2nd century AD is also discussed.

Introduction

Gilding is the application of a layer of gold onto the surface of a less precious metal, where it is attached either mechanically or physically. It is used as a method of decorating other metals (by partial gilding) and of making them appear to be made of solid gold (by all-over gilding). Because gold has always been regarded as a precious metal, the use of gold for gilding has made the available supplies of the metal go further.

As a result of the intrinsic value of gold, one of the aims of the gilder must always have been to develop techniques of achieving thinner and thinner coatings in order to economize even further on the use of the precious metal. Notable advances in the economical use of gold can be seen in the introduction of gold leaf in the second millenium BC, the introduction of fire-gilding *c.*2000 years ago and in the numerous 'dipping' processes which have emerged during the last three centuries. Most recently, the development of electroplating in the mid-19th century has meant that gilders can control very accurately the amount of gold deposited on the surface, thus achieving either very thin or much thicker layers.

The outline history of gilding is well known, having been the subject of several reviews in recent years (for example Oddy 1981, 1985, 1988, and 1990, Oddy and Cowell 1993). However, the detail of the changeover from one technique to another and the reasons why some older techniques continued in use alongside newer and more economical practices remains to be explored. From its origins in the late fourth or early third millennium BC, gilding technology gradually spread throughout the Mediterranean and the Middle East, but practice in both China and Egypt differed from the rest of the Old World in important respects.

Foil gilding

Gold is very malleable and can easily be hammered out into a thin foil. This was used for gilding by mechanical processes as early as the beginning of the third millenium BC. One often quoted example is some silver 'nails' (British Museum, WAA 127430) from Tell Brak in Syria (Oddy 1981) (Figure 15.1), and similar gilded copper 'nails' (British Museum, WAA 1919.10-11. 4862-7) have been excavated at Abu Sharain in Iraq, dating from *c*.2500 BC. A copper boss (British Museum, WAA 121435) from the site of Ur of the Chaldees can also be cited (Figures 15.2 and 15.3). These rather unprepossessing objects clearly show the mechanical attachment of the gold foil which has been pressed over the base metal surface and the edges of the foil crimped into position.

As far as these objects are concerned, the only method of attachment of the foil was

Figure 15.2. *Foil-gilt copper boss from Ur, (modern Iraq)* c.*2500* BC.

Figure 15.3. *Reverse of the boss from Ur showing the crimping of the gold foil around the edge.*

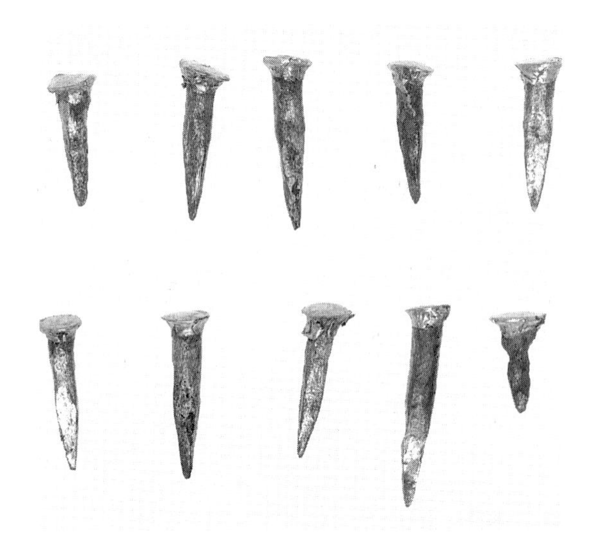

Figure 15.1. *Silver nails with foil-gilt heads from Tell Brak (modern Syria)* c.*3000* BC.

mechanical, by folding and crimping, but by the middle of the first millenium BC the Greeks were using a modified foil gilding method which allowed them to gild much larger objects. The best example is a life-size equestrian bronze statue originally erected in Athens, of which a few fragments survive. This has been identified as depicting Demetrios

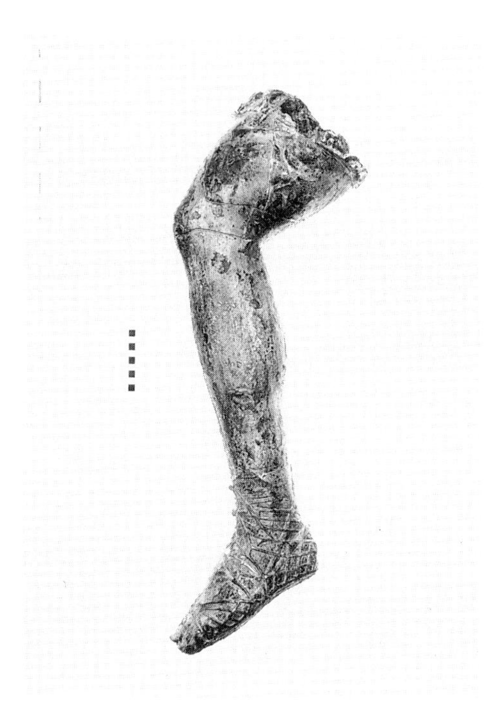

Figure 15.4. *Leg from a fragmentary equestrian statue from Athens with traces of gilding protruding from grooves on the surface of the bronze. End of 4th century* BC.

Figure 15.5. *Gilt silver disc from the Oxus Treasure. 5th/4th centuries* BC.

Poliorketes and dated to the very end of the 4th century BC (Houser 1987). On these fragments very little of the gilding remains *in situ*, but 'lines' of gold are visible in several places (Figure 15.4). Close inspection of the surviving pieces shows that grooves were cut in the surface of the bronze into which were inserted the edges of pieces of gold foil. The foil was then pressed around the surface of the bronze and the other edges similarly inserted in grooves. The gold foil was secured in position by wedging a soft material into the groove, or by hammering the groove to trap the gold. This method of gilding was mentioned by Pliny in the 1st century AD (Rackham 1968, pp. 175).

One adaptation of this technique for use on small objects is the attachment of gold foil to small areas of an object by punching round the periphery of the gilded area to drive the edges of the foil into the underlying metal. Good examples are a gilt silver disc from the Oxus Treasure (Dalton 1964, no. 24) (Figure 15.5), and a silver statuette of an Achaemenid King, also from the Oxus Treasure (Dalton 1964, no.1).

Gold foil was also used by the Celts for the gilding of counterfeit coins, very often using rather debased gold (Oddy and Cowell 1993), and similar base gold foil has been used for gilding some horse trappings in a hoard of Iron Age metalwork from Stanwick in North Yorkshire (MacGregor 1962). The method by which the gilding was applied to the Celtic gold staters is uncertain, but it seems likely that two discs of gold foil would be applied to the faces of a coin-shaped blank of silver or copper and the foil burnished down around the edge. If the surface of the silver or copper blank had been roughened, the adhesion of the gold foil would be improved by the operation of striking to emboss the designs on the two faces. Further increase in the adhesion of the gold foil to the base metal could be achieved by heating, when some interdiffusion of the gold and the copper or silver would take place. Whether heat was normally

Figure 15.6. *Gilt bronze horse harness decoration from Stanwick, North Yorkshire. Second half of 1st century AD.*

Figure 15.7. *Reverse of the Stanwick horse harness to show traces of 'adhesive' squeezed out from beneath the edge of the gold foil.*

applied is uncertain, and metallographic examination is essential to demonstrate this point, but it seems likely that coins were struck on hot blanks in the regular mints of the Roman world (Cope 1972) in order to increase the ability of the metal to 'flow' into the designs on the dies.

The gilt bronze horse trappings from Stanwick (Figure 15.6) have also not been investigated metallurgically, but the gold appears to be well bonded to the surface of the bronze. On the back of one of the trappings (MacGregor 1962, No.1) there is a black deposit adjacent to the edge of the gold foil, which looks as though it might be the remains of an adhesive which has been squeezed out from between the gold foil and the bronze casting (Figure 15.7). Although it is not possible to theorize as to exactly how the gold foil was attached, it is possible to discount a simple mechanical process. The gilding is certainly attached physically, and the use of either an adhesive or solder are possibilities, although MacGregor's statement (1962, p. 20) that the gold foil was 'soldered on to the upper surface with the aid of a flashing of copper' must be treated with reserve pending a scientific examination.

Gold foil for gilding other metals continued in use into the Roman period and has been identified in use on Roman silver as late as the 3rd century AD (Oddy 1988, nos. 32 and 39), but it was by then an obsolete and wasteful method of gilding.

Leaf gilding

Gold leaf differs from gold foil in being much thinner; so thin in fact, that it cannot support its own weight and cannot be picked up in the fingers. The thinnest gold leaf can only be made from pure (or almost pure) gold, and so it did not become available until sometime in the second millennium BC when methods for the refining of gold were developed. Because gold leaf is so weak, it cannot be attached to a surface mechanically, so physical methods had to be invented. The most simple of these is to use an adhesive to stick the gold to the surface, just as gold leaf is still applied today to wooden and stone sculpture and to leather bookbindings. However, gold leaf attached with an organic glue is not very durable because of the susceptibility of the adhesive to biodeterioration, and so it is rare to find leaf gilded objects before the Hellenistic and Roman periods. One example is a pair of copper alloy wings (Walters 1899, no. 1728) from a cult statue found at Priene in Asia Minor, where careful inspection reveals the overlapping edges of the original sheets of

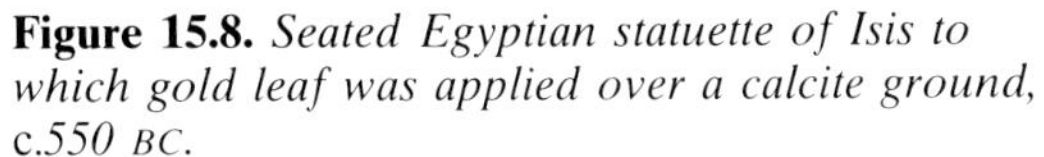

Figure 15.8. *Seated Egyptian statuette of Isis to which gold leaf was applied over a calcite ground,* c.*550* BC.

Figure 15.9. *Egyptian bronze head of Hathor to which gold leaf was applied over a gypsum ground.* c.*1000* BC.

gold leaf (Oddy 1990). Another is the arm of a life-size Roman bronze statue found in a well in France (British Museum G&R 1904.2-4.1249), where again the overlapping edges of the rectangles of gold leaf are clearly visible (Oddy 1985). In both these cases the gold leaf has been glued directly onto the metal surface.

When gold leaf is attached to wood or stone, it is usual to first apply a 'ground' which can be rubbed down to give a very smooth finish on which to stick the gold. Interestingly this technique was used on bronze sculpture in ancient Egypt (Oddy, Pearce and Green 1988).

Analyses of a number of grounds on Egyptian gilded bronzes have shown that some of them consist of gypsum and some of calcite. The gypsum presumably represents a layer of plaster of Paris which has been spread over the surface of the bronze before gilding, but the calcite is thought to be derived from ground up limestone which was mixed with an organic adhesive (possibly animal glue) to make it stick to the surface of the bronze. Conventional terminology refers to the ground as 'gesso', and typical examples of 'gesso and gold leaf' gilded Egyptian bronzes are a seated statuette of Isis dating from the Late Dynastic Period (*c*.550 BC) (British Museum, E.A. 43380), which has a calcite ground (Figure 15.8), and a head of Hathor, dating from the Late New Kingdom (*c*.1000 BC) (British Museum, E.A. 30492), which has a gypsum ground (Figure 15.9). Some of these Egyptian bronzes have been deliberately cast with a

Figure 15.10. *Elamite roundel made of gilt silver foil. 14th/13th century BC.*

Figure 15.11. *Greek ear pendant from Ephesus (modern Turkey) 7th century BC.*

rough surface to improve the keying of the gesso layer to the metal (Oddy *et al.* 1990).

Diffusion bonding

A second way of sticking gold leaf to a silver or copper surface is to lay the gold leaf on the surface and heat the object gently. This causes some interdiffusion of the gold and silver or copper to take place, and is therefore known as diffusion bonding. The gold may then be burnished, and the gilded metal may also be embossed or worked in other ways. The technique has been proved by a scientific examination to have been used on an Elamite roundel of the later second millennium BC (Oddy *et al.* 1981) (Figure 15.10), and it has been inferred on numerous items of Greek, Etruscan and Roman silver in The British Museum (Oddy 1988). Examples are an ear-pendant of the 8th/7th centuries BC, excavated at Ephesus (Marshall 1961, no. 949) (Figure 15.11), and a partially gilt silver cup of the 1st centuries BC/AD, found in Asia Minor (Corbett

Figure 15.12. *Partially gilt silver Roman drinking cup. First century BC/AD.*

and Strong 1961) (Figure 15.12). The extreme thinness of the gilding which can be achieved by diffusion bonding has misled some investigators into thinking that a more technical explanation is called for. Hence both electrolytic gilding and electrochemical gilding have been proposed, but may be rejected. (See Oddy and Cowell 1993, for references to supposed electrolytic gilding and Oddy 1988, for brief discussion of electrochemical gilding).

Fire-gilding

The most important advance in the technology of gilding to be made in antiquity was the introduction of the gold amalgam process, which is usually nowadays known in English by the name fire-gilding. This involves the spreading of a mixture of gold and mercury (a gold amalgam) onto a scrupulously cleaned metal surface (usually silver or copper). Heating then causes the mercury to evaporate, leaving behind a very well-bonded layer of gold. The details of the process are now too well known to bear repetition here (Oddy, Bimson and La Niece 1981) and suffice it to say that it is possible, by this technique, to apply a very thin layer of gold indeed, or to repeat the process numerous times so as to build up a considerable thickness of the precious metal.

Fire-gilding is very easy to detect, as the mercury is never driven off completely by the heating process, and the residual amount can often be detected by X-ray fluorescence spectrometry, and always detected by more sensitive analytical techniques such as emission spectrography.

As the positive identification of mercury is so easy, it might be expected that the origins of fire-gilding would have been chronicled with some degree of certitude. Sadly this is not so, and a number of enigmatic fire-gilded objects bedevil the issue.

The origins of fire-gilding

There seems to be little doubt that the earliest occurrence of fire-gilding is in China in the Warring States period (468–221 BC). Mercury has been detected in the gilding of belt-hooks of this period (Lins and Oddy 1975) and circumstantial evidence suggests that fire-gilding of these ornaments was common (see reference to work by Tom Chase, reported in Oddy 1985). Wang Hai-wen (1984) has argued that because fire-gilding was obviously a well-established process in China by the mid to late Warring States period, it must have been introduced somewhat earlier, perhaps near to the beginning of this period. This type of argument is dangerous, and judgement must be reserved until archaeologically datable objects of the early Warring States period have been excavated and shown by chemical analysis to be fire-gilded.

Analysis has shown that fire-gilding continued in use in China during the Han (206 BC–AD 220) and succeeding periods (Wu Kunyi 1981a and b). In seeking to trace the transmission of the technique to the West, it is interesting to note the occurrence of imported Achaemenid style silver in Chinese tombs of the 2nd century BC (Rawson, forthcoming), but conversely there is, as yet, no evidence for the knowledge of fire-gilding in Parthian Iran. Though few Parthian gilded items have been scientifically examined, those that have appear to have been gilded by the diffusion bonding process (Lins and Oddy 1975, nos 6 and 7, Oddy and Meeks 1978). It is only with the inception of Sasanian rule in Iran that suddenly there is a wealth of fire-gilded silver (Harper and Meyers 1981, p. 150, British Museum, unpublished analyses).

As far as the civilizations of the Mediterranean are concerned, recent work on silver has shown that fire-gilding first becomes widespread in the 2nd/3rd centuries AD (Oddy 1988) and that it continued as the method *par excellence* of gilding silver and copper alloys throughout the Middle Ages (Oddy 1977, Oddy, La Niece and Stratford 1986).

Further evidence that the technique was not widely used before the 2nd century AD is provided by X-ray fluorescence analyses (Cowell, unpublished analyses) of 20 gilt silver objects from Bulgarian Museums which were exhibited at the British Museum in 1976 (Venedikov 1976). These objects ranged in date from the early 4th century BC to the 2nd century AD (Venedikov nos 280, 283, 284, 288,

325, 327, 328, 330, 334, 347c, 347e, 390, 394, 400, 414, 425, 429, 430, 543, 547) and, although the absence of mercury as demonstrated by X-ray fluorescence was not tested by a more sensitive technique of analysis, the consistency of the results argues that fire-gilding was unknown, at least in the Balkans, before Roman Imperial times. Mercury was, however, detected in the gilding on some Roman gilt-silver chariot fittings of the 2nd to 3rd century AD which were found in Bulgaria (Venedikov nos 500, 501, 503, 504).

More recently, the discovery in 1986 of a large silver treasure at Rogozen, in north western Bulgaria, has stimulated interest in methods of gilding carried out by the ancient Thracians. Photographs in a recent exhibition catalogue (Fol, Nikolov and Hoddinott 1986) clearly show that of the numerous gilded items, some were definitely gilded by the application of thick gold leaf or thin foil. On some pieces, however, the gilding is thinner and, in at least one case, the catalogue identifies the gold as having been applied by the 'amalgam' process (Fol *et al.* 1986, no. 97). Analysis of the surface of several pieces from the hoard by X-ray fluorescence spectrometry failed to detect the presence of mercury (Craddock, unpublished analyses) and it seems unlikely that fire-gilding was known in the Balkans in the first half of the 4th century BC.

Nevertheless, some years ago it was reported that two objects from the so-called tomb of Philip II at Vergina in Northern Greece were fire-gilded (Assimenos 1983) and clearly the time is ripe for a thorough technological study of gilding of other metals in the Balkans during the second half of the first millennium BC.

These unconfirmed analyses add to a growing body of evidence that fire-gilding was known in Europe before the 2nd or 3rd century AD. Both Vitruvius, writing in the first century BC (*De Architectura* VII, 8, 4), and Pliny, writing in the first century AD (*Naturalis Historiae* XXXIII, 64-65, 100, 125), mention the use of mercury for gilding copper alloys and silver, but the Pliny reference in particular may possibly be interpreted as the use of mercury as an adhesive for gold leaf (Vittori 1978, 1979). This interpretation supposes that mercury was rubbed over the surface of the copper or silver and that as much mercury as possible was then removed by wiping to leave only the thinnest of films. When gold leaf is laid down on this thin mercury film it will stick to the surface, and the mercury will slowly evaporate during the course of the next few days or weeks without the application of heat. Such a cold mercury gilding technique may be inferred from Pliny, who wrote that 'The copper ... after a thorough polishing ... is able to take the gold-leaf laid on with quicksilver,' and he also wrote, 'When things made of copper are gilded a coat of quicksilver is applied underneath the gold leaf and keeps it in place with the greatest tenacity; but if the gold leaf is put on in one layer or is very thin it reveals the quicksilver by its pale colour.' It is interesting that cold mercury gilding is still used today for the gilding of copper domes on churches in Bavaria (information from Dr C J Raub).

Such a method of gilding would be difficult to distinguish chemically from true fire-gilding, as both will contain a detectable amount of mercury in the gold.

Of the many analyses which have been carried out on Hellenistic and Roman gilded silver in the British Museum, only two objects which can be dated before the 2nd century AD have revealed the presence of mercury in their gilding. One is a silver kylix dated to *c.*300 BC (Walters 1921, no. 15) (Figure 15.13). It has a very small area of gilding in the centre of the cup, (Figure 15.14), which has been shown by emission spectroscopy to contain mercury (Lins and Oddy 1975, no. 15). This gilding, however, looks like a piece of foil, and, as it has a decorative border of punch marks, it is likely that the attachment was mechanical. The other object is a silver ladle from the Arcisate Treasure (Walters 1921, no. 128) (Figure 15.15), which has a thin wash of gilding on the bird's head terminal to the handle. The ladle is dated to the first century BC, and the gilding, which contains mercury (Lins and Oddy 1975, no. 18), has every appearance of being an example of fire-gilding. If, however, the correct interpretation of Vitruvius and Pliny is the theory that true fire-gilding was preceded by a period of time when a cold mercury gilding process was used, perhaps this is an example of that process.

Figure 15.13. *Greek silver kylix with raised gilt knob in the centre of the bowl,* c.*300* BC.

Figure 15.14. *Detail of the gilt knob on the silver kylix.*

Whatever the interpretation, this object so far stands virtually alone in the Graeco-Roman world as possibly being an example of fire-gilding before the 2nd century AD.

There are, however, some classical 'bronzes' which Paul Craddock has published (1977 p. 109) as being examples of fire-gilding in the Hellenistic period. These are ten finger rings and two bracelets which, when analysed, were (uncharacteristically) found to be made of impure copper or low tin bronze. Such a composition for a gilded object made by casting immediately suggests that the low proportion of alloying elements is deliberate and is related to the fact that certain alloying elements (particularly zinc and lead) can interfere with the process of fire-gilding.

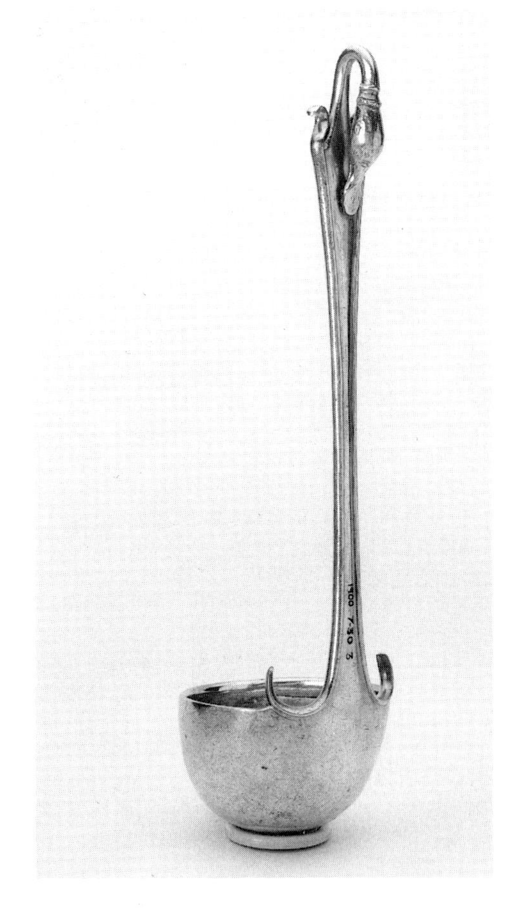

Figure 15.15. *Roman silver ladle with the end of the handle terminating in the head of a bird, which is gilded. From the Arcisate Treasure, 1st century* BC.

Now, however, emission spectroscopy has shown that three of the objects, a Hellenistic ring excavated at Naukratis (Marshall 1907, no. 1258) and two Hellenistic bangles from tomb 84 at Amathus, Cyprus (Marshall 1961, nos 1993/4), do not contain mercury in the gold. Of the remaining nine finger rings, seven of them, which were originally catalogued as Hellenistic (Marshall 1907, nos 1275, 1290, 1291-3, 1296-7), have now been reattributed to the Roman period (letter to the author from B.F. Cook, dated 17.12.1979). Only two rings remain which are stylistically pre-Roman and which contain mercury in the gilding (Marshall 1907, nos 1272 and 1274), but both these rings were acquired from the dealer Castellani in Rome in 1872. Castellani is well known as an 'improver' of antiquities and it would be unwise to build any theories involving the history of technology solely on ex-Castellani objects.

Evidence from the Classical world is, therefore, unclear about the date of the introduction of fire-gilding, with only one or two undoubtedly authentic objects and two documentary sources which indicate the use of mercury in the gilding process and which can be dated before the second century AD. Surprisingly, however, evidence is now beginning to appear for a knowledge of fire-gilding in the Celtic world, apparently by the first century BC at the latest! In all, six objects have been analysed which contain mercury in their gilding.

Some years ago Professor Zwicker (1973) published the analysis of a gilded stater of the Vendelici which he showed to contain mercury in the gold. Since then, two gilt Celtic staters have been identified in The British Museum which also contain mercury in the gold (Oddy and Cowell 1993, nos 24 and 25). These coins were, however, struck in Southern Britain at the end of the 1st century BC or in the early years of the 1st century AD by the Dobunni and the Corieltauvi respectively. The other Celtic objects which have been shown to contain mercury in the gold are a brooch of the Spanish Iron Age (Tait 1976, no. 139), which is attributed to the 2nd or 1st centuries BC, and fragments of two gilt bronze torcs of about 50 BC from Snettisham, Norfolk (personnel communication from Peter Northover). The gilding on the Spanish brooch is patchy, but does look more like a primitive attempt at fire-gilding than the result of cold mercury gilding with gold leaf.

These three coins, brooch and two torcs do, then, raise the spectre of a knowledge of fire-gilding by the Celtic tribes of Western Europe in the 1st century BC. Clearly what is needed now is an extensive survey of other Celtic gilded objects, and a start has been made by analysing 15 other gilded Celtic coins in The British Museum (Oddy and Cowell 1993), but none of them contained mercury in the gold! Spain, however, with its well-known deposits of mercury, is an obvious place to look for fire-gilded coins or other objects.

This brief survey of gilding in the Old World stops with the end of the Roman period, but developments in gilding technology in post-Medieval Europe introduced numerous techniques which need to be explored from a museum point of view. These include the use of gold powder, of organo-gold complexes and of electrochemical gilding, and their history can be followed in the numerous technical dictionaries and encyclopaedias of the 18th and 19th centuries (Oddy and Cowell 1993). These techniques have largely, but not entirely, been rendered obsolete in modern times by the introduction of commercial electro-gilding in the 1840s (Hunt 1973, Raub this volume Chapter 23).

Acknowledgements

I have been assisted in the X-ray fluorescence and emission spectrographic analysis by numerous colleagues and students over the years, among whom I am particularly grateful to Paul Craddock, Susan La Niece, Michael Cowell, Andrew Lins and Tim Padley. I would also like to acknowledge all those curators who trusted us with their objects for the purposes of examination. I am also indebted to a succession of British Museum photographers for their work on these objects, but especially to Trevor Springett who made the illustrations for this paper. Figure 15.4 was kindly provided by the American School of Classical Studies in Athens.

References

Assimenos, K. (1983) Technological and analytical research on precious metals from the chamber tomb of Philip II (Vergina). Abstract of paper given at the International Symposium on the history and technology of the precious metals, Meersburg, 25–28 April, 1983

Cope, L.H. (1972) The metallurgical analysis of Roman Imperial silver and *aes* coinage. In *Methods of Chemical and Metallurgical Investigation of Ancient Coinage* (eds E.T. Hall and D.M. Metcalf) Royal Numismatic Society Special Publication No. 8, London, 261–278

Corbett, P.E. and Strong, D.E. (1961) Three Roman silver cups. *The British Museum Quarterly*, **23**(3), 68–86

Craddock, P.T. (1977) The composition of the copper alloy used by the Greek, Etruscan and Roman Civilization: 2. The Archaic, Classical and Hellenistic Greeks. *Journal of Archaeological Science*, **4**, 103–123

Dalton, O.M. (1964) *The Treasure of the Oxus*, 3rd edition, British Museum, London

Fol, A., Nokolov, B. and Hoddinott, R.F. (1986) *The New Thracian Treasure from Rogozen, Bulgaria*, British Museum, London

Harper, P.O. and Meyers, P. (1981) *Silver Vessels of the Sasanian Period: Volume 1: Royal Imagery*, Metropolitan Museum of Art, New York

Houser, C. (1987) *Greek Monumental Bronze Sculpture of the Fifth and Fourth Centuries* BC, Garland Publishing, New York and London, pp. 255-281

Hunt, L.B. (1973) The early history of gold plating. *Gold Bulletin*, **6**(1), 16–27

Lins, P.A. and Oddy, W.A. (1975) The origins of Mercury Gilding. *Journal of Archaeological Science*, **2**, 365–373

MacGregor, M. (1962,) The Early Iron Age metalwork hoard from Stanwick, N.R.Yorks. *Proc.Prehist.Soc.* (New Series) **28**, 17–57

Marshall, F.H. (1907) *Catalogue of the Finger Rings, Greek Etruscan and Roman, in the ... British Museum*, London

Marshall, F.H. (1961) (Reprint of 1911 edition) *Catalogue of the Jewellery ... in the British Museum*, London

Oddy, W.A. (1977) Gilding and tinning in Anglo-Saxon England. In *Aspects of Early Metallurgy* (ed. W.A. Oddy) (reprinted as British Museum Occasional Paper No. 17, London, 1980)

Oddy, W.A. and Meeks, N.D. (1978) A Parthian bowl: study of the gilding technique. *MASCA Journal*, **1**, 5–6

Oddy, W.A., La Niece, S., Curtis, J.E. and Meeks, N.D. (1981) Diffusion bonding as a method of gilding in Antiquity. *MASCA Journal*, **1**(8), 239–41

Oddy, W.A., Bimson, M. and La Niece, S. (1981) Gilding Himalayan images: history, tradition and modern techniques. In *Aspects of Tibetan Metallurgy* (eds W.A. Oddy and W. Zwalf) British Museum Occasional Paper No. 15, 87–101

Oddy W.A. (1981) Gilding through the ages: an outline history of the process in the Old World. *Gold Bulletin*, **14**(2), 75–79

Oddy, W.A. (1985) Vergoldungen auf prähistorichen und klassichen Bronzen. In *Archäologische Bronzen Antike Kunst, Moderne Technik* (ed. H. Born) Staatliche Museen Preussischer Kulturbesitz, Berlin, 64–71

Oddy, W.A., La Niece, S. and Stratford, N. (1986) *Romanesque Metalwork: Copper Alloys and their Decoration*, British Museum, London

Oddy, W.A. (1988,) The gilding of Roman silver plate. In *Argenterie Romaine et Byzantine* (ed. F. Baratte) De Boccard, Paris, 9–25

Oddy, W.A, Pearce, P, and Green L (1988) An unusual gilding technique on some Egyptian bronzes. In *Conservation of Ancient Egyptian Materials* (eds S.C. Watkins and C.E. Brown) UKIC, London, 35–39

Oddy, W.A. (1990) Gilding - an outline of the technological history of the plating of gold on to silver or copper in the Old World. *Endeavour*, **15**(1), 29-33

Oddy, W.A., Cowell, M.R., Craddock, P.T. and Hook, D.R. (1990) The gilding of bronze sculpture in the Classical World. In *Small Bronze Sculpture from the Ancient World* (papers delivered at a symposium ... held at the J. Paul Getty Museum, March 16-19, 1989) Malibu, California, 103–124

Oddy, W.A. and Cowell, M.R. (1993) The technology of gilded coin forgeries illustrated by some examples in The British Museum. In *Metallurgy in Numismatics III* (eds M.M. Archibald and M.R. Cowell) Royal Numismatic Society, London

Rackham, H. (1968) *Pliny: Natural History: Vol.9*, Loeb Classical Library, London

Rawson, J. (forthcoming) Chinese gold and silver. In *Macmillan Dictionary of Art*, London

Tait, H. (ed.) (1976) *Jewellery through 7000 years*, British Museum, London

Venedikov, I. (1976) *Thracian Treasures from Bulgaria*, British Museum, London

Vittori, O. (1978) Interpreting Pliny's gilding: archaeological implications. *Rivista di Archeologia*, **2**, 71–81

Vittori, O. (1979) Pliny the Elder on gilding. *Gold Bulletin*, **12**(1), 35–39

Walters, H.B. (1899) *Catalogue of the Bronzes, Greek, Roman and Etruscan in the ... British Museum*, London

Walters, H.B. (1921) *Catalogue of the Silver Plate ... in The British Museum*, London

Wang Hai-wen (1984) Survey of gilding (in Chinese). *Gugong bowuyuan yuankan* (Journal of the Palace Museum, Beijing), pp. 50–58, 84 and pls. 6–7

Wu Kunyi (1981a) Chemical distribution of mercury in ancient Chinese gilding (in Chinese). *Zhongguo Keji Shiliao* (Material about the History of Chinese Science and Technology) **1**, 90–94

Wu Kunyi (1981b) Gilding (in Chinese). *Zhongguo Keji Shiliao* (Material about the History of Chinese Science and Technology) **1**, 157–9

Zwicker, U. (1973) Untersuchungen an goldplattierten keltischen und griechischen Münzen. *Jahrbuch für Numismatik und Geldgeschichte*, **23**, 115–117

16

Techniques of gilding and surface-enrichment in pre-Hispanic American metallurgy

Warwick Bray

Abstract

Between *c.* 1500 BC and the European conquest, native American metalworkers discovered (with no apparent influence from the Old World) almost all the techniques of gilding and plating known to pre-industrial goldsmiths in Europe and Asia. In addition, the Indians of coastal Ecuador and Colombia were making platinum jewellery by the time of Christ, and were able to produce platinum-clad surfaces on objects of copper or copper alloy.

Amerindian techniques of surface-enrichment can be grouped into two major categories: those (like foil gilding, fusion gilding and electrochemical replacement plating) which add a layer of gold or silver to a substrate of base metal, and those which depend on the principle of 'depletion gilding', employing an acid substance to remove copper from the surface of an object containing a percentage of gold and/or silver. These techniques are described, drawing on ethnohistorical evidence, laboratory analyses and replication experiments. Although Old and New Worlds had access to the same range of techniques, American metalsmiths showed a marked preference for depletion gilding. The reasons for this have more to do with cultural and symbolic values than with technology or with the properties of the metals themselves. The rationale behind Amerindian technology for the 'surface transformation' of metals has no counterpart in the Old World.

Introduction

This chapter summarizes our present knowledge of gilding and silvering practices in the New World before the European conquest and, in the final section, attempts to set this technology in its native cultural context. The metallurgical data reviewed in this essay are not my own, and my debt to two friends and colleagues – David Scott and Heather Lechtman – will be clear from the bibliography.

The Americas, from Mexico in the north to Chile and Argentina in the south, were the homeland of one of the world's great metallurgical traditions (Figure 16.1). Iron and steel were unknown until introduced by the Europeans at the time of conquest, but native Amerindians had discovered the properties of gold, silver, copper, tin, lead, platinum and their alloys. In spite of much wishful thinking, no convincing evidence has been produced for Eurasian influence. Beginning shortly before 1000 BC in South America, indigenous metalworking has an internal development of its own, with new techniques being progressively added to the repertoire. Without transatlantic or transpacific contacts, native metalsmiths discovered for themselves most of the

Plate 8.1. *Kozuka with Chinese beneath a tree.* Shakudo, *silver and gold inlay on a worn* shakudo nanako *ground. (17th-18th century).*

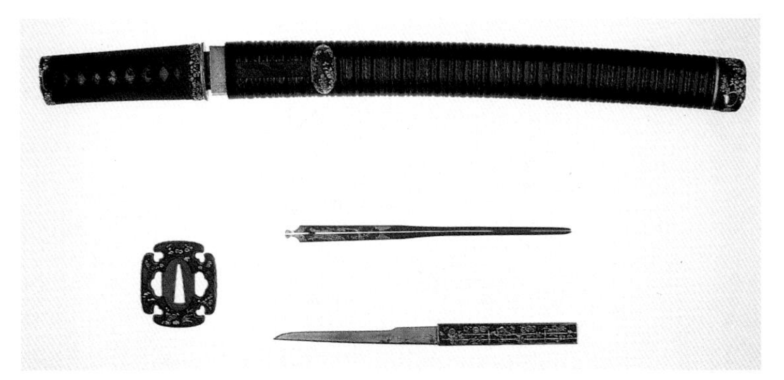

Plate 8.2. *The components of a sword mounting with metal fittings decorated with flowers in gold inlay on a* shakudo nanako *ground. (17th-18th century).*

Plate 8.3. *Kozuka with a geisha and her maid.* Shakudo, *silver, and gold inlay and* katakiri bori *on a polished* shibuichi *ground. Signed Ichijo. By Goto Ichijo (1791–1876).*

Plate 8.4. *Fuchi and kashira with the immortals Gamma Sennin, he with the three-legged toad, and Tekkai Sennin, who transports his miniature self upon his breath.* Shakudo, *copper, and gold inlay on a* shakudo nanako *ground. Signed Eisho. By Omori Eisho (d.1772).*

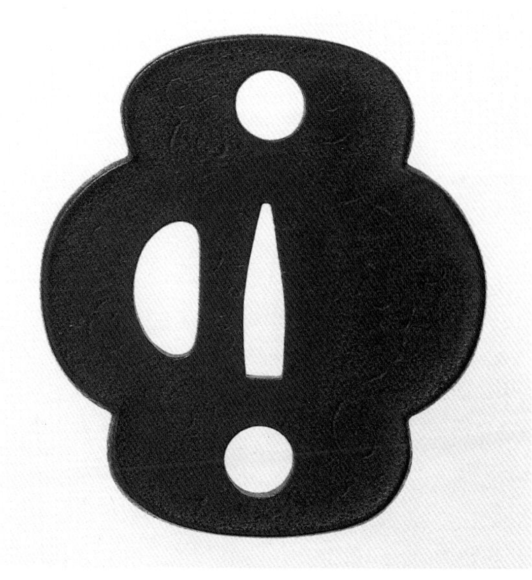

Plate 8.5. Tsuba. *Lacquered wood in imitation of patinated iron. Inscribed with the name of the 16th century maker of iron* tsuba, *Nobuie, and signed on the reverse Zeshin. By Shibata Zeshin (1807–1891).*

Plate 9.1. *A pair of Roman inkpots, G R 1853.2-18.6 and 7. They are composite, the tops are separate, with small amounts of gold in the alloy and inlaid with gold, whereas the bodies are of ordinary bronze and inlaid with silver.*

Plate 9.2. *Small Roman bronze plaque, G R 1979.12-13.1, with patinated black front surface inlaid with gold and silver, the latter now largely missing.*

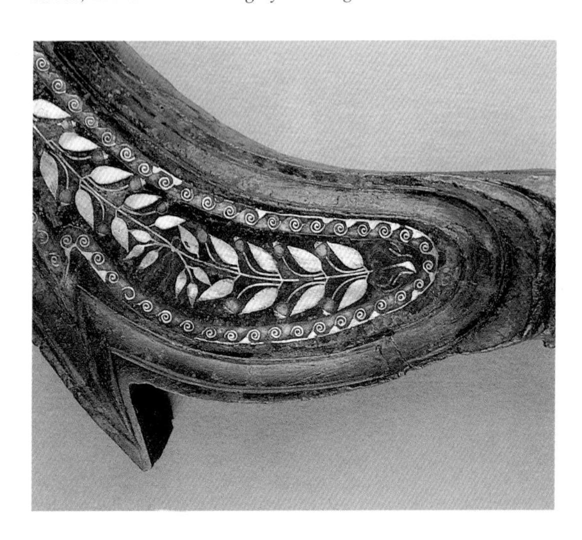

Plate 9.3. *Roman bronze couch end, G R 1784.1-31.4, with insert of black patinated bronze inlaid with copper and silver.*

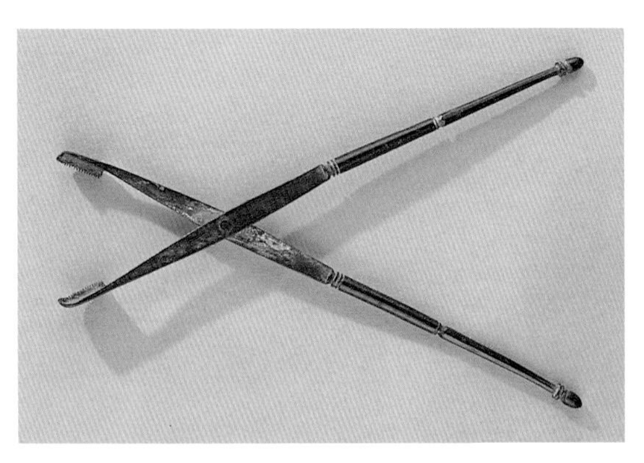

Plate 9.4. *Roman forceps, G R 1814.7-4.969. The arms are inlaid with bands of black bronze.*

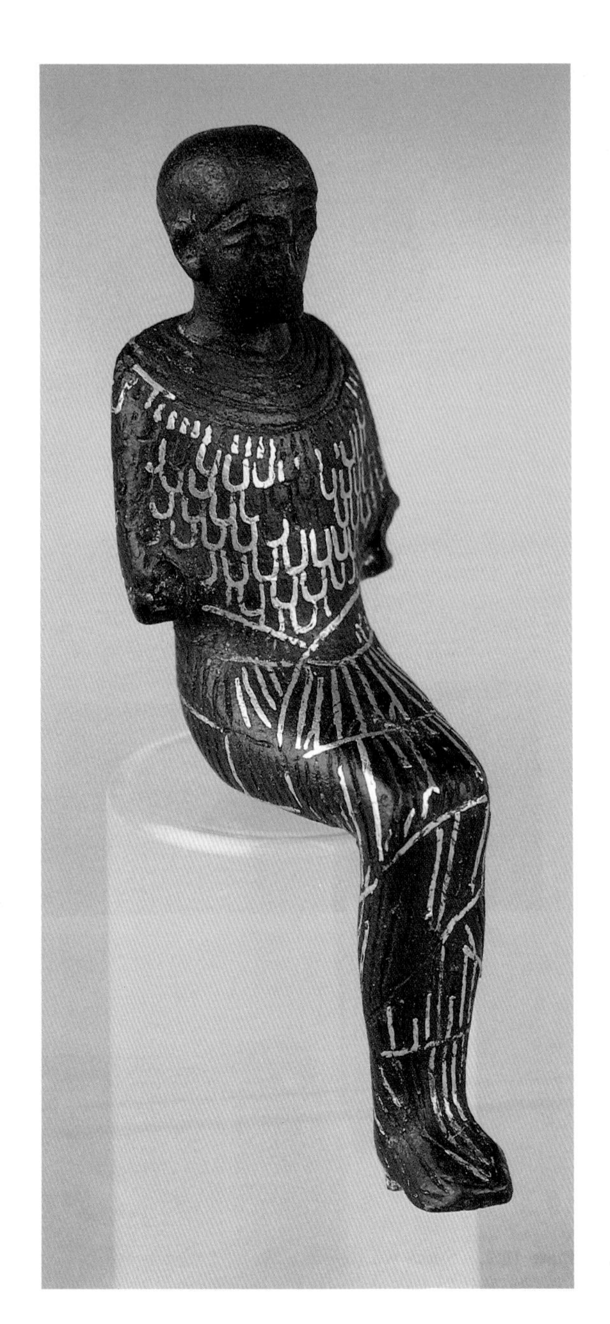

Plate 9.5. *Black patinated and inlaid figure of Ptah, EA 27363, containing 2.71% of gold, 0.45% of silver and 0.57% of arsenic.*

Plate 22.1. *Copper and brass plated iron stirrup from Seagry. (Photo courtesy of Wiltshire Archaeological and Natural History Society).*

Plate 23.1. *St Isaac's Cathedral, St Petersburg. The statues and bas-reliefs are electroforms and gilded domes are electroplated. (Photo from Novosti Press Agency).*

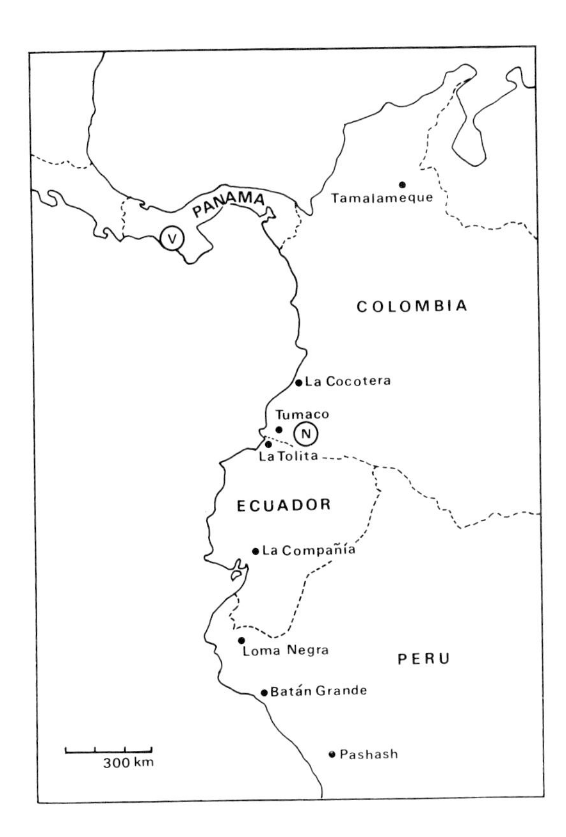

Figure 16.1. *Panama and northern South America, with archaeological sites mentioned in the text. Encircled areas are zones containing several poorly documented localities. V = Veraguas; N = Nariño.*

techniques employed in the Old World. One of the fascinations of working in American metallurgy is the opportunity to study independent, convergent, technological evolution in a cultural milieu entirely different from that of Europe or Asia.

The sections which follow show that native Americans, well before the Conquest, were practising nearly all the gilding and colouring techniques that were standard in pre-industrial Europe, with the exception of fire-gilding (amalgam gilding), even though cinnabar was used as a pigment in Peru from earliest times, and metallic mercury (probably from sources near Lake Amatitlan, in highland Guatemala) has been found in Maya tombs (Berlo 1984: I, 143, 145; Pendergast 1982).

American techniques of surface enrichment can be divided into two major categories:

1. those (e.g. foil-gilding, fusion-gilding, plating) which deposit a surface layer of the desired composition onto a metal substrate of any composition whatsoever, and
2. depletion-gilding, effective only with alloys already containing gold, which works by removing the less noble metals at the surface of the object, while leaving the gold unaffected.

Foil-gilding

This is the most straightforward of the gilding methods and consists of the mechanical attachment of thin sheets of gold to a substrate of another material, usually metal or wood, more rarely stone, bone, shell, and even pottery (Rivet and Arsandaux 1946, pp. 160–65). Gold leaf is less than *c.* one micron in thickness, anything thicker being described as foil. The natural impurities in native gold make the production of gold leaf difficult unless the metal is first purified, so it is hardly surprising that the analysed specimens from South America turn out to be foil-gilded (Scott 1986b). The disadvantage of using the thicker sheet is that the foil does not easily follow changes of contour, and therefore tends to wrinkle (Figure 16.2), in particular along the inner side of curved forms. These wrinkles, and sometimes also the presence of joined,

Figure 16.2. *Three foil-gilded nose ornaments from Colombia in the collections of the Museo del Oro, Bogotá. The example at the bottom shows the wrinkling of the foil overlay on the inside of the curve. (Photo and information from Juanita Sáenz.)*

overlapping, or multilayered sheets, are diagnostic of foil-gilding. Where these visual criteria are not present, foil-gilding can easily be confused with other forms of surface treatment, and metallographic analysis is recommended.

The simplest technique for applying gold foil is to burnish it directly on to a cleaned and roughened substrate, though this does not give a very firm bond. Adherence can be improved by diffusion-bonding, when burnishing is followed by a brief heating to a temperature well below red-heat. This heating promotes solid state diffusion between the gold and the substrate metal across their common interface, creating a thin zone of interalloyed metal (Scott 1986b, pp. 317–18).

Little attention has been given to foil-gilding in the archaeological literature, though the technique has been reported sporadically from Costa Rica, Panama, Colombia, Ecuador and Peru. Most of these items are of simple bar metal (Figure 16.2), but foil-gilded pins with elaborate heads are reported from Pashash, in Peru, dating to the early centuries after Christ (Grieder 1978), and there are rare figurines of birds and human beings from Veraguas, Panama, from the final centuries before the Spanish conquest (Root 1950, p. 93, Bonhams 1992, no. 5). Root suggested that one Panamanian item made from gold–copper alloy might have been cast directly into a foil-lined mould, but this would be impractical in most cases, and absolutely impossible for a complex lost wax casting, like the figurine in question. The only systematic studies are those by Scott (1986a, b), who surveys the evidence from Ecuador and includes metallographic data on a penannular nose-ring from an urn burial of the Milagro-Quevedo phase (AD 500–1500) at La Compañía on the central coast. The substrate metal of this item was copper with *c.* 0.5% arsenic, shaped by hammering and annealing. Over this were applied several small sheets of gold foil, varying in thickness from 90 to 35 microns. The overlapping of these foil sheets produced as many as three superimposed layers. Analysis of three separate foil sheets gave figures in the range of 69% to 86% of gold, with some silver and copper. Diffusion bonding was detected at the interface between the copper base and the adjacent (innermost) foil sheet.

Fusion-gilding

This process involves the application of molten metal to the surfaces of an object made of base metal, usually of copper or copper-rich alloy (Bergsøe 1938, Lechtman 1971, Scott 1986a, b). The gilding metal is almost always an alloy of gold with copper, which melts at a considerably lower temperature than pure gold itself. If the ornament is to be coated all over it can be dipped into a bath of molten gold; if only one surface is to be treated, the coating has to be flushed on by hand. In either case, when the molten metal runs over the heated substrate this creates a firm fusion bond caused by the 'melting' together of the metals at their interface.

The process requires precise control over alloys and their different melting points. Fusion gilding does not leave a uniform surface; drips or runs in the surface coating are sometimes visible and, in section, the coating is of variable thickness. Experience has shown, however, that it can be difficult to distinguish, by visual inspection alone, between fusion-gilding and depletion-gilding or foil-gilding.

Fusion gilding seems to be relatively uncommon in the Americas. Where the process was invented is still unsure, but it was known to the people of La Tolita, in coastal Ecuador, sometime between 300 BC and AD 800, has been recorded in Peru during the Moche-Vicús period of the early centuries after Christ, and in Nariño (southern Andes of Colombia) shortly before the Spanish conquest (Scott 1986b and personal communication).

The first archaeological studies of American material were carried out by Paul Bergsøe (1937, 1938), who examined the thick gold coatings on copper items from La Tolita, in northern Ecuador, and outlined the process of manufacture. The copper substrate was carefully cleaned and then heated to at least 850°C. Onto the prepared copper was flushed an alloy of 80% gold and 20% copper, with a melting point 200°C below that of the substrate. The coated artifact was then hammered and burnished, and, in some cases, pickled to remove unwanted copper from the surface alloy.

This pioneering study was followed up by Scott (1986a, b), who analysed a gilded copper fish hook from Bergsøe's collections in Copenhagen. The hook was basically of copper, with about 4% gold. The gold-rich surfaces had an average composition of 26% Au, 2% Ag, and 72% Cu. The metal of the hook has a melting point around 1080°C, and the surface alloy melts at *c.* 1010°C (a difference of only 70°C). After the gold had been applied, the object was further worked and annealed. As a consequence of this activity, the outer layer of the *tumbaga* (gold-copper) surface was further enriched, and proved to contain more than 40% of gold.

Scott has also carried out metallographic analyses of fusion-gilded sheet fragments from the department of Nariño (Colombia) dating from around the 9th to 12th centuries AD. One of his samples (No. AI 100) consisted of worked and annealed copper sheet coated on both faces with a gold alloy (*tumbaga*) containing 51.1% Au, 5.9% Ag, and 41.9% Cu. Electron probe analysis of the 'gold' coating showed that further surface enrichment (pickling or depletion gilding) had taken place, and that the outermost surface contained *c.* 72% Au, 8% Ag, and 20% Cu. The electroprobe scan showed (working from interior to surface) a sheet substrate of almost pure copper, then a wide region of interdiffusion (12 to 15 microns across) until the gold alloy coating is reached. There is another compositional change at the surface, where the gold layer has been depletion gilded.

Scott's second Nariño sample (AI 89) consisted of copper sheet with a thick gold alloy coating on one face only. Metallographic analysis indicated that a diffusion bond was formed between the copper substrate and the applied coating, and that the coated object was further finished by working and annealing. The *tumbaga* coating proved to be a low-silver, 14 carat gold alloy (*c.* 59% Au, 8% Ag, 33% Cu by weight) with a melting point of about 900°C, nearly 200° below that of copper (Figure 16.3).

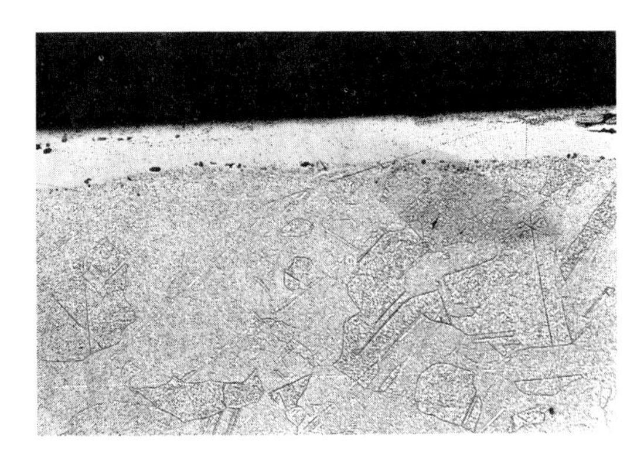

Figure 16.3. *Photomicrograph (×120 approx.) of a sheet of copper, fusion gilded on one face. From Yacuanquer, Nariño, Colombia. The cross-section has been etched in ammonium persulphate and potassium cyanide. The surface coating is an alloy of about 59% gold, 8% silver and 33% copper. The substrate is copper of a high purity. The twinned and recrystallized grains show that the object was shaped by working and annealing. (Photo and information from David Scott.)*

The same molten alloy/fusion technology can also be used to produce silvered surfaces over base metal. Silvered objects are not uncommon in archaeological collections from Peru and Ecuador, but few of them have received detailed analysis. One exception is a silver-coated nose ring from the urn burial at La Compañía, on the central coast of Ecuador, belonging to the final centuries before the Spanish conquest.

Scott (1986a, b) has been able to reconstruct the entire process of fabrication. A cast ingot of copper was worked and annealed into the shape of a penannular nose ornament, and the surfaces were carefully cleaned to remove all copper oxide scale. The coating material, consisted of an alloy of 40% silver and 60% copper, with a melting point of 920°C (165° lower than that of the substrate). The copper nose-ring was then dipped or coated with the molten silver–copper alloy. The entire object was finished by heating and hammering, and then further cleaned to remove copper oxides and to create a silvery appearance.

Electrochemical replacement plating

In the 1970s, while examining a group of metal objects from the site of Loma Negra, in the far north of Peru, researchers at the Massachusetts Institute of Technology made an unexpected discovery (Lechtman 1979, 1984b, Lechtman *et al.* 1982). The objects in question, belonging to the Moche-Vicús

Figure 16.4. *Seated man, of gilt sheet copper, Loma Negra, Peru,* c. AD *300 (Metropolitan Museum of Art, New York, No. 69.82). This object was gilded by the electrochemical replacement process. (Photo from Heather Lechtman.)*

culture of the early centuries AD, were entirely covered with green corrosion products formed during burial. Removal of the green mineral layers showed that many items were originally covered on one or both sides with extremely thin coatings of gold or silver (Figure 16.4). The coatings themselves measured only 0.5 to 2.0 microns, and they were often not visible in cross-section at a magnification of × 500. Metallographic examination showed that:

1. The metal coatings were of uniform thickness and were present on all surfaces, including the thin edges of the objects and of the holes made to receive danglers. This, with the lack of wrinkles or overlaps, rules out the possibility of gold leaf (see also pp. 183–184).
2. The presence of a solid-state diffusion zone between the gold and the underlying copper indicated that heat had been applied at some stage of the process.
3. There was no trace of mercury, or the flushing on of molten gold or silver.

The general appearance, in fact, was very like that of modern electroplating. Since the modern technique requires an external source of electrical current, this could be eliminated from consideration, but it turned the investigators' attention towards another technique of plating, electrochemical replacement, known since Roman times in the Old World (Lechtman 1984b p. 56). During the 9th century of the Christian era it was used to prepare iron for gilding, and was employed by 18th century European armourers to decorate iron and steel with gold (Smith 1720 p. 130).

In the European formula, when silver or copper were to be plated with gold, an aqueous solution of gold was prepared by dissolving the metal in a bath of aqua regia (a mixture of hydrochloric and nitric acids). Since distilled acids were unknown in prehispanic America, Lechtman and her associates attempted to replicate the process by dissolving the gold in an aqueous solution of corrosive minerals, several of which occur naturally in coastal Peru.

Lechtman (1984b p. 57) describes the results of the experiments as follows:

> The simplest and most effective method we devised was to prepare a corrosive bath by dissolving in water equal parts of potassium aluminium sulphate, potassium nitrate and sodium chloride (common salt). Gold heated gently in the mixture for two to five days dissolves readily. (The mixture, incidentally, contains among other ions those present in aqua regia.) The solution is highly acidic, however, and immediately attacks any copper immersed in it. We had to neutralize it before any gold in the solution could plate out onto the copper. The salt we found to be most effective for this purpose was sodium bicarbonate ('baking soda'), and the optimum pH (the standard measure of acidity, or alkalinity) for plating proved to be 9, or somewhat alkaline.
>
> After five minutes of gentle boiling, a sheet of copper immersed in the neutralized solution was uniformly coated with gold on all its surfaces, edges included. The gold film was approximately one micron thick. The plating process by itself, however, was rarely sufficient to bond the gold to the copper permanently. Unless the copper was extremely clean and free of oxide the gold tended to pop off.
>
> Our metallographic studies of the Loma Negra objects revealed a distinct zone of solid-state diffusion between the gold layer

and the underlying copper, indicating that the Moche smiths had heated their plated copper to achieve a strong metallurgical bond at this interface. When our laboratory-prepared samples were heated for only a few seconds in the temperature range between 650° and 800° degrees Celsius, we also achieved excellent bonding between the gold plate and copper substrate.

Our experiments were eventually successful in producing electrochemically deposited gold platings that quite closely resembled those on the Loma Negra artifacts both in their surface visual characteristics and in their microstructure.

Other Loma Negra ornaments were plated on one side only, and could not have been produced by the dipping technique. Lechtman (1979 p. 157) proved that it was possible to produce a shiny gold film by brushing a hot diluted solution of gold at pH 9 onto the copper substrate.

The procedure for silvering is identical to that for gilding:

> The same aqueous solution of salts can be used to dissolve silver, but in this case silver chloride (AgCl) precipitates out of solution. We have found it convenient to use calcium carbonate (as precipitated chalk) to neutralize the solution; and if sufficient quantities are added, a thick slurry or paste can be produced which incorporates the silver chloride crystals. Either dipping the copper sheet into this slurry or rubbing the paste onto a piece of copper immediately caused silver to plate out. (Lechtman 1979 p. 157)

Most of the gold employed by Peruvian jewellers contains silver in amounts that range up to 15% by weight. Since both these metals can be dissolved in the same aqueous salt solution, the MIT team was also able to plate copper with alloys containing gold in concentrations from 95% to 5% by weight.

Platinum cladding

Some of the world's richest platinum deposits occur in the alluvial sands and gravels of the Pacific coastal plain on either side of the modern Ecuador–Colombia frontier. Since platinum grains and nuggets are found alongside gold, in the same river gravels, the Indians of the Tumaco–Esmeraldas zone developed techniques for dealing with platinum at the same time as they were experimenting with the metallurgy of gold. Since platinum melts at *c.* 1770°C, and the indigenous blowpipe-and-crucible technology could not produce temperatures above 1200–1400°C, the metal was never cast. Large platinum nuggets were forged and hammered to shape, but more commonly small grains of platinum were 'sintered', or bound together in a fusible matrix of gold to produce a workable ingot of gold–platinum 'alloy' (Bergsøe 1937, Rivet and Arsandaux 1946, Scott and Bray 1980 and in press, Bray 1988). This is the earliest known example of powder metallurgy

The oldest dated find of gold–platinum jewellery comes from La Tolita on the north coast of Ecuador, with a radiocarbon date of AD 90 ± 60, uncalibrated (Scott and Bouchard 1988, Valdez 1987), and from the site of La Cocotera, on the Colombian side of the frontier, there is evidence of platinum cladding over a *tumbaga* (gold–copper) substrate by about AD 100 (Patiño 1988 p. 24). Among the mass of looted trinkets from La Tolita and related sites are items of gilded copper, and

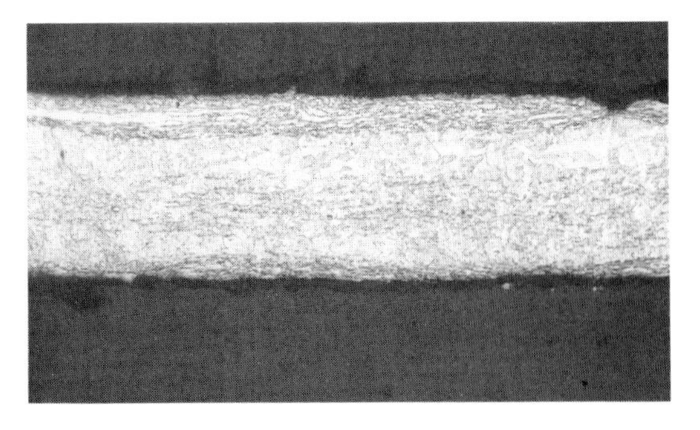

Figure 16.5. *Platinum cladding. Etched section across a small, crescentic dangler from the general area of La Tolita, Ecuador. (National Museum of Denmark 08274). On both faces, small platinum particles have been flattened out over the underlying substrate (of sintered gold-platinum) and then burnished to produce a smooth, silvery finish. (Photo from David Scott.)*

several examples of *tumbaga* clad with platinum on one or both faces. Metallographic examination of one such platinum coated ornament showed that the cladding was composed of a number of small particles flattened out by hammering, and then well burnished to give a silvery finish (Figure 16.5). In a laboratory experiment carried out by David Scott, a crude simulation of this kind of structure was achieved by using a sintered gold–platinum matrix as the substrate, onto which were hammered filings of an alloy containing 96% platinum with 4% iron (present up to 12% in native Ecuadorian platinums), followed by cycles of working and annealing at 700°C. After about 20 minutes an acceptable platinum surface was produced. After burnishing, the plated surface was dark grey, while the reverse remained golden (Scott and Bray 1980).

Depletion-gilding

The methods of gilding described in the preceding sections rely on the deposition of a layer of gold onto an item made of a different metal. Depletion gilding (*mise en couleur*) works on the diametrically opposed principle, not the addition of new material, but the removal of unwanted base metals from an object made of alloy already containing gold. This technique is widely distributed in the New World from Mexico to Peru, and is the norm in Mesoamerica, the Isthmian countries and Colombia.

In this northern sub-area, the material to be treated is almost always *tumbaga*, an alloy of copper and gold which may contain silver (up to 10% or >15% in extreme instances) as a natural impurity. For Aztec Mexico the method of depletion gilding is described in detail in a much-quoted account by the Spanish friar Bernardino de Sahagún (e.g. in Emmerich 1965 pp. 177–183), and there are similar early colonial descriptions from Caribbean Colombia (Bray 1978 p. 38). The technique is effective with alloys containing as little as 10% of gold; in certain damaged, high-copper items from the Isthmus and Colombia the interior core, below the gold-rich surfaces, has corroded to a black powdery material which is all that remains of the original alloy (Scott 1983a p. 196) (Figure 16.6).

Figure 16.6. *Objects of* tumbaga *(gold–copper alloy) treated by depletion gilding. Left: fragment of a multiple bird pendant from central Panama,* c. *AD 500–1000. Lost wax casting over a clay and charcoal core. This item is made of high-grade, gold-rich* tumbaga*; the inner face, protected by the core, shows the original reddish colour of the alloy, and the yellower outer surface is the result of pickling. Right: Frog pendant from Veraguas, Panama, AD 1000–1550. This is an open-back, lost wax casting in low-grade* tumbaga *(oxidized almost black) with depletion gilded surfaces.*

The basic principle of depletion-gilding involves the removal, by means of some acid substance, of copper and/or silver at the surface of an item made from gold alloy. The pickling agent may be of plant or mineral origin. Sahagún's description refers to the use of alum and 'gold medicine ... just like yellow earth mixed with a little salt'. In Colombia and the Isthmus the preference was for acid plant juices (Bray 1978, Scott 1983b p. 112). A document of 1555, in the Archive of the Indies (Seville), describes the practice in the Tamalameque region of Caribbean Colombia: '... the herb they brought to give it [the *tumbaga* alloy] colour was crushed on a stone and, once crushed in this way, they placed it in a small pot which they brought in, and added water and ground white salt, and stirred all together'. After polishing, heating and quenching in the solution several times, 'it took on the colour it should have'. There are similar descriptions from neighbouring zones of

Colombia by Gonzalo Fernández de Oviedo, Martin Fernández de Enciso, and Francisco López de Gómara.

Oviedo complained that the Indians would not show him the plant they used for gilding: 'They do it with a certain herb, and it is such a great secret that any goldsmith in Europe, or in any other part of Christendom, would soon become a rich man from this manner of gilding ... I have seen the herb, and the Indians have taught me about it, but I was never able, by flattery or in any other way, to get the secret from them.' The active substance may well have been oxalic acid. Present-day Indians in Ecuador still clean jewellery by heating items in a copper pot with a mixture of common salt, water, and macerated plants of the Oxalis family (Zevallos Menéndez 1965–6 pp. 72–73, Rivet and Arsandaux 1946 p. 77). Under the *Quechua* name *chullco*, Oxalis juice was used for the same purpose in Peru until recent times (Teresa del Solar, personal communication). Scott (1983b p. 113) experimented in the laboratory with solutions of organic acids (citric and oxalic). Both of these were effective, but he obtained the best results with a solution of 10% oxalic acid plus 10% sodium chloride.

Depending on the composition of the original alloy, the strength of the pickling agent, and the length of the treatment, a whole spectrum of surface colours, from red to yellow, can be produced. In some cases depletion gilding was clearly carried out for 'cosmetic' purposes, to obtain a golden colour, but we should keep in mind that the removal of some copper at the surface is an inevitable consequence of hammering and annealing an object made of copper alloy; this produces an oxide scale on the surface, which must then be cleaned off with acid, resulting in the selective depletion of copper at the surface of the item (Lechtman 1971 p. 19, 1984b p. 60).

The method of depletion-gilding also allows the goldsmith to produce bi-coloured pieces (Plate 16.1). Scott (1983b) has studied the manufacture of colour-contrasting artifacts in Nariño, south Andean Colombia, and has identified the following techniques:

1. Selective depletion by masking off certain areas of the surface so that these are not attacked by the acid (the resist agents remain unknown).
2. Selective abrasion, i.e. removal of areas of the depletion gilded surface to reveal the differently coloured alloy below.
3. Selective burnishing or etching of the depletion gilded layer to produce zones of contrasting colour and/or texture.

As one moves south from Colombia into Ecuador and Peru, silver becomes a more significant component of jewellery alloys. Recorded alloys include gold–copper, gold-silver, and ternary alloys of gold, copper and silver. All of these can be treated by depletion gilding.

Lechtman (1971, 1973, 1984b, 1988) has paid particular attention to these central Andean silver alloys. Silvered surfaces were produced on objects of copper–silver alloy in the same way as the gold surfaces on *tumbaga* – by the selective depletion of superficial copper. The process of creating a golden surface on a sheet item made of a ternary alloy, with gold as a minor constituent only, was longer and more complicated. First of all, the copper oxide scale resulting from hammering and annealing was removed with a mild acid pickle. This reduction of the copper content left a silver-coloured surface layer containing a little gold. The next stage involved the removal of more silver, to leave a gold-rich, depletion-gilded surface. This effect can be achieved (and has been replicated in the laboratory) by a process akin to cementation, for instance by heating the alloy in a crucible packed with clay and salt moistened with urine, or by using an acid bath. The ancient Peruvians may well have used a solution containing one of the corrosive minerals (e.g. ferric sulphate or cupric sulphate) which are abundant in the coastal zone. Dissolved in water, with the addition of salt or alum, these minerals are capable of removing both copper and silver, to leave a gold-rich surface.

The quality and thickness of the resultant gold layer depend on the temperature and duration of the treatment. The method works well for ternary alloys high in silver, and also for *tumbagas* of gold and copper. Laboratory experiments show that alloys with as little as 20% of gold can be gilded in this way.

A very similar technique for enriching *tumbaga*, using a hot solution containing 25% of alum, 50% of sodium nitrate and 25% of common salt, is used today by the black goldsmiths of Barbacoas in the southern Pacific lowlands of Colombia (Friedemann 1974). Whether this technique is a Colonial introduction, or was taken over from the Indians who made the gold and platinum jewellery in the Tumaco/Tolita style (see pp. 187–8) has still to be established.

Cultural considerations

So far, this paper has concentrated on the techniques that native American metalsmiths employed, rather than on the reasoning behind their choices. Lechtman's studies of Moche-Vicús metalwork have demonstrated that several methods of surface-enrichment were in use at the same time, and by the same people, in coastal Peru (Lechtman 1988), and Scott (1986b) has identified foil-gilded, fusion-gilded and depletion-gilded objects among the funerary offerings in a single urn burial from La Compañía, in Ecuador. In each case, then, the goldsmith had to decide which technique of gilding to employ out of the several available to him.

Surprisingly (at least, to most metallurgists trained in the high-tech laboratories of the industrialized world), the choices made by Amerindian goldsmiths seem to have been determined by cultural attitudes and values, rather than by purely technical considerations.

In her study of West Mexican copper bells, Hosler (1988) has demonstrated that the choice of whether to use unalloyed copper, or to add either tin or arsenic in varying proportions, was made on the basis of colour, and not primarily on the basis of the hardness of the metal or the ease of casting. Lechtman (1977) and Helms (1981) note that, in the Andean world, gold and silver (the 'sweat of the sun' and the 'tears of the moon') were monopolized by the elite and the gods who supported them, and were used primarily for display, and as indicators of status, by both the living and the dead. Alloys were selected not only for their mechanical properties, but also because they gave different colours, were charged with symbolic values, or contained within themselves some essential principle.

This argument applies, above all, to the depletion gilded objects made of gold, copper and silver alloys. Depletion gilding is not a very effective way of economizing on gold, for all the gold below the surface layer, in the core of the piece, is 'wasted', and a golden appearance can be achieved more cheaply by other means. Perhaps, as Lechtman suggests, the 'essence' of the object required gold to be present throughout.

Sixteenth-century Spanish chronicles give little information on the symbolism of metals, but they do hint at a system of values very different from our own. Szaszdi Nagy (1982–3 pp. 11–21) has brought together much of the Caribbean data. Fray Bartolome de las Casas makes it clear that, in the Antilles, gold-copper alloy (locally called guanín gold) was preferred to pure gold 'for the smell which they perceive in it, or for some virtue which they believed there was in it'. Whatever the reason, the value of guanín gold among the native peoples was one hundred, or even two hundred, times that of high carat gold, and this was so even under Spanish dominion (Szaszdi 1982–3 pp. 18, 20). A depletion gilded guanín item has a yellow surface (and one that resists corrosion) but retains its coppery smell.

Rather similar attitudes towards metals were observed by Reichel-Dolmatoff (1981) among the modern Tukanoan Indians of Colombian Amazonas. In this case the sexual and cosmic symbolism is explicit. Like the Caribbean islanders, the Desana appreciate the smell of copper and *tumbaga*, comparing it to that of a certain frog with female and sexual connotations. In addition,

> The process of intrauterine embryonic development is imagined as one of 'cooking', of the embryo being transformed in a fiery furnace, or crucible. The color combination yellow/red stands, therefore, for male/female fertility and fecundity ... Ideally the sun fertilizes a brilliant New Moon which, at first Quarter, proves to be pregnant. Moon then passes through a sequence of yellowish, reddish and copper-colored phases which are compared to the

menstrual cycle and the process of embryonic development. At the same time this process is said to be a model of metallurgical combinations. (Reichel-Dolmatoff 1981 p. 21)

Attitudes and beliefs of this kind, belonging more to alchemy than to our kind of science, help to explain the passion of American metalsmiths for alloys of all kinds, and for transformations by which a metal of one composition is given the surface appearance of another. What is known to be there is as important as what can be seen.

This psychology could also account for the custom (which is common in Peru and occurs sporadically elsewhere) of covering objects of precious metal with paint and other substances. Black or red pigment was applied to sheet gold ornaments by the Chavín period (first millennium BC) and the habit continued until the Spanish conquest. In a volume dedicated to surface colouring of metals, I will end this contribution with a description, taken from Carcedo and Shimada (1985 p. 65), of Middle Sicán mummy masks from Batán Grande in the Lambayeque valley, dated between AD 800 and 1100:

> On most masks, part of the face – except around the nose, eyes and below the mouth – was painted red (with a thick coat of cinnabar) or other color(s). [One mask] still retains green and white paint on the eyes, below the nose and on the ears ... At times, the unpainted areas of masks show traces of carefully pasted feathers of varied colors that once formed colorful mosaics. Also, miniscule feathers were individually pasted onto pieces of leather, which were fastened to the lateral projections of the mask corresponding to the ears and ear ornaments.

Having demonstrated their technical virtuosity by producing gilded and silvered surfaces on objects made from various alloys, the Sicán jewellers then hid most of the metal from sight.

With these examples in mind, we must begin to ask what technology is for, not merely how it works.

Acknowledgements

I am grateful to David Scott, Juanita Sáenz, Heather Lechtman and John Merkel for their help with the preparation of this paper.

References

Berlo, J. C. (1984) *Teotihuacan Art Abroad. A Study of Metropolitan Style and Provincial Transformation in Incensario Workshops*, British Archaeological Reports International Series 199, Oxford

Bergsøe, P. (1937) The Metallurgy and Technology of Gold and Platinum among the Pre-Columbian Indians. *Ingeniørvidenskabelige Skrifter*, A44, Copenhagen

Bergsøe, P. (1938) The gilding process and the metallurgy of copper and lead among the Pre-Columbian Indians. *Ingeniørvidenskabelige Skrifter*, A46, Copenhagen

Bonhams (1992) *Tribal Art* (Sale catalogue), London, 23 June 1992

Bray, W. (1978) *The Gold of El Dorado*, Times Books and the Royal Academy, London

Bray, W. (1988) Resolving the platinum paradox. *Americas*, **40**(6), 45–49

Carcedo Muro, P., and Shimada I. (1985) Behind the Golden Mask: Sicán gold artifacts from Batán Grande, Peru. In *The Art of Precolumbian Gold: The Jan Mitchell Collection* (ed. J. Jones) London, Weidenfeld and Nicolson, pp. 61–75

Emmerich, A. (1965) *Sweat of the Sun and Tears of the Moon*, University of Washington Press, Seattle

Friedemann, N. S. de (1974) *Minería, Descendencia y Orfebrería Artesanal, Litoral Pacífico, Colombia*, Universidad Nacional, Facultad de Ciencias Humanas, Bogotá

Grieder, T. (1978) *The Art and Archaeology of Pashash*, University of Texas Press, Austin and London

Helms, M.M. (1981) Precious metals and politics: style and ideology in the Intermediate Area and Peru. *Journal of Latin American Lore*, **7**(2), 215–238

Hosler, D. (1988) The metallurgy of ancient West Mexico. In *The Beginning of the Use of Metals and Alloys* (ed. R. Maddin) Cambridge, MIT Press, pp. 328–343

Lechtman, H. (1971) Ancient methods of gilding silver: examples from the Old and New Worlds. In *Science and Archaeology* (ed. R.H. Brill) Cambridge, MIT Press, pp. 2–30

Lechtman, H. (1973) The gilding of metals in Pre-Columbian Peru. In *Application of Science in Examination of Works of Art* (ed. W. J. Young) Boston, Museum of Fine Arts, pp. 38–52

Lechtman, H. (1977) Style in technology – some early thoughts. In *Material Culture: Styles, Organization, and Dynamics of Technology* (eds H. Lechtman and R.S. Merrill) 1975 Proceedings of the American Ethnological Society, pp. 3–20

Lechtman, H. (1979) A pre-Columbian technique for electrochemical replacement plating of gold and silver on copper objects. *Journal of Metals*, **31**(12), 154–160

Lechtman, H. (1984a) Andean value systems and the development of prehistoric metallurgy. *Technology and Culture*, **25**(1), 1–36

Lechtman, H. (1984b) Pre-Columbian surface metallurgy. *Scientific American*, **250**(6), 56–63

Lechtman, H. (1988) Traditions and styles in Central Andean Metallurgy. In *The Beginning of the Use of Metals and Alloys* (ed. R. Maddin) Cambridge, MIT Press, pp. 344–378

Lechtman, H., Erlij, A. and Barry B.J.Jr. (1982) New perspectives on Moche metallurgy: techniques of gilding copper at Loma Negra, northern Peru. *American Antiquity*, **47**(1), 3–30

Patiño Castaño, D. (1988) Orfebrería prehispánica en la costa Pacífica de Colombia y Ecuador. Tumaco-La Tolita. *Boletín del Museo del Oro*, **22**, 17–31 Bogotá

Pendergast, D.M. (1982) Ancient Maya mercury. *Science*, **217**, 533–534

Reichel-Dolmatoff, G. (1981) Things of beauty replete with meaning. In *Sweat of the Sun, Tears of the Moon* (exhibition catalogue) Natural History Museum of Los Angeles County, Los Angeles, pp. 17–38

Rivet, P., and Arsandaux H. (1946) La métallurgie en Amérique précolombienne (Travaux et Mémoires de l'Institut d'Ethnologie XXXIX), Musée de l'Homme, Paris, pp. 113–115

Root, W.C. (1950) A report on the metal objects from Veraguas. In Samuel Kirkland Lothrop, *Archaeology of Southern Veraguas, Panama* Memoirs of the Peabody Museum of Archaeology and Ethnology, Harvard University, Vol. IX, No. 3, pp. 93–96

Scott, D. A. (1982) *Prehispanic Colombian metallurgy : studies of some gold and platinum alloys.* Unpublished PhD dissertation. Institute of Archaeology, University College, London

Scott, D. A. (1983a) The deterioration of gold alloys and some aspects of their conservation. *Studies in Conservation*, **28**, 194–203

Scott, D. A. (1983b) Depletion gilding and surface treatment of gold alloys from the Nariño area of ancient Colombia. *Journal of the Historical Metallurgy Society*, **17**(2), 99–115

Scott, D. A. (1986a) Gold and silver alloy coatings over copper: an examination of some artifacts from Ecuador and Colombia. *Archaeometry*, **28**(1), 33 50

Scott, D. A. (1986b) Fusion gilding and foil gilding in pre-Hispanic Colombia and Ecuador. In *Metalurgia de América Precolombina/Precolumbian American Metallurgy* (ed. C. Plazas) Bogotá: Banco de la República, pp. 283–325

Scott, D. A. and Bouchard J.F. (1988) Orfebrería prehispánica de las llanuras del Pacífico de Ecuador y Colombia. *Boletín del Museo del Oro*, **22**, 3–16, Bogotã

Scott, D. A., and Bray W. (1980) Ancient platinum technology in South America. *Platinum Metals Review*, **24**(1), 147–157

Scott, D. A., and Bray W. (in press) Prehispanic platinum alloys: their composition and utilisation in Ecuador and Colombia. In Pre-Colombian Archaeometry (ed. D.A. Scott) Getty Conservation Institute, Marina del Rey, California

Smith, G. (1720) *The Laboratory or School of Art*, London

Szaszdi Nagy, A. (1982–3) Las rutas del comercio prehispánico de metales. *Cuadernos Prehispánicos*, **10**, 5–127, Valladolid

Valdez, F. (1987) *Proyecto Arqueólogico La Tolita (1983–1986)*, Quito, Editorial Luz de América, Museo del Banco Central del Ecuador

Zevallos Menéndez, C. (1965–6) Estudio regional de la orfebrería precolombina de Ecuador y su posible relación con las áreas vecinas. *Revista del Museo Nacional*, **34**, 68–81, Lima

17

A study of the gilding of Chinese Buddhist bronzes

Paul Jett

Abstract

Seven gilt Chinese Buddhist bronzes, dating from the early 6th to early 10th centuries AD, were examined to determine the method of gilding employed. Chemical analysis and metallographic examination of samples taken from the sculptures demonstrate the use of fire-gilding (amalgam gilding) on a leaded tin bronze substrate. A test of mercury fire-gilding on a modern bronze sample is also described.

Introduction

The earliest gilt bronze Buddhist image from China for which an historical account is available is thought to date to AD 190 and was probably made in Kiangsu Province (Soper 1959 p. 4). A passage from the history of the Three Kingdoms describes the patron's dedication of the image: 'He erected a buddha shrine, making a human figure of bronze whose body he coated with gold and clad in brocades' (Soper 1959 p. 4). Thus, from nearly the introduction of Buddhism to China until the late dynasties, gilded bronze Buddhist images were widespread, although the artistic highpoint for such works is considered to have occurred during the latter half of the first millennium AD (Mizuno 1960 p. 7). In spite of the great number of such bronzes, there is little technical information published about these images. Only isolated analyses of Chinese Buddhist bronze figures are found (Uhlig 1979 pp. 67, 220, Beguin and Liszak-Hours 1982 pp. 30–34, 58–60). Also, while there are some articles in Chinese describing so-called traditional Chinese gilding methods (Wu 1981), only one has been found that discusses the technical study of particular gilded objects (Wang 1984). The art historical literature on Buddhist bronzes makes different claims about the nature of the gilding techniques used. Some authors state that fire-gilding, was used (Soper 1959 p. 254), while others state that gold leaf was applied using mercury (Munsterberg 1967 pp. 20–21, Deydier 1980 p. 141).

In order to better understand the nature of the copper alloys and gilding techniques used during the later periods in China, a technical study was begun of the later Chinese bronzes in the collection of the Freer Gallery of Art. One hundred and six works of sculpture dating from the 5th through the 19th century were chosen for study. About 80% of these date to the Tang Dynasty (AD 618–907) or earlier. Also, almost 80% of the total group of objects are gilded. The following describes the technical study of the nature and methods used in the gilding of some of these bronzes.

Examination

Visual examination of a number of the Freer Gallery bronzes revealed attributes commonly noted as indicating fire-gilding. The presence of a pasty, granular layer of gold was often found lying in the recesses of engraved lines. The gilding was also often found on the undersides of the feet of the bases, a feature one would expect with fire-gilding due to its tendency to flow as it becomes more fluid during the heating which drives off the mercury. Visual evidence of leaf gilding, such as overlapping borders of leaves, or loose, ragged edges of leaves, was not found.

While the microscopic features of fire-gilding are often distinctive, these features may be unclear due to wear, corrosion, later surface treatments that alter the gilding surface, or the original application of gilding. Surface analysis which indicates the presence of mercury in combination with good visual evidence is a strong argument for the use of fire-gilding. But in the absence of visual evidence, surface analysis for mercury would not differentiate between the use of leaf or fire-gilding. Leaf gilding may be done employing mercury (Vittori 1978), or the presence of mercury might be due to reasons other than the gilding method used. Thus it was decided to remove metallographic samples so that the method of gilding could be amply studied and determined. Seven works of sculpture, dating between the early 6th and late 10th centuries AD, were sampled. The criteria for choosing those bronzes to be sampled were their apparent authenticity, the presence of inscriptions, and the presence of broken edges which allowed for sampling with a minimum of disfigurement to the objects. The samples were studied using light microscopy, scanning electron microscopy, and electron microprobe analysis. Three of the samples are illustrated here and are representative of the general features noted in the gilding of all seven samples.

The earliest work of sculpture sampled, dated by its inscription to AD 516, is a representation of two Buddhas, Sakyamuni and Prabhutaratna (Figure 17.1). Microscopic examination of the surface shows features that are typical of fire-gilding, such as a grainy texture in the recesses of the engraved lines (Figure 17.2). The metallographic sample shows a typical cast bronze structure with the

Figure 17.1. *Gilded Buddhist bronze, Northern Wei dynasty, Freer Gallery of Art accession no. 11.130.*

Figure 17.2. *Recess of engraved line on bronze no. 11.130.*

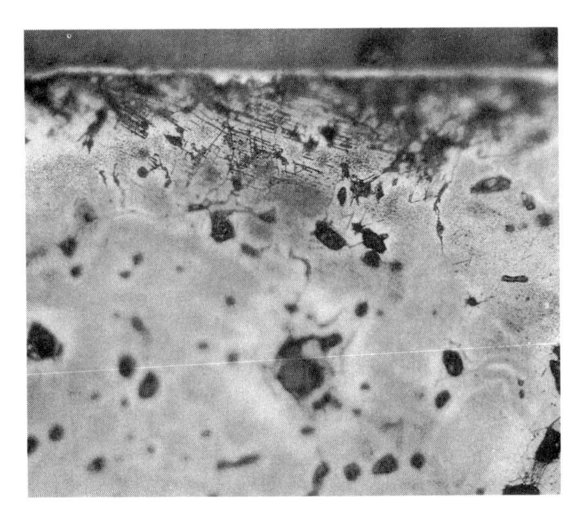

Figure 17.3. *Metallographic sample from gilt bronze no. 11.130.*

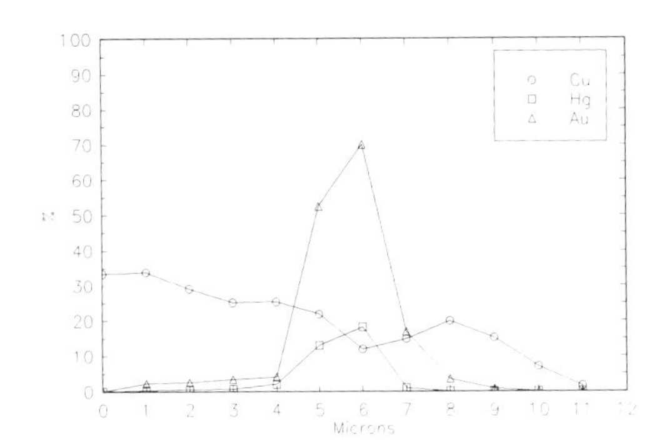

Figure 17.4. *Electron microprobe scan of sample from gilt bronze no. 11.130.*

lead lying in small globules throughout the bronze (Figure 17.3). A microprobe scan for gold, mercury, and copper from the interior of the bronze through the gilding layer reveals that mercury is present in the gold layer and is roughly consistent in its relative amount to the gold present (Figure 17.4). The rise in the copper content at the outer surface is due to the presence of corrosion products.

The second example is a Tang dynasty sarcophagus (Figure 17.5) composed of many separate pieces. Analyses of one of the figures shows the composition to be 74.1% copper, 14.2% tin and 8.1% lead. A scanning electron

Figure 17.5. *Gilt bronze sarcophagus, Tang dynasty, Freer Gallery of Art accession no. 15.106.*

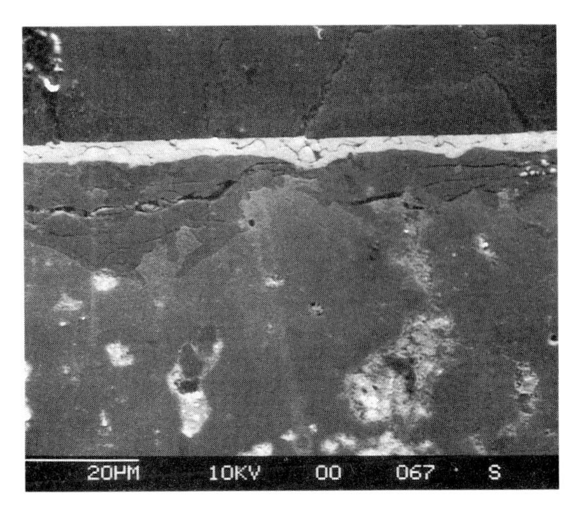

Figure 17.6. *SEM image of metallographic sample from bronze no. 15.106 showing granular gilding layer.*

Figure 17.7. *Gilt bronze guardian figure, Tang dynasty, Freer Gallery of Art accession no. 16.249.*

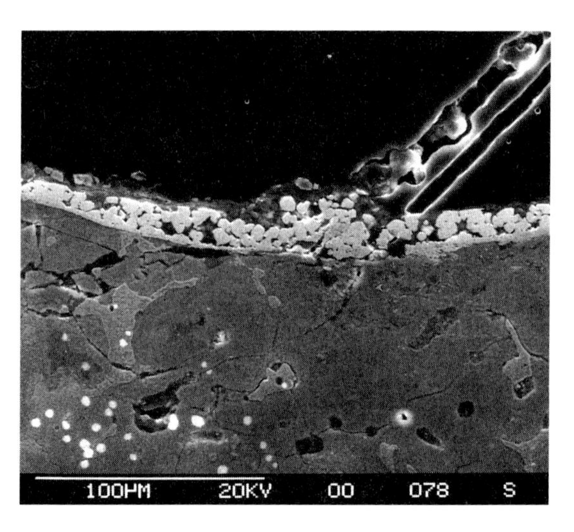

Figure 17.8. *SEM image of metallographic sample and gilding layer from bronze no. 16.249.*

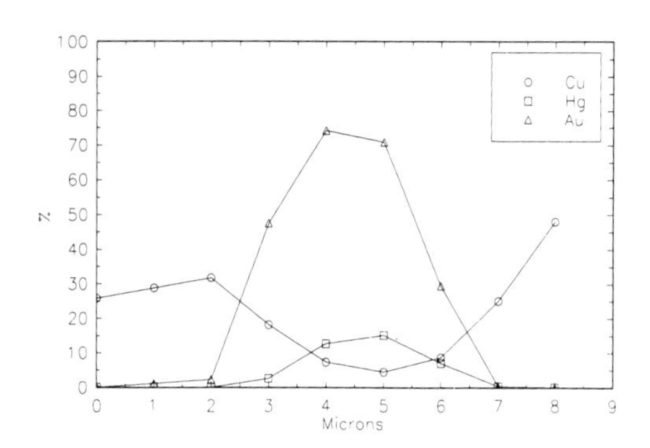

Figure 17.9. *Electron microprobe scan through gilding layer of metallographic sample from bronze no. 16.249.*

microscope image of a sample removed from one of the figures reveals individual grains in the gilding layer which reflect the physical structure of gold–mercury amalgam, and which did not entirely lose their form in the course of heating and burnishing (Figure 17.6). The microprobe scan shows the same features noted for the previous sample.

The third illustrated example is the metallographic section taken from a small Tang dynasty guardian figure which stands about 11 cm in height (Figure 17.7). The composition of this figure is 74.9% copper, 11.1% tin, and 13.2% lead. Here again the granular structure of the gilding is apparent in recesses on the surface and in cross-section; this feature is best illustrated using scanning electron microscopy (Figure 17.8). The microprobe scan across the surface shows the same pattern found in the other samples (Figure 17.9).

As noted earlier, the four other samples show many of the same features. Many of the cross-sections reveal areas where the gilding flowed into cavities in the surface (Figure 17.10). One sample, taken from a figure which

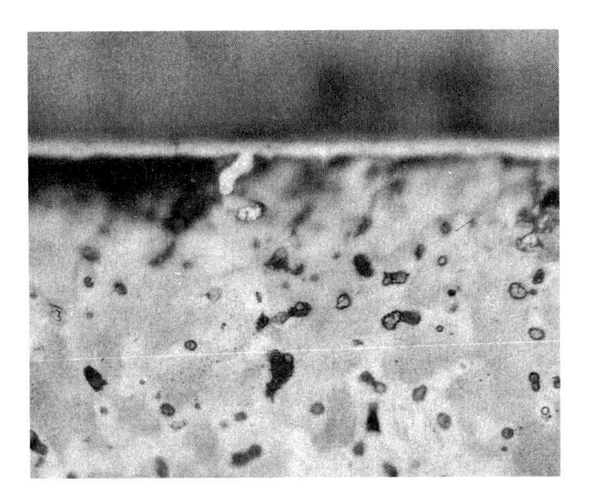

Figure 17.10. *Metallographic sample from gilt bronze, Freer Gallery of Art accession no. 11.132.*

dates to AD 561, displayed very large lead globules within the bronze matrix indicating a relatively high lead content. The thin gilding layer on this sample also has the granular structure one often sees in cross-sections of fire-gilded objects. In addition, the electron microprobe scans for these bronzes also showed a consistent relationship between the distribution and relative amounts of gold and mercury in the gilding layers very similar to that shown in Figure 17.4.

As seen in the photomicrographs and the scanning electron microscope images, the gilding often has a granular structure and is irregular in thickness; both features indicate the use of fire-gilding. The microprobe scans also suggest the use of fire-gilding as opposed to leaf gilding with mercury because the mercury does not appear to be concentrated at the interface between the gilding and the substrate. If mercury had first been applied to the substrate and then leaf applied, it seems that one would see a higher concentration of mercury at the interface. A study by Lechtman (1971) demonstrated that in the fire-gilding of silver, an increase in the mercury content occurs at the interface due to the diffusion of the mercury into the silver. This does not appear to occur with fire-gilded copper alloys however, and it has been suggested that this is because of the insolubility of mercury in copper as opposed to the relatively high solubility of mercury in silver (Wu 1981). It also was shown in the photomicrographs that the bronze substrates appear simply to be cast.

It thus seems quite clear that the Chinese were successfully applying mercury amalgam gilding to early Buddhist bronzes which contain relatively large amounts of tin and lead. This fact appears to stand in contrast to what we know about the use of mercury amalgam gilding during other periods and in other cultures. The use of mercury amalgam gilding by the Chinese dates back to at least the Warring States period (480–221 BC), and one author suggests a slightly earlier date, just prior to the Warring States period, for its introduction (Wang 1984). Previous studies have confirmed the presence of fire-gilding on various copper alloys from China during the first centuries prior to the modern era (Lins and Oddy 1979). In Chase's study of early Chinese belt hooks, fire-gilding was found on copper alloy substrates with low levels of tin and lead, while those belt hooks composed of leaded tin bronze were not fire-gilded (Oddy 1985). This finding agrees with studies of gilded copper alloys from the West. Craddock (1977), in his study of Greek copper alloys, found that alloys low in tin and lead were used for fire-gilding. Oddy's study (1981) of Roman gilded copper alloys demonstrates the same pattern. When discussing the reason given for the preference for low tin and lead contents in the alloys to be mercury amalgam gilded, authors often cite Theophilus (1979 pp. 145–146) and his statement about white spots forming during the gilding of brass if the substrate is not free of lead or well alloyed with the calamine. How fire-gilding can be used for leaded tin bronze, given the apparent problems it entails, became the next focus of this study.

Experimentation

An attempt was made to gild a modern bronze sample with a composition roughly similar to that of the two Tang dynasty bronzes just described, and composed of 76.1% copper, 13.1% tin, and 9.2% lead. A gold-mercury amalgam was prepared using five parts liquid mercury and one part powdered gold. The gilding procedure used was basically the same as one described as a traditional method by some Chinese authors (Wang 1984, Wu 1981). This method is as follows. First, the bronze surface is cleaned and then washed with an acid solution. The amalgam is then applied to the surface using a 'gold rod' or 'gold stick' (Wang 1984, Wu 1981), which is a copper burnisher, apparently similar to that described by Theophilus (1979 p. 113). The bronze is then heated in an oven or over charcoal. During the heating process the surface is rubbed with the copper burnisher; this procedure, as experimentation later showed, serves to smooth the gilding as the mercury is volatilized. Following heating, the surface is cleaned with a solution made from honey locust pods and then burnished. The process is repeated, as many as seven times, until the gilding layer is judged acceptable.

This process was followed in my experiments with the exception that a torch was used to heat the samples rather than using an oven or a charcoal fire. Using a torch probably caused the surface to heat much more quickly than in an oven, and the rapid volatilization of the mercury that resulted may have caused the gilding layer to be rougher than would occur during a slow heating process. Three applications of the amalgam followed by heating were carried out. During the first heating, the surface was burnished with the copper burnisher, and after the volatilization of the mercury, the surface was silver in appearance and slightly rough. The second application and heating left the surface a mottled silver and gold colour; during this heating, the surface was not rubbed with the copper burnisher and the resulting gilding layer was very rough. The last application of amalgam was, after heating, purely golden in colour, though after burnishing it had an undulating surface, not the smooth, flat surface that one would want on a gilded object. It was then found that the surface could be smoothed with fine emery paper and easily reburnished to give a relatively flat, coherent gilding layer with a good appearance to the unaided eye. Final burnishing of the gilding layer was not found to be difficult and the gilding was not found to be 'brittle' as has been observed in experiments with the cold mercury leaf gilding of copper (Vittori 1978 pp. 76–79). That the handling properties of leaf and amalgam are different is not surprising given the fact, among others, that their physical structures are not in any way analogous.

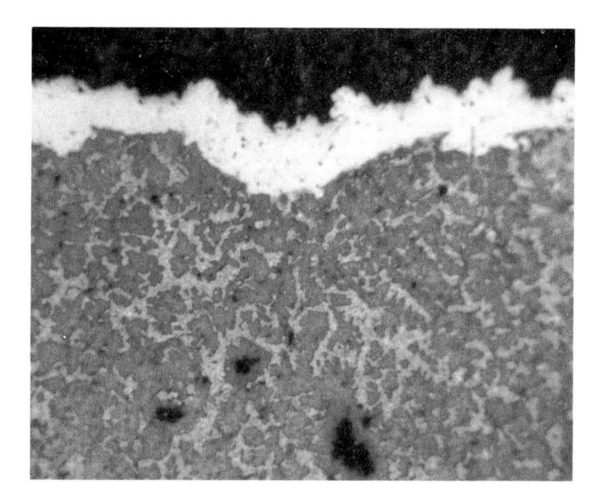

Figure 17.11. *Metallographic sample from prepared sample of fire-gilded leaded tin bronze.*

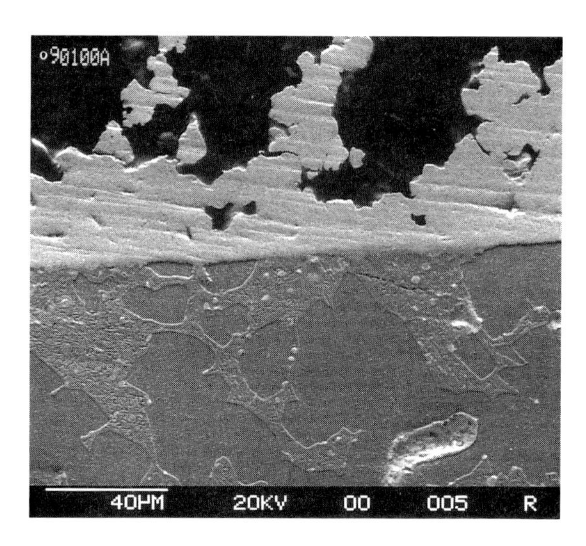

Figure 17.12. *SEM image of metallographic sample shown in Figure 17.11.*

The surface appears rather coarse in cross-sections of the prepared gilded bronze (Figure 17.11). I believe this is more due to my

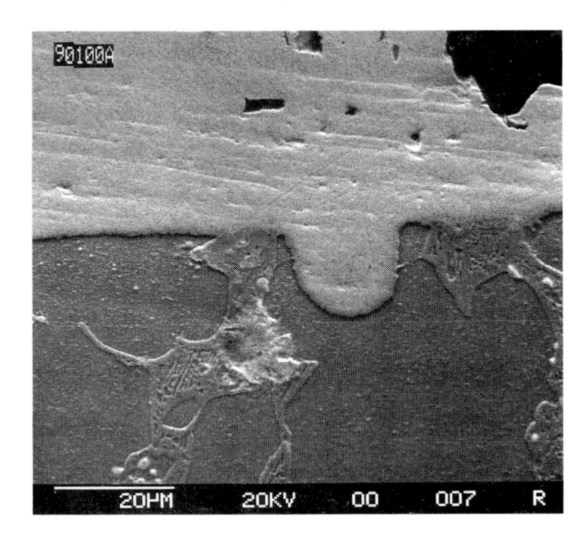

Figure 17.13. *SEM image of metallographic sample shown in Figure 17.11.*

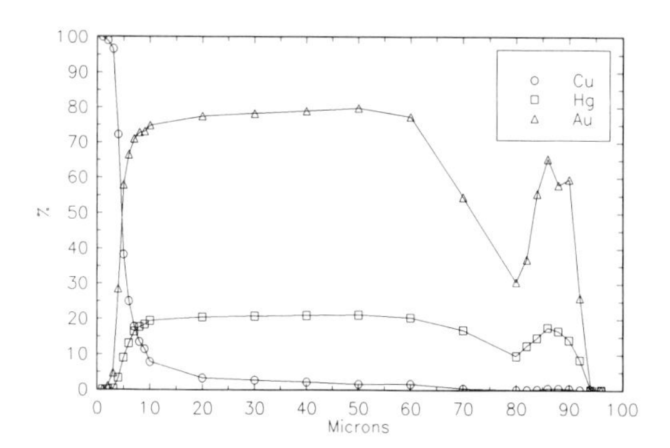

Figure 17.14. *Electron microprobe trace through gilding layer of metallographic sample shown in Figure 17.11.*

inexperience with the process rather than a fault in the basic technique. Other aspects of the cross-section from the experimental gilded bronze resemble those found in the samples from the Buddhist bronzes. The gilding layer still bears some remnants of a granular structure (Figure 17.12). Under higher magnification, one sees a rather sharp division between the gilding and the copper rich areas of the surface and what appears to be some interdiffusion between the gilding and the tin rich phase (Figure 17.13). A microprobe trace across the gilding layer of the prepared sample, moving out from the copper rich phase, shows the same pattern as seen in the traces from the Buddhist bronzes (Figure 17.14). The thickness of the gilding on the prepared sample, however, is between 40 to 95 microns, much greater than the thickest gilding layer found on the Buddhist bronzes which measured about 20 microns. Most of the gilding layers studied measured between 5 and 10 microns. Achieving thinner gilding layers may be the product of experience rather than a different technical approach. It is also interesting to note the consistent and seemingly high level of mercury in the gilding layers from the prepared sample and those samples from objects. The maximum amount of mercury found is roughly in the area of the maximum solid solubility of mercury in gold, around 19.5% (Okamoto and Massalski 1989 p. 53).

The fire-gilding of leaded tin bronze is thus quite feasible, but to achieve an acceptable surface, three things seem necessary. First, the application and heating of the amalgam must be repeated a number of times. Second, the surface must be burnished during the heating process to achieve the smoothest possible surface. Third, polishing prior to burnishing to remove irregularities in the gilding layer due to the volatilization of mercury must also be done to smooth the surface.

Conclusion

Theophilus (1979 p. 145) warns in his instructions for gilding brass that: 'The amalgamation of brass must be done more scrupulously and carefully and it must be gilded more thickly and washed more often and dried for a longer time'. His advice seems equally relevant to working with leaded tin bronze; it is, for example, an altogether more laborious task than the fire-gilding of silver substrates. It spite of this, by the early centuries of the first millennium AD, Chinese craftsmen chose to use leaded bronze alloys as substrates for fire-gilding.

Acknowledgments

Scanning electron microscopy was performed by Melanie Feather and Walter Brown of the Smithsonian Institution. Many thanks are due to Eugene Jarosewich and Joe Nelen of the Smithsonian Institution for their collaboration in performing the electron microprobe analyses. The compositions of bronzes nos. 15.106 and 16.249 were determined using gravimetric analyses performed by I.V. Bene. Part of this study was conducted under the auspices of a collaborative research effort by the Smithsonian Institution and the Institute for Cultural Affairs, Japan.

References

Beguin, G. and Liszak-Hours, J. (1982) Objets himalayens en metal du musée Guimet: Etude en laboratoire. *Annales du Laboratoire de Recherche des Musées de France.* Reunion des Musees de France, Paris, pp. 28–82

Craddock, P.T. (1977) The composition of the copper alloys used by the Greek, Etruscan, and Roman civilizations. 2. The Archaic, Classical, and Hellenistic Greeks. *Journal of Archaeological Science*, **4**(2), 103–123

Deydier, C. (1980) *Chinese Bronzes*, Rizzoli Publishing Co., New York

Lechtman, H.N. (1971) Ancient methods of gilding silver: examples from the Old and the New Worlds. In *Science and Archaeology* (ed. R.H. Brill) MIT press, Cambridge, Massachusetts, pp. 2–32

Lins, P.A., and Oddy, W.A. (1979) The origins of mercury gilding. *Journal of Archaeological Science*, **12**(1), 35–39

Mizuno S. (1960) *Bronze and Stone Sculpture of China*, Nihon Keizai Shimbun Sha, Tokyo

Munsterberg, H. (1967) *Chinese Buddhist Bronzes*, C.E. Tuttle Co., Rutland, Vermont

Oddy, W.A. (1981) Gilding through the ages. *Gold Bulletin*, **14**(2), 75–79

Oddy, W.A. (1985) Vergoldungen auf prehistorischen und klassischen Bronzen. In *Archaologische Bronzen, Antike Kunst, Moderne Technik* (ed. H. Born) Dietrich Reimer Verlag, Berlin, pp. 64–70

Okamoto, H. and Massalski, T.B. (1989) The Au-Hg (gold-mercury) system. *Bulletin of Alloy Phase Diagrams*, **10**(1), 50–58, American Society for Metals, Ohio

Soper, A. (1959) *Literary Evidence for Early Buddhist Art in China*, Artibus Asiae, Ascona, Switzerland

Theophilus (1979) *On Divers Arts* (translated by John G. Hawthorne and Cyril Stanley Smith) Dover Publishing Co., New York

Uhlig, H. (1979) *Das bild des Buddha*, Safari Verlag, Berlin

Vittori, O. (1978) Interpreting Pliny's gilding archaeological implications. *Revista di Archeologia*, **2**, 71–81

Wang Haiwen (1984) A study of gold plating technology. *Gugong Bowuyuan Yuankan, (Palace Museum Journal)*, **2**, 50–58

Wu Kunyi (1981) Chemical distribution of mercury in ancient Chinese gold gilt. *Zhongguo keji Shiliao, (China Historical Materials of Science and Technology)* **1**, 90–94.

18

Silvering

Susan La Niece

Abstract

Cladding base metal objects with a thin layer of silver originates almost as far back as the earliest use of silver. At first, silver foil was attached by simple mechanical methods and with non-metallic adhesives. The use of silver–copper solders for silvering, at least as early as the 5th century BC, improved the permanence of foil silvering and the method continued to be used for small, flat items. Soft solders (tin and/or lead) were commonly used by the Romans to attach silver foil to more complex shapes such as cups and figurines, by a method which centuries later became known as close-plating. Self soldering, known in the 18th and 19th centuries as Sheffield plating, is said to occur centuries earlier but conclusive evidence of this is lacking. Small items were plated with silver alloys applied molten at least as early as the 5th century BC but the use of mercury–silver amalgam does not seem to have become an established silvering technique until the 13th century AD in Europe, in spite of the widespread use of mercury for gilding many centuries earlier. In China, however, mercury silvering was known at least as early as the 1st century AD. Examples of mercury silvering start occurring in substantial numbers from the 13th century AD in Europe, nowhere near as early as mercury gilding. It is not known when silvering pastes and solutions were discovered, but in Europe they are documented from the 16th century. All these methods were used at some time by contemporary coin forgers, but since the 1840s electroplating has superseded other silvering methods for most purposes.

Introduction

The history of silvering has been much less studied than that of gilding. This is partly because silvering is not as common in the archaeological record, probably because of its poorer corrosion resistance rather than its original popularity. It is difficult to distinguish between silvering and tinning from superficial appearances; many of the items labelled as silvered in museums and catalogues are, in fact, tinned.

Plating with any precious metal, whether for decorative purposes or for fraud, is a means of maximizing display for minimum cost. This chapter reviews the major developments in silvering from its origins, up to the introduction of Sheffield plate in the mid-18th century.

A technical study of the silvering used on Roman coin forgeries can be found in Chapter 20. The Sheffield plating industry is discussed fully in Chapter 19, and Chapters 23 and 24 are devoted to the history of electroplating and its modern applications.

Foil silvering

Mechanical attachment of foil

The earliest methods of applying a layer of silver were purely mechanical. From at least as early as the 3rd millennium BC, silver foil was attached by hammering into keying grooves, crimping the edges or securing them with pins. These methods do not provide a very durable join and none of these mechanical methods of attaching silver foil to metal can produce a convincing illusion of solid silver, although they continued in use, especially for inlay work and for silvering items such as furniture and doors.

Adhesives in silvering

Adhesives rarely survive from antiquity, but an example of silver foil attached with a adhesive has been tentatively identified on an Illyrian helmet dated to the later 6th century BC (Hockey *et al.* 1992). The bronze helmet has badly corroded repoussé silver decoration on each side of the brow and on the cheek-pieces (Figure 18.1). A layer of calcite ($CaCO_3$) appears to be the only means of attachment or support for the silver repoussé decoration. The calcite is still firmly fixed to the bronze and retains the shape of the decoration. Gesso of powdered calcite bonded with an organic adhesive is well known as a ground for gilding wood or stone with leaf, and it has now been identified as the adhesive used to gild Egyptian bronzes (see Chapter 15), so it need not be too suprising to find it used for silvering bronze. There is no firm evidence for an organic adhesive in the calcite on the helmet, but an alternative explanation is that the calcite itself, in the form of a lime plaster, was the adhesive. Lime plaster was well known in antiquity (Kingery *et al.* 1988), although it is not normally considered to be a metalworkers' adhesive. However, experiments have proved that it can firmly stick a metal foil to a smooth sheet of brass or copper (Hockey *et al.* 1992). Other examples of Greek helmets with applied repoussé silver decorative motifs are known (Kunz 1967) and examination of these may provide more information about the adhesives used by the Greeks to attach small pieces of silver foil to bronze.

Figure 18.1. *A bronze Illyrian helmet with applied silver foil decoration of horsemen, rearing animals, palmettes and volutes on brow and cheek pieces (British Museum GR 1914.4–8.1).*

Soldered foil

Solders are far more commonly used than adhesives to join metals. A solder is a metal, or more usually an alloy, with a lower melting point than the metal(s) which it joins together. The first evidence for the use of silver–copper solder for attaching silver foil is on early coin forgeries. There are plated coins going back to at least the fifth century BC imitating Greek types (Campbell 1933). They typically consist of a copper or copper alloy disc which was encased with silver foil and bonded to the disc by silver–copper solder. The silvered disc was then struck with a coin die and was passed off as a solid silver coin (see Figures 20.9, 20.11, 20.16, etc. of silver plated Roman coins). The silver-copper solder alloy is close to the eutectic composition (72% silver : 28% copper). This is the composition of the alloy with the lowest melting point (780°C), well below that of the silver foil (about 950°C) or the copper core (1083°C for pure copper). Examples of the microstructure produced by experiments in soldering silver foil to copper in this way are seen in Figures 18.2(a) and (b). As can be seen from these plates, the length of time and temperature of heating will affect the microstructure seen.

The use of soldered foil plating was not confined to contemporary coin forgeries.

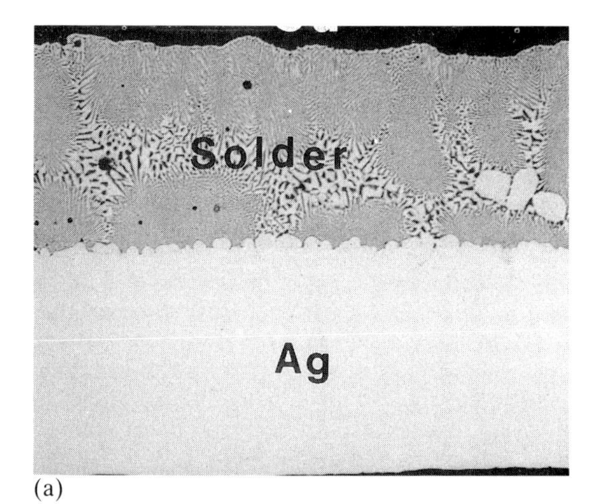

(a)

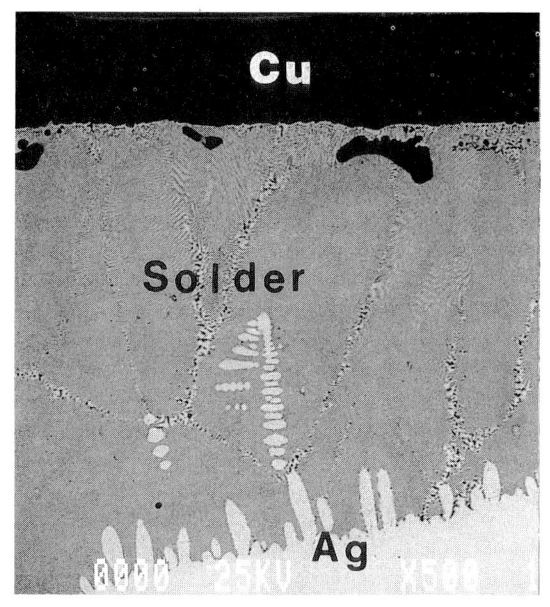

(b)

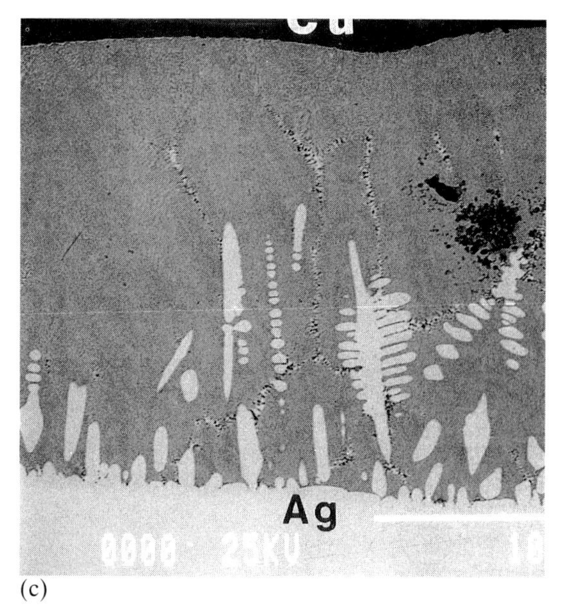

(c)

Figure 18.2 *(a) Cross-section of silver foil (approx. 0.75 mm thick) soldered to copper sheet with a silver–copper alloy solder. (b) As Figure 18.2(a) but heated for a further 5 minutes. Note the white, silver-rich, dendritic structures in the solder region. (c) Cross-section of experimental 'Sheffield' plating of silver onto copper. No solder was used, but note the similarity to Figure 18.2(b).*

Figure 18.3. *11th century* AD *brass pendant with chased decoration and fire-gilding on the silvered upper surface (approx. 7 cm long). British Museum ML 1921,1–1,189.*

Examples of soldered foil have been found on jewellery from the Viking period in Gotland. Brooches and pendants like the one illustrated in Figure 18.3 were made in the 9th and 10th centuries AD. These were generally cast in brass, with 4–10% of zinc and little or no tin. Silver foil, about 0.1 mm thick, was soldered on with an alloy which is approximately 5 parts silver, 3 parts copper and 2 parts tin, with a small amount of lead in some cases, but no zinc. It is, of course, impossible to be certain of the exact composition of the original solder alloy because this will be altered as a result of diffusion from the metals being soldered, depending on the length of time and temperature of the soldering process and the elements present (Lang and Hughes 1984).

Figure 18.4. *Roman silvered bronze goat (ht. 6 cm). British Museum PRB 1856,7–1,24. The darker areas are where the silver foil has peeled off, revealing the solder layer.*

Tin or tin–lead solders were commonly used to attach silver foil to bronze or brass items in antiquity. These 'soft' solders melt at temperatures as low as 183°C, much lower than silver solders. In the Roman period, soft solders were used to attach foil to small, decorative items silvered all over in imitation of solid silver (Figure 18.4), and for tableware and military parade items such as horse-trappings, but it is never found on coin forgeries. The silver foil was bonded by a continuous layer of solder, not just patches at key points. This continuous layer was obtained by coating the object with molten tin or tin-lead solder and allowing it to cool. Silver foil was then worked to fit the object closely, including over any details of decoration. With the foil in close contact, the object was heated gently until the solder flowed. To plate all surfaces, more than one piece of silver foil was needed. The joins were usually positioned at an unobtrusive place or sharp change of angle such as a rim or footring where the edges of the foil were butted together or overlap slightly. This method produces a good imitation of solid silver, but the solder is not strong enough to withstand working after the silver foil is applied, therefore it was unsuitable for the manufacture of plated coin blanks which have to be struck with a die.

This type of plating is easily recognized because the silver foil frequently has peeled away, to reveal an even, dark grey, metallic coating of solder underneath. Where the foil has been lost altogether, the object will appear to be tinned. Examples of soft-soldered foil silvering are very common from the Roman period, and are found throughout the medieval period. The method was eventually patented in 1779 and named 'Close Plating' by Richard Ellis, a London goldsmith (see pp. 211–12).

'Sheffield' plating

Silver can be bonded to copper by simply heating the two metals in close contact with each other with no intermediate joining material or solder. This method formed the basis of the Sheffield plate industry in the mid-18th century (see Chapter 19). When silver and copper are heated in close contact with each other, limited diffusion between them will form a low melting-point alloy at the interface. This will melt at a temperature as low as 780°C, well below the melting point of either of the two pure metals. The typical microstructure is shown in Figure 18.2(c). The resulting bimetal block can be worked and shaped as if it was solid silver, a property which made it far more verasatile and durable than silver-plating attached with soft solder or adhesives.

The earliest published example of this type of silvering is on a rivet from a mid-2nd millennium BC Minoan dagger (Charles 1968). The rivets were found to be made of high purity copper with a cap of silver about 0.25 mm thick fused onto the top. The silver–copper eutectic microstructure of the join is typical of Sheffield plate. A number of contemporary coin forgeries from the Greek city states and the Roman Republican period are also said to be silvered with foil over a copper core by this method (Campbell 1933 and Zwicker *et al.* 1968). However, the eutectic microstucture expected from 'Sheffield' plating can also be produced by a silver–copper solder. It can be seen by comparing Figures 18.2(b) and 18.2(c) that there is a very real problem in distinguishing between the two joining methods.

It has been suggested that the band of eutectic will be more regular if the joint is soldered (Campbell 1933 p. 144), though this can depend on how evenly the flux and the heat were applied. Where there is an element present in the bonding layer which is not present in either of the two metals being joined, it can be proved that the join was soldered, not made by the Sheffield plating method. This was the case in the example of the Gotlandic jewellery discussed above, where zinc was present in the alloy of the object but not in the bonding layer, and tin was found in significant quantities in the bonding layer but none was detected in the brass or its silver plating. Unfortunately the components of the earliest silver solders are silver and copper, the very metals that they join together.

The conclusion, from this study of many plated coins and other items, is that it is not possible to prove the use of the 'Sheffield' plating technique in antiquity, but that it cannot be ruled out.

French plating

An intermediate stage between mechanical attachment of foil and methods which relied on the action of heat discussed above, was a silvering method which became known as French plating in the 18th century (see p. 211). It was in use considerably earlier and was described by Theophilus as a method for silvering iron spurs (Hawthorn and Smith 1979 pp. 184–185). The object is scored with fine grooves to provide keying, then the silver foil is applied by rubbing and heating, until it is pushed well into the grooves. This method of plating may be recognized by the keying, which frequently retains traces of silver even when it is lost from the rest of the surface (Figure 18.5). The heating is not enough to melt the metal and, because of this, French plating was used in the Sheffield plating industry to repair small bubbles and blemishes in the silver without damaging the rest of the plated block by overheating.

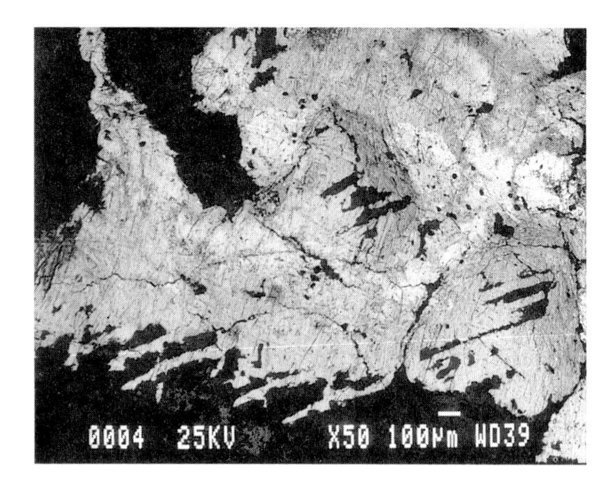

Figure 18.5. *Magnified detail of French plating on iron, from a late medieval rod (Devizes Museum 1988.216.3). The silver (white in photograph) is preserved in the diagonal lines of the keying.*

Silver coatings applied molten

A coating of molten metal is ideally suited to plating small objects with an uneven shape. Gilding by this method is sometimes known as fusion gilding. It has been suggested that objects with this type of silver coating were plated by dipping into a bath of molten silver or silver–copper alloy (Cope 1967). Amongst the practical objections which have been made to this explanation are the difficulty of keeping the temperature of the bath high enough to keep the silver molten when the item is immersed, but low enough to prevent the copper core from melting. There is little economic sense in having a bath of silver which will become progressively more contaminated by the copper dissolved from items dipped into it. An alternative method of applying a molten metal skin is to coat the copper core with a flux, apply powdered silver–copper alloy, and heat until it flows; in effect plating the object with hard solder. The microstructure of this type of plating is of a cast silver–copper alloy (Figure 18.6). Where the heating of foil plating continued for so long that the silver foil melted and dissolved some of the copper it was plating, a similar microstructure might also be found (see pp. 228–29).

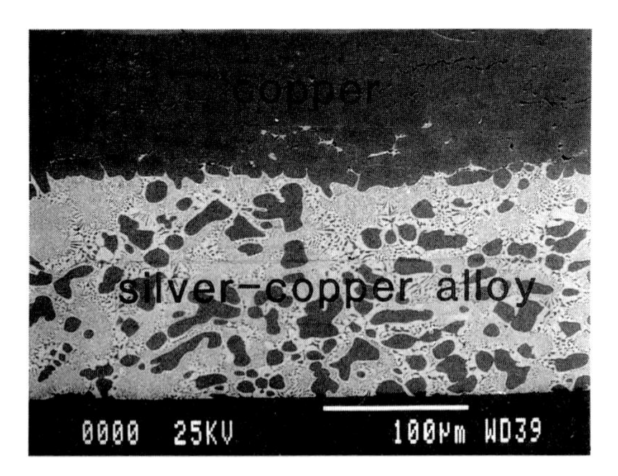

Figure 18.6. *Cross-section showing the microstructure of a silver plating layer (approx. 0.13 mm thick) which was applied molten.*

Plating with molten silver has been found in the New World (p. 185), but it is better documented on small items such as plated forgeries of Roman (see Chapter 20) and Celtic coin types (La Niece 1992). Wires forming a Celtic torc from Bawsey, Norfolk, have also been identified as being plated with a molten silver–copper alloy (Northover and Salter 1990). An early example of molten silver plating is on a coin forgery of a type issued in mid-5th century BC Macedonia (Campbell 1933 pp. 87–93).

Without examination of a section through the silver surface, or specific gravity testing, objects plated by this method could be mistaken for cast base-silver throughout, and it is possible that there are many unrecognized examples of plating with molten silver. This plating is always of base silver for the practical reason that it melts at a lower temperature than pure silver. The colour of the silver alloy is less white than pure silver because of the copper content, so it is not surprising that several items plated with molten silver alloy have been found to be also depletion silvered, to improve the surface appearance (Scott 1986, Northover and Salter 1990).

Depletion silvering

Depletion silvering is not strictly a plating technique, although the end result of a silver surface on a baser alloy is much the same. In the case of depletion silvering, the baser alloy must contain some silver. The method involves the removal of copper from the surface, either chemically, or by heating to oxidize the copper, followed by its chemical removal. The silver-rich components of the alloy will be left on the surface but it will be matt and require burnishing to give a bright finish.

The technique was used by the pre-Hispanic cultures of the New World, largely for 'gilding' base gold alloys but also for 'silvering' base silver (p. 189). It could be exploited for larger-scale production than would be possible with many of the other silvering techniques. The clearest example of the technique outside the New World is in the Roman coinage of AD 63–*c*.260. Cope (1972) has suggested that the blanks, with as little as 12–18% silver in the alloy were successfully treated to produce a bright silvery surface. It is thought that they were pickled in citrous fruit acids or vinegar to remove the copper oxides before the final striking, which consolidates the surface as an almost continuous thin skin of silver.

The typical microstructure of depletion silvering is of a base silver–copper alloy with a thin surface layer composed of silver-rich lenses, which will appear smeared together if the object has been burnished, or struck with dies as in the case of the coins (Figure 18.7). Where no such consolidated skin of silver is found, it is unlikely that the depletion was deliberate rather than an accidental result of corrosion during burial.

Mercury silvering

The discovery that mercury could be used to plate a thin layer of silver onto copper or a copper alloy was a major advance in plating technology. The method could be used to plate complex shapes without the joins that go with all the methods of plating with foil. It could be used to plate large objects, or to silver parts of a decorative design. The thickness of the plating depends on the number of layers applied, so the method could be very economical in the use of silver. An additional advantage of mercury silvering was that the plating could penetrate into the recesses of a decora-

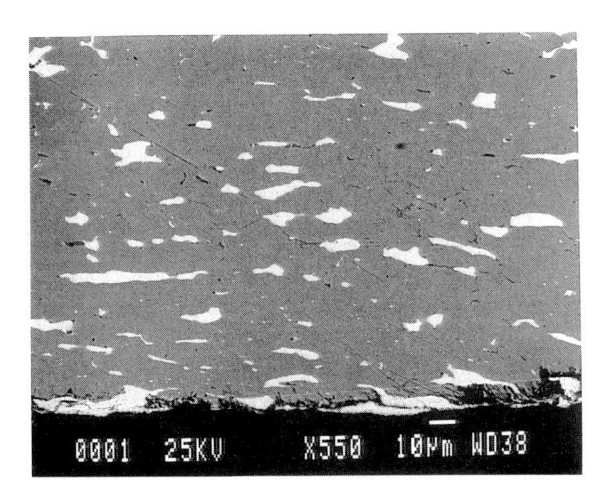

Figure 18.7. *Cross-section of a debased antoninianus of Gallienus. The white lenses are silver-rich. Although the overall silver content of the coin is only about 8%, the concentration of the lenses at the surface (running parallel to the bottom of the photograph) give the coin a surface silver composition of about 85%. (White scale bar = 0.01 mm).*

tive design, preserving the sharpness of detail, so the method was readily adopted by, amongst others, coin forgers.

The method is the same as that used for fire-gilding. Silver forms an amalgam with mercury. This amalgam is applied to a well-prepared copper or copper alloy surface and is heated gently to drive off the mercury. The surface is finished by burnishing. Like fire-gilding, the method can be positively identified by analysis because the mercury is never completely driven off by the heating process. Enough is usually left to be detected non-destructively by surface analysis by X-ray fluorescence spectrometry, and it can always be detected by more sensitive analytical techniques such as emission spectrography. Nevertheless, the extent to which mercury silvering was used in the past has only recently become apparent.

It is not certain whether there was an equivalent process to cold mercury gilding, with silver leaf or foil being stuck down by a thin film of mercury without any heat being applied (see p. 178), as it is impossible to distinguish chemically from true amalgam silvering. Only where the foil retains its worked structure and mercury is detected can a firm identification of cold mercury silvering be made.

Like fire-gilding, the earliest use of mercury for silvering appears to be in China. Analysis has identified mercury silvering and gilding on several small ornamental items dating to the Han period (1st centuries BC and AD) (La Niece 1990 p.108), perhaps the result of the contemporary alchemical interest in mercury and amalgams. Fire-gilding continued in use in China and, soon after, became widespread in the Roman world. There is at present no evidence at all for mercury silvering in China after the Han period but it seems unlikely that the technique was lost, and examples may be found in the future.

Equally surprisingly, there is no evidence for mercury silvering from the Roman world, even though the method was commonly used for gilding from the 2nd century AD. Silver plating was used by the Romans for small ornaments and tableware, as well as the ubiquitous contemporary coin forgeries, but it is generally of the soldered foil type. The earliest published example of mercury silvering from Europe is an 8th century AD forgery of a coin of Pepin (Metcalfe and Merrick 1967), and even this is an isolated example. It is not until the 13th century that mercury silvering appears to have been widely used, first for contemporary coin forgeries, and later for belt fittings and other small items.

Mercury silvering was particularly suited to silvering items like heraldic pendants, which combined silvering with gilding and enamel work to give a polychrome effect. Only a handful of 13th- and 14th-century pendants have been identified so far (Figure 18.8), but this type of object is frequently damaged and the silvering tarnished, so it is not immediately obvious. More examples might be expected to come to light as the low temperature required for mercury silvering make it ideal for combining with other decorative techniques like enamelling.

Another advantage of mercury silvering over foil silvering was its ability to cover the whole of an object to give a convincing imitation of solid silver. By the 13th century the coin forgers were exploiting this method which avoided the tell-tale joins of foil silvering and could be applied over the design without

Figure 18.8. *Heraldic pendant with enamelling and mercury silvering (length 3.5 cm). British Museum ML 1868, 7–9, 74.*

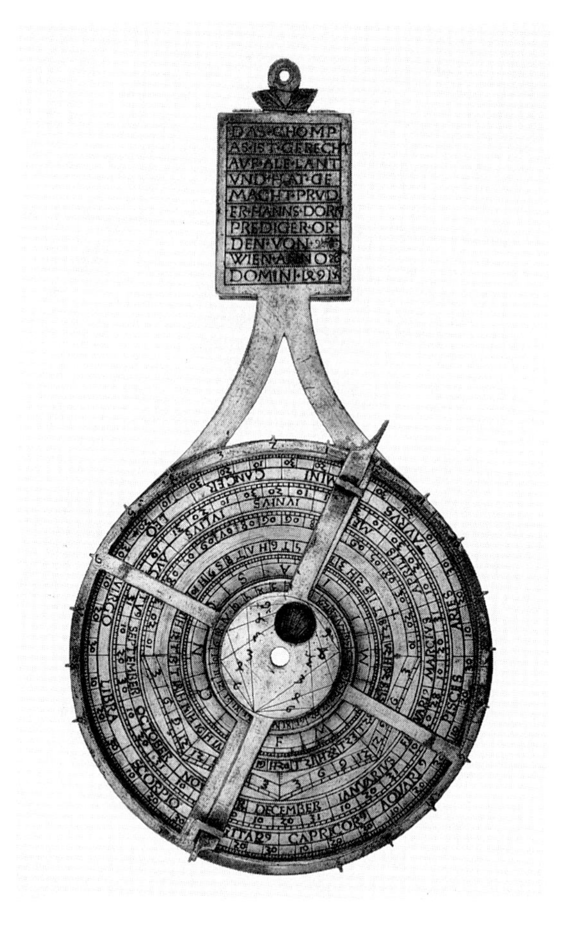

Figure 18.9. *Mercury silvered brass nocturnal made in Vienna in 1491 (length 22 cm). British Museum ML 1894, 6–15, 1.*

reducing the sharpness of detail. A further advantage over other silvering methods was the thinness of the layer of silver needed, increasing the profit margins. Not all imitations of solid silver are fraudulent; they also find a ready market amongst those who cannot afford the real thing. A distinctive group of 14th/15th-century lyre-shaped belt chapes and buckles, decorated with black letter script, which imitated solid silver examples of the same style, have been found to be made of mercury silvered brass (Hook *et al.* 1988).

By at least the 15th century the technique was being used on larger items, notably scientific instruments. Analysis has identified mercury in the all-over silvering of a fine mechanical celestial globe (now on loan to the British Museum) made by Wolff Meyer of Nuremberg for William IV, Landgrave of Hess. A letter, dated 25th April, 1575, instructing Meyer to mercury silver it is still in existence in the Hessisches Staatsarchiv, Marburg (4a31 Nr.11, personal communication J. Leopold). Other examples of mercury silvered brass scientific instruments include a nocturnal (Figure 18.9) (used for finding the time by night from observations on fixed stars) made by Hans Dorn in Vienna in 1491 and a polyhedral dial, made by Hans Tucher in Nuremberg in 1582, showing the hours, length of day according to the season, with a volvelle for interconverting solar and lunar times according to the moon's age and with a compass in the centre (British Museum ML 1857, 5–23,2) (Ward 1981). These pieces are of extremely fine workmanship, and were made for patrons who could well afford solid silver, but presumably the precision required from these instruments demanded a tougher metal, hence the use of brass, with the mercury silvering to give the appearance of precious metal.

Silver does not form an amalgam as easily as gold, but whether this is the only reason for the relatively late introduction of mercury silvering is not entirely obvious. It is interesting to note that brass as well as copper was silvered by this method with equal success.

Silvering pastes and solutions

Modern pastes are generally made of silver in solution with a nitrate or ammonium salt. The solution is thickened by evaporation and the addition of a natural salt such as alum or cream of tartar. It is rubbed onto a clean matal surface to produce an extremely thin (less than 0.005 mm), fine-grained deposit of silver with no distinguishing structure. The mechanism is electrochemical and occurs when a metal at the negative end of the electromotive series, such as copper, is placed in an electrolyte containing ions of a more noble metal, from the positive end of the series. The chemical reaction involved is a simple replacement reaction. No external source of electric current is required, instead the copper provides both anodic (electrically positive) and cathodic (negative) areas. Disruption and removal of the oxide film from the metal to be plated is essential and most recipes include an acid etchant or abrasive material. Silvering solutions, used for immersion plating, operate on the same principles.

It is not known when such pastes and solutions were first used. The extreme fragility of the thin plating limits its chances of survival. Very thin coatings of silver and gold (0.0005–0.002 mm thick) have been reported on copper objects of the Mochica culture of Peru (pp. 185–87). A 3rd century BC bronze coin, overstruck with the die of a contemporary silver Carthaginian coin, has been found to have a very thin, structureless plating of silver (Carradice and La Niece 1988) and may be an early example of this type of silvering. It has also been suggested as the method used in the manufacture of the official coinage of the so-called silver-washed debased coinage of the late Roman Empire. Certainly the ingredients and technology required could have been available in antiquity.

From the 16th century, silvering pastes are known to have been applied to the breastplates of coffins and also fittings on furniture, clocks and instruments. Recipes are given in workshop handbooks (Stalker and Parker 1960, Hooper 1958) and such pastes are available commercially today for the restoration of worn and damaged plated wares.

The history of silvering could be said to be a search for better ways of applying thinner layers, though the difficulties of cleaning tarnished silver has always put a practical limit on the thinness of plating. Economy was not the only driving force; the shape of the object and the uses to which it was to be put were also factors in the choice of silvering technique. Some methods were more labour intensive than others, and in an age of mass-production, electroplating has relegated most other silvering methods into the history of technology.

References

Campbell, W. (1933) *Greek and Roman Plated Coins*, The American Numismatic Society monograph 57, New York

Carradice, I. and La Niece, S. (1988) The Libyan War and coinage: a new hoard and the evidence of metal analysis. *Numismatic Chronicle*, **48**, 33–52

Charles, J.A. (1968) The first Sheffield plate. *Antiquity*, **42**, 278–284

Cope, L.H. (1967) A silvered bronze false antoninianus ascribed to the Roman Emperor Gordian III AD 238–244. *Metallurgia*, Jan., 15–20

Cope, L.H. (1972) Surface-silvered ancient coins. In *Methods of Chemical and Metallurgical Investigation of Ancient Coinage* (eds E.T. Hall and D.M. Metcalf) Royal Numismatic Society Special Publication 8, London, pp. 261–278

Hawthorne, J.G. and Smith, C.S., (trans.) (1979) *Theophilus. On Divers Arts.* Dover, New York

Hockey, M., Johnston, A., La Niece, S., Middleton,

A. and Swaddling, J. (1992) An Illyrian helmet in the British Museum. *The Annual of the British School of Archaeology at Athens*, **87**, 281–291, plates 19–27

Hook, D., La Niece, S. and Cherry, J. (1988) A fifteenth century mercury-silvered buckle from Hillington, Norfolk. *The Antiquaries Journal*, **68**, 301–305

Hooper, P.R. (ed.) (1958) *The Shop Records of Daniel Burnap, Clockmaker 1759–1838*, Connecticut Historical Society

Kingery, W.D., Vandiver, P.B. and Prickett, M. (1988) The beginnings of pyrotechnology, part II: production and use of lime and gypsum plaster in the pre-pottery Neolithic Near East. *Journal of Field Archaeology*, **15**, 219–244

Kunze, E. (1967) *Bericht uber die ausgrabungen in Olympia*, Berlin, helmets B5316 (44) and B5105 (43)

Lang, J. and Hughes, M.J. (1984) Soldering Roman silver plate. *Oxford Journal of Archaeology*, **3**(3), 77–107

La Niece, S. (1990) Silver plating on copper, bronze and brass. *The Antiquaries Journal*, **70**(1), 102–114

La Niece, S. (1993) The technology of silver plated coin forgeries. In *Metallurgy in Numismatics III* (eds M. Archibald and M. Cowell) Royal Numismatic Society Special Publication No. 23, London, pp. 227–236

Metcalfe, D.M. and Merrick, J.M. (1967) Studies in the composition of early medieval coins. *Numismatic Chronicle*, seventh series, vol VII, 172–175

Northover, J.P. and Salter, C.J. (1990) Decorative metallurgy of the Celts. *Materials Characterization*, **25**, 109–123

Scott, D.A. (1986) Gold and silver alloy coatings over copper: an examination of some artefacts from Ecuador and Columbia. *Archaeometry*, **28**, 33–50

Stalker, J. and Parker, G. (1960) *A Treatise of Japanning and Varnishing*, Quadrangle Books, Chicago, Illinois, pp. 66–67

Zwicker, U., Hedrich, D., Kalsch, E. and Stahl, B. (1968) Untersuchungen über Plattierungen antiker Münzen mit Hilfe der Metallographie, der Spektralund Mikroröntgenfluoreszenzanalyse, **43**, 371–380

Ward, F.A.B. (1981) *Catalogue of European Scientific Instruments*, Catalogue No. 65, p. 32, British Museum, London

19

Silver plating in the 18th century

Eric Turner

Abstract

The introduction of Sheffield plate revolutionized the silver-plating industry for it offered an effective, relatively cheap, supremely versatile and durable plating technique which could convincingly imitate solid silver. This paper examines the early history, manufacturing techniques and markings of Sheffield plate and also the disputes with the London Goldsmiths' Company. After the introduction of electroplating, the use of Sheffield plate rapidly declined, though it was retained for some specialized uses, where its durability was needed, well into the 20th century.

Introduction

It is important for the student of Sheffield plate to remember that there were several options available to the plater before Boulsover's discovery in the early 1740s. Two of the most popular 18th-century techniques which continued to be used throughout the history of the Sheffield plating industry were French plating and close plating. In the early 1700s, a method of attaching leaves of pure silver to brass and copper was evolved in France which continued in use until the commercial introduction of electroplating in the middle of the 19th century when the plating industry was entirely changed forever. Known in England as French plating, the procedure can be briefly described as follows.

The object, made of either brass or copper, was first smoothed and polished and then thoroughly de-greased by being heated until almost red hot before being plunged into a bath of nitric acid. The actual silvering, known as charging, required a progressive application of four to six silver leaves at a time by a variety of steel burnishing tools. Before the plater applied the leaf to the object, it was heated until almost red hot and then applied, quickly and under great pressure, to the surface of the object itself. As many as 50–60 leaves of silver leaf were fixed in this way before the surface was finally smoothed to provide a uniform, finished surface.

Close plating was a simplified variant of the above method and was patented by the Englishman, Richard Ellis, in 1779 although it was a further 25 years before the Sheffield and Birmingham metalworking trades adopted his method extensively. His procedure required the same degree of thorough cleansing as the French platers. Then the object was first heated until red hot, plunged into a solution of sal ammoniac followed by molten tin. The plater then wiped the surface with hemp until only a thin film of tin remained. While still hot, the surface was covered with silver foil, cut to shape and pressed onto the surface to ensure complete contact with the tin. A soldering iron was then passed swiftly over the entire surface of the foil which remelted the tin beneath and so ensured that the silver foil was effectively soldered to the base metal. Subsequent burnishing with a blood stone or agate made the joins invisible and the silver surface convincingly resembled silver plate.

Close plating was frequently used by the cutlery trades to coat steel blades. While steel might be unrivalled for cutting meat, close plated blades proved useful for dessert knives where lightly flavoured fruits could be ruined by the use of acid contaminated base metal. In the Sheffield plate industry itself, taps for urns and other complicated cast fittings were often firstly produced in brass and then subsequently close plated.

Both techniques, however, had their limitations. They were slow, laborious and costly methods of imitating the appearance of solid silver and moreover, were ultimately, not very durable. French plating in particular was only successful when it was applied to surfaces which were not susceptible to excessive wear and tear. The introduction of Sheffield plate revolutionized the plating industry for it offered an effective, relatively cheap, supremely versatile and durable plating technique which at times directly threatened the silver industry itself as we shall see.

The discovery

Various accounts exist for the discovery of Sheffield Plate. All are unsatisfactory in that they rely more on romantic appeal than factual accuracy but the common ground they share is this. In 1742, a minor Sheffield cutler, Thomas Boulsover (1705–1788), was repairing a knife handle composed of both copper and silver when he accidentally overheated it, causing the silver to melt and spread on to the copper surface (Hatfield and Hatfield 1974). After this, the story varies but the preferred version is that in his ensuing panic, he put the knife haft through some rollers in order to force it back in to its original shape and was intrigued to discover that instead of the two metals acting independently of each other, they spread in unison at a uniform rate. This last operation was the truly revolutionary aspect of this accidental discovery. Copper and silver in unequal amounts, when fused expanded together indefinitely at a uniform rate under mechanical pressure.

Boulsover did not patent his discovery. It is not clear why. One can only suppose he was either ignorant of the procedure or deterred by the cumbersome nature of it. Nor did he proceed to exploit it on the industrial scale that others soon did. For a short while he managed to keep the secret to himself. He produced a range of small goods including a series of military and livery buttons for which the new material proved ideally suited. Boulsover prospered but the industrial potential was developed by others. One of the earliest entrepreneurs was Joseph Hancock (1711–1790), also a cutler and a distant relative of Boulsover's by marriage. The process he developed for the large-scale production of fused plate differed little throughout the course of the industry and it is worth describing in some detail here (see also Bradbury 1912, Hughes 1970 and Bambery 1988 for good technical descriptions).

An ingot of copper, alloyed with a small percentage of lead and zinc was covered with a thin sheet of sterling silver. (Figure 19.1) These ingots were usually approximately 1½ to 1¾ in thick and 2½ in wide by 8 in long. This

Figure 19.1. *Two ingots during preparation for Sheffield Plate production. The block on the right-hand side of the photograph is ready for firing in the furnace, while that on the left has been fused and is ready to roll into sheet form. (Photo courtesy of the Board of Trustees of the Victoria and Albert Museum.)*

could vary according to the weight and size of the plated sheet that was required to be made. Generally speaking however, the thickness of the silver sheet was 1/40 that of the copper block which meant that 10–12 oz of silver was used for every 8 lb of copper. The surface of the ingot was planed to remove any inequalities from the casting, filed and then scraped to produce a smooth, clean surface. The sheet of silver was cut to the size of the face of the ingot and the surface similarly prepared. The two surfaces were carefully placed together, great care being taken that no dirt or moisture could intervene. They were then firmly pressed together. At first, this meant a workman holding a piece of iron, called a 'bedder' and weighing at least 20 lb, over the surface of the metal while another struck the iron block with a mallet. This served to expel all air and moisture remaining between the layers of metal, flattening both surfaces and embedding the silver sheet in the surface of the copper block. Later, this action was done by a powerful hydraulic press. Then a copper, or sometimes iron plate, covered in a chalk paste (to prevent it fusing to the silver sheet), was placed on top to protect the silver surface from fire damage in the furnace. The three pieces of metal were then tied together by iron wire at regular intervals along the ingot and the exposed edges, where the silver and copper came into contact, were swabbed with a borax solution. The block was now ready for firing.

After about 1760, it became the practice to plate two sides of the copper ingot so that the resulting sheet was plated with silver on both sides. In 1830, Samuel Roberts (1763–1849) patented a variation (no. 5963, July 1830) whereby a sheet of German silver, an alloy of copper, zinc and nickel, was inserted between the silver and the copper block. This produced a laminate of far greater durability. An improved version of Robert's development was patented (no. 7018) by Anthony Merry, a Birmingham metal dealer in March, 1836, which dispensed with the copper and brass substrates altogether; the silver being plated onto an improved-quality German silver where the base metal alloy was whiter than formerly and less liable to split under pressure. But when Merry tried to introduce his samples of plated German silver to Sheffield manufacturers, he was dismayed to discover that his invention had already been fully anticipated by Thomas Nicholson, a partner in the Sheffield firm of Gainsford & Nicholson. His patent was therefore invalidated and made universally available to the trade. British plate, as Merry's invention became known, largely superseded the traditional form of Sheffield plate, even though the German silver had to be entirely imported from Hamburg. This too was soon largely supplanted by the introduction of electroplating.

However, whatever base metal was used for the core of the laminate, the technique of fusing the silver to the base metal substrate did not vary. Fusion of silver and copper relies on the differing melting points of the two metals. Silver melts at 960°C whereas copper melts at the slightly higher temperature of 1083°C. The 'sandwich' of copper, topped by silver and iron plates, was placed into a coke fired furnace which had a spyhole so that the workman could monitor the firing closely. The block was extracted as soon as the exposed silver edges started to melt. This was indicated by the silver beginning to ooze out from between the copper sheets, a process known as 'weeping'. Great skill needed to be exercised at this stage. If the block was removed too soon before the silver sheet had properly started to melt, then fusion did not successfully take place. If, on the other hand, the block was left in the furnace a moment too long, then the silver simply melted and ran off into the fire. Moreover removal of the block from the furnace presented its own hazards. It had to be carefully removed with tongs, grasping the sides and keeping the block both level and steady. Any pressure on the upper surface or careless tilting would again lose the silver. Once the block was removed, it was set aside to cool and when it had done so sufficiently, it was fed through a graduated series of rollers, with occasional annealing, which progressively flattened the ingot into a laminated sheet of copper and silver.

Techniques

As perhaps has already become clear, the unique property of Sheffield plated wares which sets them apart from all other forms of

Figure 19.2. *Sheffield plate cake basket, English c.1780–1790. A good example of pierced work using a fly punch, and die stamping for the embossed swags and medallions. (Photo courtesy of the Board of Trustees of the Victoria and Albert Museum.)*

plated articles, is that the plating takes place before the article itself is fashioned. After the laminated plate was produced, it could be worked in much the same way as sterling silver, with one important and obvious exception; it could not be cast. The techniques used to overcome this problem relied on parallel developments in the steel industry.

Benjamin Huntsman's (1704–1776) invention of crucible or cast steel had a decisive influence on the Sheffield plate industry. An instrument maker, he started experiments to produce a more uniform steel for the springs and pendulums of clocks. After many failures, he eventually found a satisfactory method of doing so. Bars of blister steel, mixed with fluxes were fused in closed clay crucibles under intense heat within a coke fired furnace. His method produced a steel that was more homogeneous in composition and freer from impurities than any produced previously. It became the standard for the cutlery trade and for the production of precision tools. It was crucial for the cutting of accurate dies used in the plating industry.

The techniques used in the industry were a combination of the traditional silversmithing skills such as raising hollow ware from flat sheet on a stake, soldering, planishing, burnishing etc. and several new procedures which were developed to overcome the limitations of the material itself. I shall only describe a selection which were developed specifically for the plating trade.

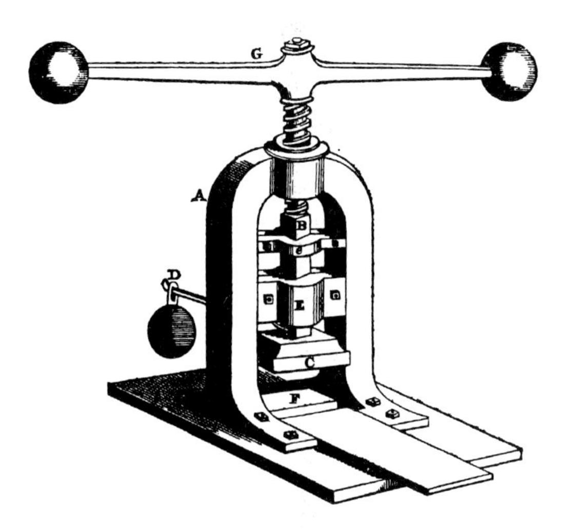

Figure 19.3. *An illustration of a fly press taken from* A Treatise on the Improvement and Present State of the Manufactures in Metal, *Vol II, Rev. D. Lardner London 1833 (p.196).*

Figure 19.4. *Sheffield plate mustard pot, English c.1780. The piercing on the frieze has been made using a fly press and embellished by bright cut engraving. (Photo courtesy of the Board of Trustees of the Victoria and Albert Museum.)*

Silver pierced work was enormously popular in the mid-18th century, when the Sheffield plating industry was just getting under way (Figure 19.2). The traditional silversmith's technique for executing this type of work was with a fretsaw which proved quite unsuitable for Sheffield plate. The teeth of the saw ripped the silver surface, tearing it away from the copper core and leaving a jagged edge. The copper itself was exposed to view at the same time. The solution adopted to overcome this was to use a fly punch (Figure 19.3). This consisted of a large screw, secured in a frame which had a cross-bar at the top with weights at either end. This machine works on the same principle as an early printing press, serving to translate a horizontal movement into a vertical downward thrust. The lower end drove a hardened steel cutting tool, shaped to the pattern required, directly through the surface of the metal in one operation causing the uppermost silver skin to be dragged over the copper core and so at the same time concealing it. The underside of the object being pierced is protected by a bedding tool, thus preventing any possible distortion.

Engraving equally presented difficulties because the tool cut through the upper surface of the laminate, exposing the copper beneath. This problem was overcome by a variety of measures. At first, the silver coating over the copper core was sufficiently thick for an experienced workman to be able to engrave the surface of the metal without cutting through to the copper beneath. Bright cut engraving, a form of faceted ornament, became fashionable about 1790 (Figure 19.4). The engraving tool was used to gouge out in a single deep stroke, a small section of silver leaving behind a bright, scooped surface. However to be able to do so with impunity, the silver surface had to be extraordinarily thick. Samuel Roberts, who is credited with introducing this form of decoration on to Sheffield plate, is recorded as using 24 oz of silver to every 8 lb of copper alloy, an extremely expensive ratio to maintain.

However, engraving, particularly for coats of arms for which the technique was commonly used, was usually only required on certain areas of the object and so various means were developed to thicken the silver surface, only in those parts which required it. Thus sterling silver shields or more commonly extra heavily plated sections of Sheffield plate

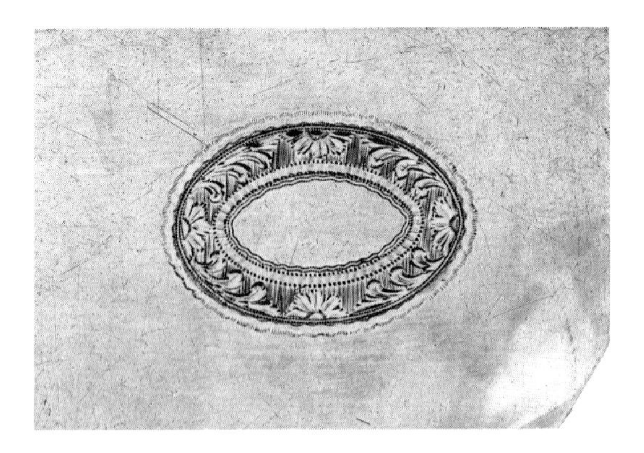

Figure 19.5. *Detail of a Sheffield plate teapot stand, English c.1790 (M.215–1920). The oval panel in the centre with engraved decoration has been 'let in'. (Photo courtesy of the Board of Trustees of the Victoria and Albert Museum.)*

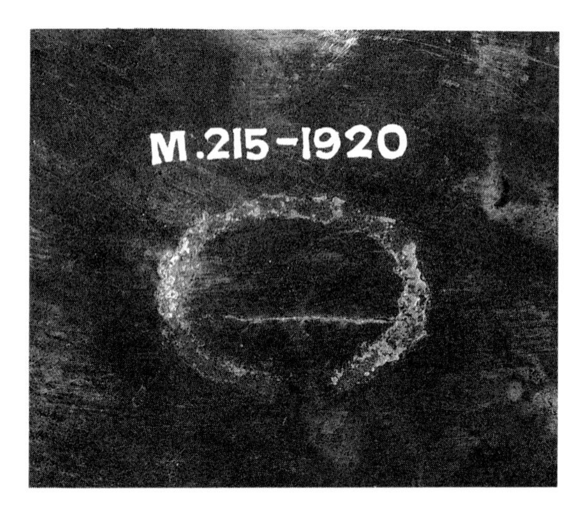

Figure 19.6. *The underside of the teapot stand (Figure 19.5) which shows clearly the soldered outline. (Photo courtesy of the Board of Trustees of the Victoria and Albert Museum.)*

were soldered in, or 'let in' as the process came to be termed (Figures 19.5 and 19.6). A round or oval piece of metal was cut out of the article. The metal to be inserted was carefully cut to the same shape and hard-soldered in to the gap. After any superfluous solder was removed, the whole section was hammered and burnished. Often, a small wavy line was lightly engraved over the join on the uppermost surface in order to distract the eye from the almost invisible join.

After about 1800, the plating thickness of the silver of Sheffield plate became progressively thinner and another method of letting in was adopted. This method took place before the article was fully fashioned. A very fine gauge piece of pure silver was placed over the area to be engraved, the metal heated to a dull red heat until the silver section was adhering to the surface of the blank at which point it was removed from the flame and the workman started to burnish the shield. Once that operation was complete, the blank was cleaned with acid, washed in water and subsequently hammered until the silver shield was level with the surface surrounding it. The silver shield had been literally forced into the plated surface.

Other techniques developed, particularly for the requirements of the plating industry, were spinning, swaging and the disguising of exposed edges with silver thread or ribbon. Spinning was the forming of a piece of hollow ware from sheet metal on a lathe, using a variety of shaped wooden chucks. Swages were used extensively by the trade when it was considered unnecessary to go to the expense of cutting dies. The swage itself consisted of two steel hammers (an upper and a lower) hinged in much the same way as a pair of fire tongs. Their faces were cut according to the cross-section of the required design, one face convex, the other concave, each fitting perfectly on the other. The top jaw was struck repeatedly with a hammer to force the metal into the required shape. Swaging was used particularly for the shaping of borders for salvers and shallow hollow ware such as meat dishes and decanter stands.

The use of silver wire edging rather than rolling the edge over to conceal the core was first used by Matthew Boulton (1728–1809) at his Soho manufactory in Birmingham during the 1780s. Later, the Sheffield plater Samuel Roberts perfected a technique of extending the silver thread beyond the ornamental border by filing it down to the plated edge and burnishing the junction so that the join became virtually invisible. By such means, a plated article could appear, at a superficial glance, to be indistinguishable from its silver duplicate.

Figure 19.7. *Three die-stamped candlesticks with identical bases and different columns and capitals. Dies, once cut, were interchanged to achieve considerable variety. The candlestick on the left is by John Winter, and hallmarked for London 1774, while the other two are of Sheffield Plate. (Sheffield City Museums.)*

All the techniques described so far were developed to overcome the limitations of the material. With the exception of engraving which really relied on traditional craft skills, the skills that were used in the plating workshops were emerging industrial techniques which changed the whole character of the silver industry itself. The investment in dies alone was substantial. Lardner (1834) offered the following comparison;

> sets (of dies) for a candlestick which sells for about three guineas the pair, often cost upwards of £50; and if the candlestick have [*sic*] branches from seventy to one hundred guineas. Nor is there in the latter case a separate die for every part; as it will be obvious to any person looking at the finished article, that many of the embossed portions of a candlestick are duplicates. The parts are, however more numerous than would be supposed; and an ordinary plated bed candlestick, with extinguisher, is often made up of more than twenty pieces. (Lardner 1834, Vol III, 363)

Duplication of die-stamped sections, considering the initial investment in the dies themselves, was common. The illustration of three die-stamped candlesticks with identical bases but different columns and capitals, demonstrates that dies were interchanged to achieve variety (Figure 19.7). What is also of interest, is that the candlestick on the left is of silver, hallmarked for Sheffield 1774 with the maker's mark of John Winter (active *c*.1765–1783), while the two to the right of it are of Sheffield Plate. Dies once cut had to be used as economically as possible and it seems that not only were the same dies used for both silver and plate production within the same firm but were exchanged with other firms so that they could be used and re-used in as many different combinations as possible.

Bradbury (1912 pp. 38–42) in his standard work on the subject, mentions that several different firms shared the same partners. John Winter, for instance, headed a firm specializing in the manufacture of candlesticks but was also a partner in the firms of Richard Morton & Co. and Samuel Roberts & Co. All these factors help to explain the extraordinary homogeneity of designs, particularly in the production of candlesticks by Sheffield manufacturers in the latter part of the 18th century.

This industrial structure was very different from the London trade and perhaps helps to explain the growing antipathy the Londoners felt for their Sheffield colleagues in particular. Until the advent of the Sheffield plating industry, there was no silver trade in Sheffield. Its development was closely dependent on the plating industry. Not only was the basic material of the plating industry significantly cheaper than sterling silver with which it directly competed, but the techniques of manufacture quickly developed along much more economical lines. Thus the Sheffield silversmiths could, by using techniques learnt from their colleagues in the plating industry, produce silver candlesticks in a variety of designs which greatly appealed to the market and which significantly undercut their London rivals. This caused considerable resentment as shall be seen.

Design

If the manufacturing techniques were developed within Sheffield itself, the designs were inevitably determined by the fashionable tastes of London society. Again, there is the temptation to quote Lardner (1834 pp. 363–4): 'Persons who have visited the British Museum...will have observed how extensively the various forms and especially the raised ornaments of our silver and plated wares have their counterparts in the designs and workmanship of the ancient artists.'

The vast collection of Attic pottery, given to the Museum by Sir William Hamilton had an enormous impact on public taste. The neoclassical style which favoured objects having a two-dimensional structure could be perfectly adapted to the requirements of mechanical production. In this of course the Sheffield manufacturers excelled and the period between 1770 and 1790 is often regarded as the time in which the most elegant forms of Sheffield plate were produced. After 1800, when a taste for massive forms of silver, requiring increasingly complicated cast work became fashionable, the Sheffield plating industry was at something of a disadvantage since they had to produce the effect by a complicated system of die stamping and cast lead backing but the industry soon proved equal to the challenge.

Marking

From the earliest days of the industry, manufacturers frequently marked their wares, sometimes with their names in full but more often with a monogram only, containing their initials. During the first 20 years or so that the trade flourished, there is clear evidence that some makers adroitly contrived to stamp their wares in such a way as to give a passing resemblance of the assay marks on sterling silver. The cup and cover illustrated (Figure 19.8) is one of the more flagrant examples of this practice. Initials (Figure 19.9), so elaborately entwined within a cartouche as to be unrecognizable have been stamped four times in a row to resemble, at least at a superficial glance, the marks denoting town of origin, date letter, lion passant and maker which were, and are still, required on silver.

This practice continued until 1773 when it was abruptly stopped. This came about

Figure 19.8. *Sheffield plate cup and cover. English c.1760 stamped with a series of false marks (M.199–1920). (Photo courtesy of the Board of Trustees of the Victoria and Albert Museum.)*

Figure 19.9. *Detail of the marks, stamped just below the rim of the cup and cover (Figure 19.8). (Photo courtesy of the Board of Trustees of the Victoria and Albert Museum.)*

directly as a result of the establishment of the Sheffield and Birmingham assay offices. Before then, the provincial silver manufacturers were obliged to have their wares hallmarked in offices well away from their own centres. Chester was preferred by the Birmingham silversmiths while London was favoured by Sheffield.

This gave rise to a number of grievances. Matthew Boulton who led the Birmingham petition accused the local Chester silversmiths of pirating his designs. This was not of such concern to the Sheffield trade since they were pretty adept at this practice themselves. What was of grave concern to them both, however, was the extreme inconvenience of having their wares marked away from their own direct supervision. However carefully the goods were packed for the long journey to London, the same care was seldom taken in packing the goods for the return so that many Sheffield silversmiths found their goods irretrievably damaged, to the point of unsaleability, by the time they received them back. Moreover, by the 1760s, the Sheffield trade had become one of the major suppliers of silver and silver plated candlesticks. It was the practice of some London retailers, most notably amongst them John Carter (active *c.*1735–77) to add or overstrike the Sheffield silver wares with their own maker's mark, thus passing them off as their own and grabbing the lion's share of the subsequent profit.

The London Goldsmiths' Company was naturally anxious not to lose the lucrative business which the provincial manufacturers afforded them and quickly mounted their own counter-petition. Parliament had no option but to appoint a select committee to investigate the matter. The evidence submitted by both sides has all the indications of a pretty squalid and acrimonious row. The London Goldsmiths' Company reported the following to the House of Commons Committee;

> That the Artificers are now arrived at so great a perfection in plating with Silver the Goods made of base Metal that they very much resemble solid Silver; and if the Practice which has been introduced, of putting Marks upon them somewhat reassembling those used at the assay offices, shall not be restrained, many Frauds and Impositions may be committed upon the Public. (Bradbury 1912 p. 426)

In so far as we can see from the cup and cover mentioned earlier, the Goldsmiths' Company had a valid point. The Sheffield trade countered by accusing some of the officers at the Hall of corruption. W. Hancock said; 'that his work had been injured by scraping; but that he went to the Hall and gave some drink to the Assay Master and Scraper, after which his plate had been less damaged', and another, a Mr Spilsbury; 'had several times treated the workmen with drink; and thought it of consequence to be on good terms with the Scrapers, for when plate had been objected to, he had known of difficulties being removed by giving liquor at the Hall'. (Jackson 1905 p. 255)

These accusations of venality were passed uncontested but it was left to Matthew Boulton who by now was coordinating the campaigns of both towns to play the trump card as described by Rowe (1965). He boldly accused the London assay office of deliberately and consistently marking substandard wares as meeting the sterling standard. This had to be either proved or disproved and the committee immediately ordered that 22 pieces of London-made plate be purchased at 'public shops' and assayed by both the assay master at the Royal Mint and the head assayer at Goldsmiths' Hall. The complete sample was gathered together in one afternoon from shops close to the Hall itself. All but one were found to be markedly below standard and the Goldsmiths' Company's case collapsed. By May, 1773 legislation was passed allowing assay offices to be established in both Birmingham and Sheffield which, however, included a clause forbidding the striking of any letter, or letters, on articles 'made of metal, plated or covered with silver, or upon any metal vessel or other thing made to look like silver' (Bradbury 1912 p. 426). The penalty for flouting this regulation was a fine of one hundred pounds which proved a sufficient deterrent to prevent any further practice of this sort.

In 1784, Sheffield successfully obtained a modification to the Act which allowed platers to mark their goods with their 'surname or partnership name, together with any mark, figure or device...such figure not being an assay office device for sterling silver, or in imitation thereof' (Bradbury 1912 p. 426). Makers were not legally obliged to mark their goods, unlike silversmiths, but if they wished to do so, they could only use a device which had the approval of, and had been previously registered at, the Sheffield Assay Office. Many platers found these conditions irksome and continued, as they had by now become accustomed, not to mark their goods at all.

The London Goldsmiths' Company only once again referred to the matter of marking Sheffield Plate wares. This is fully described by Prideaux (1897). In 1797, when threatened with a further increase in the duty payable on wrought plate, the Court of Assistants of the Company received the following petition from its members.

> Your petitioners beg leave further to represent that the advantage and encouragement derived by plated manufacturers have animated them to exertions and improvements, by which they have been able to produce articles of the highest elegance and fashion, many of which are now made with solid silver – borders, shields and ornaments, finished in exact resemblance of real plate – and which do material injury to your petitioners by curtailing the sale of wrought plate, and consequently by withholding a very considerable quantity of work from Goldsmiths' Hall, depriving the Revenue of vast duty and being the means of waisting [*sic*] a great quantity of silver, which is thereby lost to the public. (Prideaux 1897, 285)

The petition continued by explaining, in detail, their proposed measures for imposing a system of tax on finished goods for the platers, similar to that which had applied to the silversmiths themselves since 1719. Plated goods would be subject to a duty of threepence per ounce and those that have 'silver edges, bands, shields, or any other parts silver, either for use or ornament' (Prideaux 1897 p. 286) would be subject to a higher tax of sixpence per ounce.

Furthermore, all plated goods were to be marked at the assay office. Every plate manufacturer was to strike the initials of his name on every separate piece (this again was very similar to the regulations pertaining to the silversmiths themselves) and the marks to be used at the Hall or Office shall be the (word) 'Plated' and the Sheffield Arms alongside the manufacturer's own mark. In order to safeguard against any fraud, the object was to be weighed, with the weight in figures stamped under the word 'Plated' before being filled or loaded with lead or pitch, a common practice with candlesticks. Finally, any breach of these regulations was to be punishable by the same penalties instituted for the regulation of the manufacture of gold and silver.

On receipt of this petition the Court of the Goldsmiths' Company approached the Secretary of the Treasury, George Rose, with a view to obtaining an audience with William Pitt, the Chancellor of the Exchequer, which was duly granted to them on 29 May. The result was not a success. Pitt, who was shrewd enough to see through the Goldsmiths' proposals went on record as saying 'that he thought plated goods were a fair object of taxation, but that he, nevertheless, was of the opinion that plate [i.e. wrought silver] would bear some additional tax'. (Prideaux 1897 p. 287) The Company was wise enough not to press their case and not a word about a duty on plated articles was ever officially mentioned again.

Postscript

The industry continued to flourish until the commercial introduction of electroplating by the Elkington Brothers in 1840. Thereafter, the use of Sheffield plate in decorative arts manufacture rapidly declined. But specialized uses for Sheffield plate are found well into this century (Bambery 1988). Fused plate had a surface of greater durability than electroplate and it was retained in the production of military and livery buttons (the products which Boulsover manufactured on first making his discovery), carriage and lighthouse lamp reflectors and pub tankards. In the early years of this century, it was frequently used as a base for enamelled decoration, particularly for the backs of hairbrushes and hand mirrors and from the 1920s, it was used in specialized industrial applications such as electrical contacts in electromechanical switching gear. The following account of another such application is related in detail in Hatfield (1974). Briefly, the circumstances were as follows.

Almost two centuries after Thomas Boulsover's discovery, Britain, in 1940, was involved in a full-scale war with Germany. The Air Ministry urgently required a modified and improved form of fighter aircraft, particularly

Figure 19.10. *A Rolls-Royce Merlin engine incorporating an intercooler made of fused (i.e. Sheffield) plate. The intercooler is housed within the jacket at the top of the picture. (Rolls-Royce Heritage Trust.)*

to serve as a long-range escort for bombers. Fighter pilots had two principal requirements from their aircraft; high altitude which they needed to gain as quickly as possible which in turn depended on an engine supplying maximum thrust. The two most successful fighter aircraft in the Battle of Britain proved, on the British side, to be the Hawker Hurricane and the Vickers Supermarine Spitfire. Both were powered by an early form of the Rolls-Royce Merlin engine which incorporated a supercharger. This compressed the air and petrol mixture which in turn boosted the thrust of the engine. In order to increase the thrust still further, it was necessary to install a second supercharger. The problem was that gas under pressure expands and under expansion increases in temperature. Hot air entering the piston chamber drastically reduced the engine's efficiency and so to overcome this, an intercooler, which is a form of heat exchanger, was inserted between the two stages of compression.

The design of the intercooler, using the cross-flow principle, superficially resembled a Rolls-Royce car radiator without the traditional 'Parthenon' pediment (Figure 19.10). Alternate, but separate, passages for air and glycol were set at right angles to each other. Originally copper was tried for its construction using soft solder (a lead, tin alloy) but under the severe operating conditions which included excessive fluctuations in temperature, pressure and vibration, the unit failed, causing the engine to disintegrate. An American design using aluminium was tried but that had a tolerated leakage rate which for a liquid cooled unit was hopeless.

All technicians concerned with the aircraft industry were alerted to the problem. The Air Ministry, in desperation, were prepared to try anything within reason. The solution offered which was to prove successful came about almost as accidentally as Boulsover's original discovery of Sheffield Plate. John Coltman, one of the Senior Development Engineers for Marston Excelsior Ltd. a Wolverhampton subsidiary of the ICI Metals Division had been shown the manufacture of fused plate by a friend, Sidney Chatwin of Johnson Matthey in Birmingham. Although his superiors were initially sceptical, Coltman succeeded in obtaining enough of the fused plate to construct a prototype. His theory was that while the copper provided the necessary structural strength, the silver already strongly bonded to the base metal would provide a soldered joint which could withstand all the extraordinary stresses subjected to it. In a controlled, reduced atmosphere, an experimental unit, made up of strips of fused plate, was heated to 800°C, forming a silver copper eutectic, i.e. a mixture whose constituents are in such proportions as to melt and solidify at one temperature. One unit alone required over six thousand lineal feet of joints which had to be formed in one operation. The result worked and proved so successful that by 1943, the design was in full production and became standard for all Rolls-Royce high performance aircraft engines for the remainder of the war.

References

Bambery, A. (1988) *Old Sheffield Plate*, Shire, Princes Risborough

Bradbury, F. (1912) *History of Old Sheffield Plate*, Macmillan, London

Hatfield, J. & Hatfield J.J. (1974) *The Oldest Sheffield Plater*, The Advertiser, Huddersfield

Hughes, B., 1970) *Antique Sheffield Plate*, Batsford, London

Jackson, C.J. (1905) *English Goldsmiths and Their Marks*, Macmillan, London

Lardner, D. (1834) *Manufactures in Metal*, Longman, London

Prideaux, W.S. (1897) *Memorials of the Goldsmiths' Company*, Eyre and Spottiswoode, London

Rowe, R. (1965) *Adam Silver 1765–1795*, Faber, London

20

Roman techniques of manufacturing silver-plated coins

Ulrich Zwicker, Andrew Oddy and Susan La Niece

Abstract

Almost since the invention of coinage, imitation silver coins have been manufactured by attaching a silver layer, with the same composition as the contemporary official coinage, on to the surface of a base metal core. In the case of copper cores, this process could be carried out by coating the surface of the core with a layer of solder consisting of a silver–copper alloy, or by attaching a layer of silver foil to the surface, either by soldering or by heating to form a layer of eutectic at the interface of the silver and copper. This latter technique is, in effect, a self-soldering process which is known today as Sheffield plating. In the case of iron cores, however, the self-soldering process could not be used, and silver plating had to be carried out either by using hard (i.e. silver–copper) solder alone or by soldering silver foil to the surface of the iron.

In the Republican and early Imperial periods, a common technique of manufacturing silver plated forgeries was by the application of silver foil, but after the debasement of the official silver coinage, which began at the end of the 1st century AD, plated coins with a high purity silver layer on the surface are not usually found. Instead, the forgers used hard solder coatings for the blanks.

Introduction

Caius Plinius Secundus reports in his *Historia Naturalis* that *'Livius Drusus in tribunatu plebei octavam partem aeris argento miscuit'* (XXXIII, 46). This means that Livius Drusus mixed one-eighth part of copper in the silver, and is usually taken to be a reference to the debasement of the silver coinage. Later on, Pliny writes *'Miscuit denario triumvir Antonius ferrum, miscent aera falsae monetae, alii et ponderi subtrahunt, cum sit iustum LXXXIIII e libris signari. Igitur ars facta denarios probare, ...'* (XXXIII, 132) This translates as: the triumvir Antony adulterated the denarius with iron, and false coin contains an admixture of copper. Others make light-weight coins, although the statutory weight is eighty-four to the pound. Denarius testing was therefore established ... (Bailey 1929 p. 131).

In view of the large number of denarii consisting of silver-plated copper (or more rarely iron) cores which have survived from the time of Pliny, this passage is generally taken to refer to the manufacture of these plated coins. Pliny clearly thought that some, at least, of the plated coins were made in the official mints, but this is not accepted by numismatists today (Crawford 1968, Crawford 1974), and the fact that the testing was established by law rather suggests that the coins were always regarded as forgeries, whatever their source. Pliny does not record the nature of the test, but it may have been a simple test of density by which a suspect coin was placed on a scale and balanced with the same weight of *silver* on the other pan. If the pans are then lowered into a vat of water, the pan with the coin on it will rise if the coin is a plated forgery. The discovery of this method of testing the purity of a precious metal is attributed to Archimedes (d. 212 BC).

Other ways of testing whether coins were plated were to make a test cut or punchmark on the surface, or to see if the coin 'rings' like a normal one when dropped on a stone surface. The latter method leaves no indication on the coin, but the results of test cuts and punches are to be seen on many surviving ancient silver (and gold) coins, including some included in this study. It seems unlikely, however, that an official coin tester would use a disfiguring test like cutting or punching.

Many silver-plated denarii of the Roman Republic were published by Bahrfeldt (1884) and his results show that the most common years for which plated coins exist are 111/110, 106, 105, 104, 102, 92, 83/82 and 79 BC. Four of these types (Crawford 1974, 313.1c, 314.1c, 364.1e and 384.1 issued in the years 106, 105, 83/82 and 79) have serrated edges, a shape at one time thought to have been devised to prevent the illegal manufacture of plated forgeries, although this theory is no longer regarded as probable (Crawford 1974 pp. 299 and 581).

It is important to remember that, in the case of an ancient forgery, the 'date' of the coin is not necessarily the date of its manufacture. Indeed, there is evidence that some of the forgeries of Republican coins were made long after their prototypes were struck officially. For instance, the St Swithin's Lane hoard of forgeries (discussed below) contains plated coins whose prototypes range in date from the late 2nd century BC to AD 52 and it was presumably buried in Roman London soon after the latter date. The fact that there are numerous die links within the hoard suggests that most, if not all, of the coins must have been manufactured not long before burial, possibly in London itself.

In the past 60 years, several silver-plated Roman denarii have been investigated metallographically (Dahl 1931, Dahl and Schwarz 1933, Campbell 1933, Kalsch and Zwicker 1968, Fried and Hammer 1974, Hammer 1984, La Niece 1990 and 1992). Several methods for carrying out the plating, both with and without the addition of a silver–copper eutectic solder, have been suggested.

The purpose of this paper is to document more examples of the metallographic investigation of silver-plated-on-copper Roman coins, and to survey their occurrence during the Roman period. It also records the rarer occurrence of silver-plated-on-iron coins. The influence of plating on the specific gravity of the coins is investigated.

The investigation of the specific gravity of Republican silver coins

Measurement of the specific gravity of a large number of silver coins of the Roman Republic gives values which are usually greater than 9.8. The slight drop from the value for pure silver (i.e. 10.5) is not surprising as coins were always struck in slightly debased silver which is harder and more resistant to wear than the pure metal. Walker (1980), for instance, reports analyses usually in the range 93–98% for the Rome mint in the period 169–40 BC. Lower specific gravities were, however, found for those coins which are visibly corroded and/or porous. Specific gravity measurements on plated coins, on the other hand, give values which are typically in the range 6.3 to 9.0 when copper cores have been used, and 3.5 to 4.9 for silver plating on iron (see Table 20.1).

The specific gravity results for 316 normal Republican denarii and 82 plated ones are plotted as a histogram in Figure 20.1, which demonstrates that specific gravity is a very useful indication of whether a silver-looking coin is plated or not. Some non-plated coins which had a specific gravity below 10 were visibly corroded or porous, and their results are excluded from Figure 20.1. However, it must be remembered that if a silver coin is debased with copper, the specific gravity will also be reduced. Debased coins could, therefore, be confused with plated coins by the specific gravity test.

Table 20.1 Specific gravity measurements on silver-plated coins of the Roman Republic,

Moneyer and date of prototype and figure number	*Catalogue reference***	*Weight (g)*	*Specific gravity*	*Comments*
Anonymous 225–212 BC (Figure 20.4.1)	C. 28.3	5.35(Qa)	8.64	Foil layer *c.* 95% Ag, intermediate layer is Ag-Cu eutectic
'Rostrum Tridens' 206–195 BC (Figure 20.4.2)	C.114.1	3.38	8.55	Foil layer *c.*96% Ag
SAFRA 150 BC	C.206.1	2.80	8.13	Foil layer >95% Ag
PINARIVS NATTA 149 BC (Figure 20.4.3)	C.208.1	2.10	7.63	Weakly magnetic. Copper core contains small amount of iron
SEXTVS POMPEIVS 137 BC (Figure 20.4.4)	C.235.1a	3.23	8.48	Foil layer is 94.8% Ag, intermediate layer is 67.9% Ag, copper core contains 2.3% Ag. All by ESMA
M.BAEBIVS 137 BC	C.236.1a	3.32	8.55	Full layer > 80% Ag, intermediate layer is Ag–Cu eutectic
L.LICINIVS and CN.DOMITIVS with M.AVRELIVS SCAVRVS 118 BC (Figure 20.4.5)	C.282.1	3.03	8.45	Surface layer is Ag-Cu alloy
L.LICINIVS and CN.DOMITIVS with L.PORCIVS LICINVS 118 BC (Figure 20.4.6)	C.282.5	3.30	8.27	Surface layer is Ag-Cu alloy, copper core contains 1% Ag
L.LICINVS and CN.DOMITIVS with L.PORCIVS LICINVS 118 BC (Figure 20.4.7)	C.282.5	2.92	8.32	Surface layer contains 75% Ag, copper core contains 2% Ag
L.CAESIVS 112/1 BC	C.289.1	2.30	8.46	Foil layer is > 95% Ag, intermediate layer is Ag–Cu eutectic
L.SCIPIO ASIATICVS 106 BC	C.311.1a	3.05	7.87	Foil layer is >95% Ag
L.MEMMI GAL 106 BC	C.313.1c	2.94	7.83	Surface layer is Ag–Cu alloy
Q.THERM M.F 103 BC	C.319.1	3.23	8.47	Foil layer >95% Ag
C.EGNATVLENS C.F Q 97 BC	C.333.1	1.39(Q)	5.80	Surface layer *c.*80% Ag
D.SILANVS L.F 91 BC	C.337.1a	3.02	8.20	Foil layer >95% Ag
L.PISO FRVGI 90 BC (Figure 20.4.8)	C.340.1	2.68	7.36	Foil layer *c.*97% Ag
L.PISO FRVGI 90 BC	C.340.2e	1.47(Q)	8.39	Surface layer *c.*80% Ag
C.VIBIVS C.F PANSA 90 BC	C.342.5b	2.60	7.37	Foil layer >95% Ag
L.TITVRIVS L.F SABINVS 89 BC	C.344.1a	2.78	8.46	Surface layer *c.*70% Ag

Table 20.1 ***continued***

Moneyer and date of prototype and figure number	*Catalogue reference***	*Weight (g)*	*Specific gravity*	*Comments*
CN.LENTVLVS 88 BC	C.345.2	1.52(Q)	6.91	Surface layer 58% Ag by ESMA
P.FOVRIVS CRASSIPES 84 BC	C.356.1a	2.48	7.00	Foil layer 97.5% Ag by ESMA
C.NORBANVS 83 BC	C.357.1b	3.18	8.40	Surface layer *c.* 50% Ag
L.SVLLA IMPE L.MANILIVS PROQ 82 BC	C.367.3	1.95	7.81	Eutectic at rim *c.*70% Ag
Anonymous 81 BC	C.373.1b	1.93(Q)	8.93	Surface layer *c.*71% Ag by ESMA
L.RVTILIVS FLAC 77 BC	C.387.1	2.79	8.44	Surface layer *c.* 50% Ag
C.EGNATIVS CN.F CN.N MAXSVMVS 75 BC	C.391.1	3.34	8.07	Eutectic at rim *c.*60% Ag
L.AXSIVS L.F NASO 71 BC	C.400.1	2.69	7.91	Eutectic at rim *c.*50% Ag
C.HOSIDIVS C.F GETA 68 BC (Figure 20.4.9)	C.407.1	2.93	8.14	Foil layer >95% Ag
M.PISO M.F FRVGI 61 BC	C.418.2a	2.71	7.65	Eutectic at rim *c.*75% Ag
M.SCAVRVS and P.HVPSAEVS 58 BC (Figure 20.4.15)	C.422.1	3.37	6.38	
M.IVNIVS BRVTVS 54 BC	C.433.1	2.79	8.28	Surface layer *c.*80% Ag
M.IVNIVS BRVTVS 54 BC	C.433.1	2.64	8.17	Foil layer *c.*95% Ag
C.COELIVS CALDVS 51 BC	C.437.1a	2.84	8.45	Foil layer > 95% Ag
Q.SICINIVS 49 BC (Figure 20.4.10)	C.440.1	3.12	8.66	Surface layer is Ag-Cu alloy
MN.ACILIVS 49 BC	C.442.1a	4.95	8.91	Core 3.1% Ag and surface layer 42.5% Ag by ESMA
CAESAR 49–48 BC	C.443.1	3.19	8.73	Foil layer >90% Ag
CAESAR 49–48 BC	C.443.1	2.61	8.33	Eutectic at rim *c.*50% Ag
Q.SICINIVS and C.COPONIVS 49 BC	C.444.1b	2.96	8.65	Only Cu core survives
CAESAR 47–46 BC	C.458.1	3.49	8.81	Surface layer *c.*60% Ag
Q.METELL.SCIPIO IMP with EPPIVS LEG.F.C. 47–46 BC (Figure 20.4.11)	C.461.1	3.22	8.76	Foil layer *c.*95% Ag
CAESAR 46–45 BC	C.468.1	2.14	6.68	Foil layer >90% Ag

Table 20.1 ***continued***

Moneyer and date of prototype and figure number	*Catalogue reference***	*Weight (g)*	*Specific gravity*	*Comments*
M.ANTONIVS and M.BARBATIVS 42 BC (Figure 20.4.12)	C.517.2.	2.63	8.22	Ag–Cu layer on surface has been enriched
IMP.CAESAR DIVI F 37 BC (Figure 20.4.13)	C.538.1	3.13	8.26	Foil layer *c*.95% Ag
M.ANTONIVS (LEG.IV) 32–31 BC	C.544.17	3.07	8.54	Foil layer *c*.90% Ag
M.ANTONIVS (LEG.VII) 32–31 BC	C.544.20	3.05	8.39	Surface layer *c*.80% Ag
M.ANTONIVS (LEG.IX) 32–31 BC	C.544.23	2.34	7.90	Surface layer *c*.85% Ag
M.ANTONIVS (LEG.X) 32–31 BC	C.544.24	2.99	7.73	Surface layer *c*.40% Ag

Qa = quadrigatus Q = quinarius All other coins are denarii C = Crawford 1974
** As these coins are forgeries, the catalogue number does not strictly apply, and the dates are actually meaningless, being only those of the *original* types. In some cases, the coins vary slightly from the catalogue description.
ESMA = electron scanning micro analysis

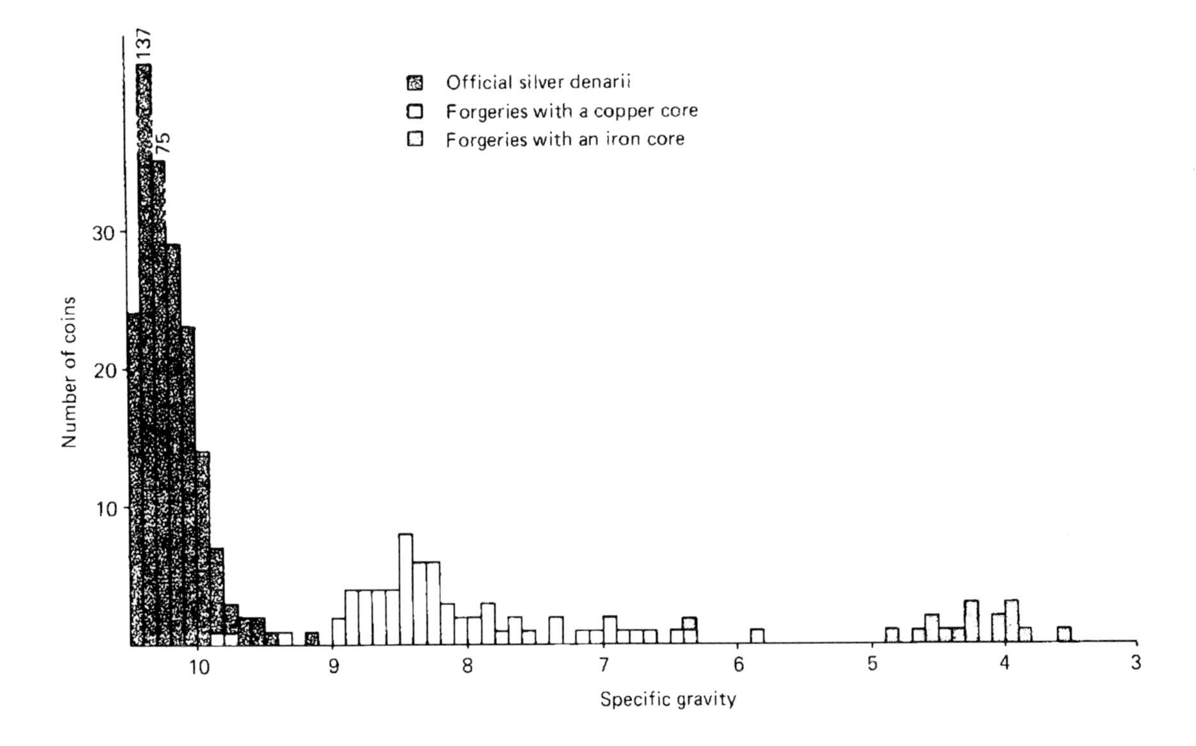

Figure 20.1. *Histogram of specific gravities of Roman Republican denarii and silver-plated forgeries.*

Methods for plating silver on to a base metal core

The addition of a substantial layer of silver to the surface of a base metal core can be carried out in two main ways:

1. by attaching an envelope of silver foil to the core; and
2. by covering the core with a layer of hard (i.e. silver–copper) solder.

In the first case the surface of the plated coin will have a high silver content (i.e. that of the original silver foil), but in the second case it will have a lower silver content, usually around the eutectic composition of 72% silver and 28% copper. Both these methods of silver plating can be carried out by two different techniques.

Silver foil can be attached to a copper or iron core by using an appropriate flux and a hard solder with a melting point below that of silver. However, in the case of copper cores, the silver foil can also be attached by a process known today as Sheffield plating in which the clean copper is covered with silver foil and then heated to a temperature well below the melting point of either metal. Under the influence of the heat, the silver and copper start to interdiffuse, forming a low melting point alloy (with a eutectic composition of 72% silver and 28% copper) at the interface between the two metals. When the temperature reaches the melting point of this alloy it melts, becoming, in effect, a layer of solder between the silver foil and the copper core. The secret of success with this method of applying silver is to stop the heating *after* the eutectic has formed but *before* all the silver foil has time to dissolve in the molten alloy. If the foil does dissolve, the coin appears, on metallurgical examination, to have been silver-plated with solder rather than with silver foil. Iron cores cannot be silver-plated by the self-soldering process as silver and iron do not form a low melting point alloy.

Silver solder can be applied to a base metal core either by wrapping the fluxed core in an envelope of a thin foil made of solder and heating to melting point, or by fluxing the core and melting granules of solder on to the surface. These two techniques are indistinguishable by scientific examination.

Experimental work on plating

One of the difficulties with the addition of a silver surface layer to a base metal core is the successful covering of the rim. It has been shown experimentally that this can be achieved by pressing the core and a piece of silver foil (thickness < 0.2 mm) into a soft material, for example a thick sheet of lead. The silver foil is thus forced up around the rim of the core (Darmstädter 1929).

In order to attempt to reproduce the metallographic structures observed on silver-plated Roman forgeries, five samples of copper, each weighing 3 grams, were individually melted in a small crucible under charcoal. The resulting small copper ingots were then reheated to 800°C and hammered to a coin-like shape. These blanks were placed upon a 0.1 mm thick disc of silver foil having a slightly larger diameter than the blank itself and then pressed into a 15 mm thick sheet of lead. The resulting 'saucer' of silver foil was trimmed so that the rim was about 1 mm deep. The process was repeated with another disc of silver foil to cover the other side of the coin. The cores with their covering of silver were then put into a crucible under charcoal and heated at 800°C for various times. They were then placed on an iron anvil and struck with a punch (copying the die of a denarius of Augustus). Figure 20.2 shows three of the resulting 'coins' which had been heated for 10, 5 and 3 minutes respectively. In all cases there has been a reaction between the silver (melting point 960°C) and the copper (melting point 1083°C) to form the eutectic (melting point 778°C), and, except for the 'coin' with the shortest heating time of 3 minutes (Figure 20.3), the eutectic has consumed most of the silver layer. The eutectic consists of copper-rich and silver-rich phases, and the former becomes oxidized after removal of the 'coin' from under the charcoal for the striking process to take place. The rather dull appearance of the eutectic on the surface can be improved by dissolving away the oxidized copper-rich component with a suitable acid, and this process could have been carried out by the Romans.

Two more blanks with their discs of silver were heated under charcoal at 800°C for 15 and 20 minutes respectively, and in these cases

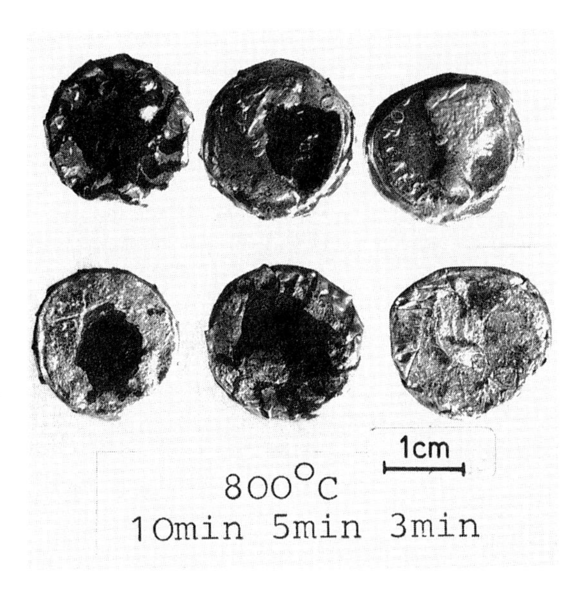

Figure 20.2. *Modern replicas of silver-plated forgeries made by enclosing copper blanks in silver foil and heating to 800°C under charcoal for various times. The silvered discs were then struck with replica dies.*

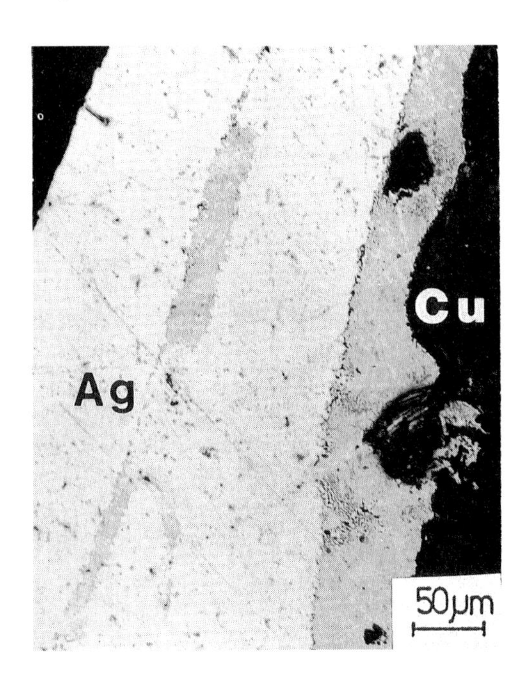

Figure 20.3. *Section at the edge of an experimental plated coin made by wrapping silver foil round a copper core and heating under charcoal for three minutes at 800°C. The section shows the copper core, the layer of eutectic and the silver foil.*

all the silver coating was absorbed into the eutectic. Analysis has shown that the surface layer contains about 40% of copper, rather more than the eutectic composition of 28%. At 800°C, the equilibrium concentration of copper in the copper-silver melt is 40%, so no more copper will dissolve out of the core into the liquid phase on the surface at this temperature. These experiments show that successful self-soldering of coin forgeries requires heating to 800°C, or even a few degrees less, for only very short times.

Other experiments on the application of a layer of silver on to the surface of copper have shown that self-soldering and the use of a separate hard solder with a silver foil are indistinguishable metallographically (La Niece 1990 and this volume Figures 18.2(b) and (c)). Plating with a hard solder and self-soldering which has been overheated so that all the silver foil has 'dissolved' would also be indistinguishable metallographically as both would have a cast silver alloy structure.

Silver-plating-on-iron cores during the Roman Republican and Early Imperial Periods

Although silver-plated-on-copper cores are the most common types of plated coins, forgeries made by plating silver on iron cores are known. The first to be recognized, because it was magnetic, was a Legionary denarius (LEG.VI) of Mark Antony (Pinkerton 1789), which led Akerman (1843–4) to equate it with the account in Pliny of Antony adding iron to the silver coinage. Akerman realized, however, that silver-plating on iron was uncommon as he 'tried with a magnet many hundreds of the legionary denarii of Antony without discovering a second example containing iron'. A recent examination with a magnet of the cabinet of plated Republican coins in the British Museum found that none of the coins were strongly magnetic and only one was weakly magnetic; a denarius of Pinarius Natta (Crawford 1974 p. 208) (Figure 20.4.3). However, analysis of the core of this coin by energy dispersive X-ray analysis in a scanning electron microscope has shown that it consists

Table 20.2 The St Swithin's Lane hoard of forged Denarii

Moneyer/ruler, date of prototype and figure number	*BM inv. number*	*Catalogue reference***	*Weight (g)*	*Specific gravity*	*Magnetic Yes*	*No*
Late 2C type	R.10910		1.908	4.50	x	
Q.THERM M.F 103 BC	R.10933 (Same dies as R.10934)	C.319.1	1.6455	4.86	x	
Q.THERM M.F 103 BC (Figure 20.5.14)	R.10934 (Same dies as R.10933)	C.319.1	2.417	4.58	x	
D.SILANVS L.F 91 BC	R.10911 (Same dies as R.10912)	C.337.3	1.635	3.94	x	
D.SILANVS L.F 91 BC	R.10912 (Same dies as R.10911)	C.337.3	Fragmentary		x	
L.PISO FRVGI 90 BC	R.10913	C.340.1	2.869	8.59		x
L.IVLIVS BVRSIO 85 BC	R.10914 (Same dies as R.10915–6)	C.352.1a	Fragmentary			x
L.IVLIVS BVRSIO 85 BC	R.10915 (Same dies as R.10914–6)	C.352.1a	3.202	8.72		x
L.IVLIVS BVRSIO 85 BC	R.10916 (Same dies as R.10914–5)	C.352.1a	2.475	6.72		x
P.CREPVSIVS 82 BC	R.10918	C.361.1	1.994	4.24	x	
CN.LENTVLVS Q 76–75 BC	R.10935	C.393.1a	2.555	6.49		x
LIBO 62 BC	R.10919	C.416.1	1.516	4.61	x	
C.(?) PISO L.F FRVGI 67(?)	R.10920	C.408(?)	2.135	3.94	x	
MN.ACILIVS 49 BC	R.10921 (Same dies as R.10922–4)	C.442.1a	3.016	8.39		x
MN.ACILIVS 49 BC	R.10922 (Same dies as R.10921–4)	C.442.1a	3.175	8.66		x
MN.ACILIVS 49 BC (Figure 20.5.15)	R.10923 (Same dies as R.10921–4)	C.442.1a	2.727	8.01		x
MN.ACILIVS 49 BC	R.10924 (Same dies as R.10921–3)	C.442.1a	3.441	8.25		x
P.ACCOLEIVS LARISCOLVS 43 BC	R.10925	C.486.1	2.201	7.16		x
L.MVSSIDIVS T.F LONGVS 42 BC	R.10926 (Same dies as R.10927)	C.494.42	1.202	3.80	x	
L.MVSSIDIVS T.F LONGVS 42 BC	R.10927 (Same dies as R.10926)	C.494.42	Fragmentary		x	
P.CLODIVS 42 BC	R.10928	C.494.23	1.657	4.08	Non-magnetic but has totally corroded iron core (XRF)	
P.CLODIVS 42 BC	R.10929	C.494.23	1.668	3.91	x	
P.CLODIVS 42 BC	R.10930	C.494.23	Fragmentary		x	

Table 20.2 *continued*

Moneyer/ruler, date of prototype and figure number	*BM inv. number*	*Catalogue reference***	*Weight (g)*	*Specific gravity*	*Magnetic Yes*	*No*
P.CLODIVS 42 BC	R.10931	C.494.23	1.481	4.27	x	
C.VIBIVS VARVS 42 BC	R.10932	C.494.38	3.227	8.89		x
M.ANTONIVS and M.BARBATIVS 41 BC	R.10936 (Same dies as R.10937)	C.517.2	Fragmentary		x	
M.ANTONIVS and M.BARBATIVS 41 BC	R.10937 (Same dies as R.10936)	C.517.2	1.824	4.47	x	
M.ANTONIVS and M.BARBATIVS 41 BC	R.10938 (Same dies as R.10939–42)	C.517.2	3.282	8.83		x
M.ANTONIVS and M.BARBATIVS 41 BC	R.10939 (Same dies as R.10938–42)	C.517.2	2.976	8.27		x
M.ANTONIVS and M.BARBATIVS 41 BC	R.10940 (Same dies as R.10938–42)	C.517.2	3.107	8.80		x
M.ANTONIVS and M.BARBATIVS 41 BC	R.10941 (Same dies as R.10938–42)	C.517.2	3.093	8.78		x
M.ANTONIVS and M.BARBATIVS 41 BC	R.10942 (Same dies as R.10938–41)	C.517.2	2.646	7.54		x
M.ANTONIVS (LEG.V) 32–31 BC	R.10943 (Same dies as R.10944 and R.10948)	C.544.18	Fragmentary		x	
M.ANTONIVS (LEG.V) 32–321 BC	R.10944 (Same dies as R.10943 and R.10948)	C.544.18	2.120	3.54	x	
M.ANTONIVS (LEG.V) 32–31 BC	R.10948 (Same dies as R.10943–4)	C.544.18	1.988	4.05	x	
M.ANTONIVS (LEG.VII) 32–31 BC	R.10945	C.544.20	2.502	6.91		x
M.ANTONIVS (LEG.VII) 32–31 BC	R.10946	C.544.20	3.112	9.30		x
M.ANTONIVS (LEG.?) 32–31 BC	R.10947 (?Same dies as R.10949)	C.544	2.213	4.33	x	
M.ANTONIVS (LEG.?) 32–31 BC	R.10949 (?Same dies as R.10947)	C.544	Fragmentary		Non-magnetic, but has totally corroded iron core (XRF)	
M.ANTONIVS (LEG.X) 32–31 BC	R.10950	C.544	Fragmentary		x	
M.ANTONIVS (LEG.?) 32–31 BC	R.10951	C.544	Fragmentary		x	
AVGVSTVS (C L CAESARES)	R.10952 (Same dies as R.10953–66)	RIC.207	3.144	8.84		x
AVGVSTVS (C L CAESARES)	R.10953 (Same dies as R.10952–66)	RIC.207	2.507	7.70		x
AVGVSTVS (C L CAESARES)	R.10954 (Same dies as R.10952–66)	RIC.207	3.060	8.64		x

Table 20.2 ***continued***

Moneyer/ruler, date of prototype and figure number	*BM inv. number*	*Catalogue reference***	*Weight (g)*	*Specific gravity*	*Magnetic Yes*	*No*
AVGVSTVS (C L CAESARES)	R.10955 (Same dies as R.10952–66)	RIC.207	3.178	8.59		x
AVGVSTVS (C L CAESARES)	R.10956 (Same dies as R.10952–66)	RIC.207	3.048	8.11		x
AVGVSTVS (C L CAESARES)	R.10957 (Same dies as R.10952–66)	RIC.207	2.818	8.22		x
AVGVSTVS (C L CAESARES)	R.10958 (Same dies as R.10952–66)	RIC.207	2.235	7.47		x
AVGVSTVS (C L CAESARES)	R.10959 (Same dies as R.10952–66)	RIC.207	3.137	8.16		x
AVGVSTVS (C L CAESARES)	R.10960 (Same dies as R.10952–66)	RIC.207	3.231	8.69		x
AVGVSTVS (C L CAESARES)	R.10961 (Same dies as R.10952–66)	RIC.207	3.176	8.87		x
AVGVSTVS (C L CAESARES)	R.10962 (Same dies as R.10952–66)	RIC.207	2.927	8.49		x
AVGVSTVS (C L CAESARES)	R.10963 (Same dies as R.10952–66)	RIC.207	2.847	8.82		x
AVGVSTVS (C L CAESARES)	R.10964 (Same dies as R.10952–66)	RIC.207	3.015	8.74		x
AVGVSTVS (C L CAESARES)	R.10965 (Same dies as R.10952–66)	RIC.207	3.385	8.97		x
AVGVSTVS (C L CAESARES)	R.10966 (Same dies as R.10952–65)	RIC.207	3.099	8.75		x
TIBERIVS (PONTIF MAXIM)	R.10967 (Same dies as R.10968–91)	RIC.25	3.465	8.97		x
TIBERIVS (PONTIF MAXIM)	R.10968 (Same dies as R.10967–91)	RIC.25	2.966	8.58		x
TIBERIVS (PONTIF MAXIM)	R.10969 (Same dies as R.10967–91)	RIC.25	3.485	8.95		x
TIBERIVS (PONTIF MAXIM)	R.10970 (Same dies as R.10967–91)	RIC.25	3.214	8.28		x
TIBERIVS (PONTIF MAXIM)	R.10971 (Same dies as R.10967–91)	RIC.25	3.365	8.67		x
TIBERIVS (PONTIF MAXIM)	R.10972 (Same dies as R.10967–91)	RIC.25	3.187	8.87		x
TIBERIVS (PONTIF MAXIM)	R.10973 (Same dies as R.10967–91)	RIC.25	3.345	8.83		x
TIBERIVS (PONTIF MAXIM)	R.10974 (Same dies as R.10967–91)	RIC.25	3.010	8.54		x
TIBERIVS (PONTIF MAXIM)	R.10975 (Same dies as R.10967–91)	RIC.25	2.820	8.09		x
TIBERIVS (PONTIF MAXIM)	R.10976 (Same dies as R.10967–91)	RIC.25	3.418	9.01		x
TIBERIVS (PONTIF MAXIM)	R.10977 (Same dies as R.10967–91)	RIC.25	3.273	8.70		x
TIBERIVS (PONTIF MAXIM)	R.10978 (Same dies as R.10967–91)	RIC.25	2.209	6.03		x
TIBERIVS (PONTIF MAXIM)	R.10979 (Same dies as R.10967–91)	RIC.25	3.061	8.70		x

Table 20.2 ***continued***

Moneyer/ruler, date of prototype and figure number	*BM inv. number*	*Catalogue reference***	*Weight (g)*	*Specific gravity*	*Magnetic Yes*	*No*
TIBERIVS (PONTIF MAXIM)	R.10980 (Same dies as R.10967–91)	RIC.25	3.102	8.40		x
TIBERIVS (PONTIF MAXIM)	R.10981 (Same dies as R.10967–91)	RIC.25	3.210	8.74		x
TIBERIVS (PONTIF MAXIM)	R.10982 (Same dies as R.10967–91)	RIC.25	3.111	8.73		x
TIBERIVS (PONTIF MAXIM)	R.10983 (Same dies as R.10967–91)	RIC.25	2.826	8.21		x
TIBERIVS (PONTIF MAXIM)	R.10984 (Same dies as R.10967–91)	RIC.25	2.648	7.26		x
TIBERIVS (PONTIF MAXIM)	R.10985 (Same dies as R.10967–91)	RIC.25	3.271	8.52		x
TIBERIVS (PONTIF MAXIM)	R.10986 (Same dies as R.10967–91)	RIC.25	2.854	8.38		x
TIBERIVS (PONTIF MAXIM)	R.10987 (Same dies as R.10967–91)	RIC.25	3.365	8.84		x
TIBERIVS (PONTIF MAXIM)	R.10988 (Same dies as R.10967–91)	RIC.25	3.107	9.07		x
TIBERIVS (PONTIF MAXIM)	R.10989 (Same dies as R.10967–91)	RIC.25	3.229	8.92		x
TIBERIVS (PONTIF MAXIM)	R.10990 (Same dies as R.10967–91)	RIC.25	3.253	8.72		x
TIBERIVS (PONTIF MAXIM)	R.10991 (Same dies as R.10967–90)	RIC.25	3.243	8.91		x
CALIGVLA	R.10992 (Same dies as R.10993)	RIC.18	3.101	8.67		x
CALIGVLA	R.10993 (Same dies as R.10992)	RIC.18	3.127	8.66		x
CLAVDIVS (TRP XI) AD 51–52 (Figure 20.5.5)	R.10994	RIC.62	2.634	7.44		x

** As these coins are forgeries, the Crawford and RIC numbers do not strictly apply, and the dates are actually meaningless, being only those of the *original* types. In some cases, the coins vary slightly from the description given in Crawford or RIC.
XRF = X-ray fluorescence analysis.

mainly of copper containing only a few per cent of iron. Two other plated denarii, of Julia Mamea (RIC 362) and Severus Alexander (RIC 160) in the Zwicker collection at Erlangen University, were also found to be magnetic, and metallographic investigation of the cores similarly showed that they consisted of copper containing about 10% iron.

The British Museum does, however, contain a hoard of denarii found before 1856 in St Swithin's Lane, King William Street, in the City of London (Lawrence 1940). Lawrence thought that the whole hoard consisted of silver-plated-on-copper coins, but testing with a magnet has shown that some of the coins are magnetic, and were, therefore, originally made by plating silver on to iron cores.

Lawrence stated that the hoard contained 89 coins (1940 p. 188), but his detailed list only included 88 coins (1940 pp. 186–7). The trays holding the hoard now contain only 84 coins, and, as a recent study by Andrew Burnett has resulted in some changes to the identifications, the hoard is listed again in Table 20.2.

The latest coin type in the hoard is a plated denarius of Claudius, indicating burial of the

Table 20.3 Forgeries made by plating silver on an iron core

Moneyer and date of prototype and figure number	*Location and inventory number*	*Catalogue reference*	*Provenance*
L.PROCILIVS F 80 BC (Figure 20.4.14)	RLM 7232	C.379.2	Novaesium (Neuss), Germany (Zedelius 1988, no.1)
M.SCAVRUS and P.HVPSAEVS 58 BC (Figure 20.4.15)	BM 1988.9–17.1	C.422.1	Found in a hoard of 217 denarii buried at Sutton, Suffolk, in *c.* AD 41 (Bland 1992 p.29, no.64) Wt: 3.371g SG: 6.38
Q.SICINIVS and C.COPONIVS 49 BC (Figure 20.5.1)	RLM 4687	C.444.1a	Martberg (Pommern a.d. Mosel), Germany (Zedelius 1988, no.2)
M.ANTONIVS (LEG.VI) 32–31 BC (Figure 20.5.2)	AM	C.544.19	Unknown (Akerman 1843–4 p. 68)
M.ANTONIVS (LEG.VI[) 32–31 BC (Figure 20.5.3)	PC	C.544.19–22	Said to have been found at the Roman site of Great Chesterford, Cambridgeshire Wt: 2.063g SG: 4.21
M.ANTONIVS (LEG.XII) 32–31 BC	RLM 86.0126	C.544.26	Vetera (Xanten), Germany (Zedelius 1988, no.3; Zedelius 1985)
AVGVSTVS* 29–27 BC	RLM 89.0006	BMC.647	Alflen (Cochem, Mosel), Germany
VITELLIVS GALBA AD 69 } hybrid	RLM 86.0147	Ob:BMC.112 Rv:BMC.190	Vettweis (Düren), Germany (Zedelius 1988, no. 9)

* = quinarius. All other coins are denarii.
RLM = Rheinisches Landesmuseum, Bonn
AM = Ashmolean Museum, Oxford
BM = British Museum, London
PC = Private collection
C = Crawford 1974
BMC = British Museum Catalogue

hoard in the early years of the conquest of Britain. The fact that there are numerous die duplicates in this hoard indicates that most, if not all, of the forgeries must be more or less contemporary with the date of burial as it is improbable that so many groups of 'coins' made at the same time would have stayed together for many years. Therefore, it is unlikely that the plated Legionary denarii of Mark Antony were made in the lifetime of Mark Antony himself and Pliny was probably in error when he castigated Antony for adulterating the coinage with iron. However, the plated Legionary denarii of 32–31 BC were clearly current in some numbers when Pliny wrote his *Natural History* in the mid-lst century AD, and Pliny therefore assumed that they had been made by the authority of Antony.

Apart from these 23 examples of forgeries with iron cores in the St Swithin's Lane hoard, a few others have come to light in recent years, and are listed in Table 20.3. Five of the coins have been investigated metallographically by polishing a section on the rim of the coin. This minimizes the visible damage to the coin, yet enables the interface between the plated layer and the core to be examined in detail. The

Figure 20.4. *For key to coins see Table 20.5.*

Figure 20.5. *For key to coins see Table 20.5.*

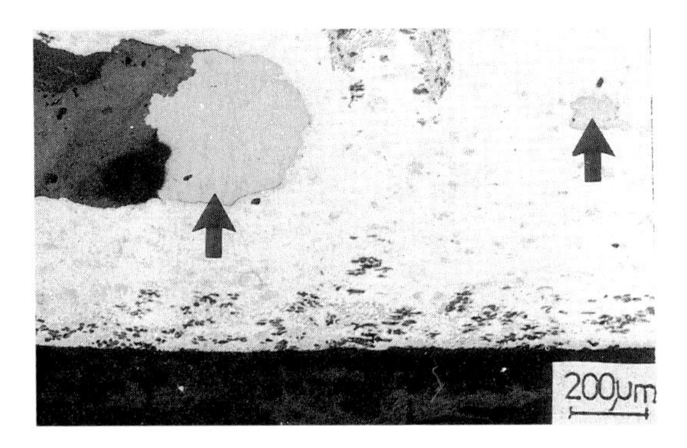

Figure 20.6. *Section at the rim of a forged denarius copying Procilius (Figure 20.4.14) showing the silver–copper alloy plating with the tips of two 'teeth' (arrowed) of the iron core showing through.*

denarius copying Procilius (Zedelius 1988, no.1), of a type struck in 80 BC (Crawford 1974 379.2) (Figure 20.4.14), has a chisel cut on the obverse, showing that it was suspect in antiquity. It is interesting that it has a serrated edge and the micrograph (Figure 20.6) shows the tips of two of the 'teeth' (arrowed). The surface layer is a silver–copper alloy which originally contained *c*.70% silver. It has suffered some corrosion of the copper-rich phase near the surface.

Another denarius copying Sicinius and Coponius (Zedelius 1988, no. 2), of a type struck in 49 BC (Crawford 1974, 444.1a) (Figure 20.5.1), also has a cut mark on the surface, this time made with a square punch. Examination

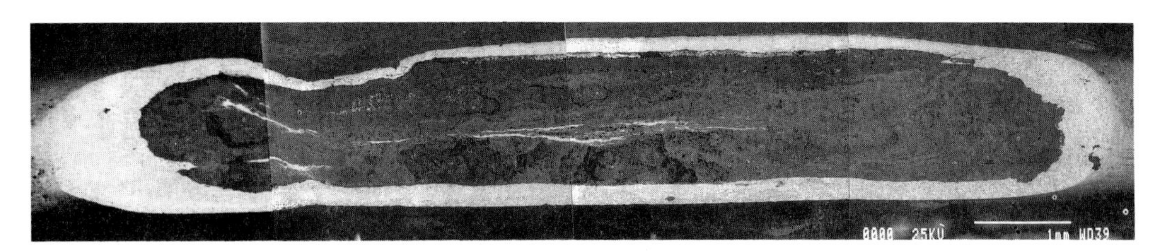

Figure 20.7. *Cross-section at the rim of a forged legionary denarius of Mark Antony (Figure 20.5.3) to show the silver–copper alloy plating surrounding the corroded iron core.*

Figure 20.8. *Detail of the section shown in Figure 20.7.*

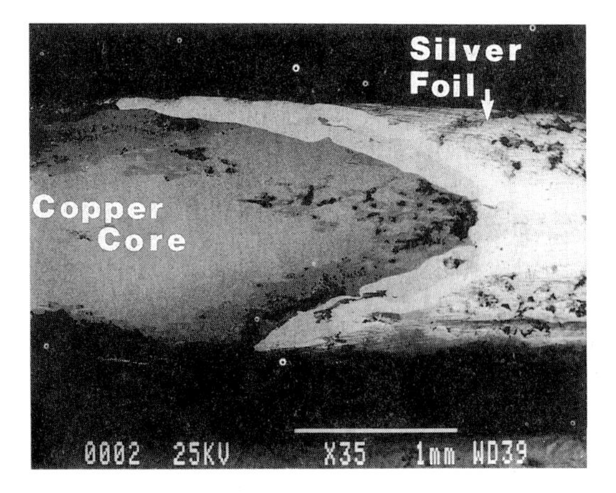

Figure 20.9. *Section through the rim of a forged anonymous denarius (Figure 20.4.2), of a type issued in 206–195 BC, showing the copper core, silver foil on surface and intermediate layer of eutectic.*

of an area polished on the rim into the silver layer again indicated that it consists of a silver–copper alloy containing *c.*70% silver.

The legionary denarius (LEG.VI) of Mark Antony (Crawford 1974, 544. 19–22), reputedly found at Great Chesterford (Figure 20.5.3), has a similar structure for the surface layer which originally contained *c.*70% silver (Figure 20.7). The copper-rich phase has been lost by corrosion. The metallographic structure is illustrated in Figures 20.7 and 20.8.

Two other silver-plating-on-iron coins have been examined metallographically, another legionary denarius (LEG.XII) (Zedelius 1988, no. 3; Zedelius 1985) of Mark Antony (Crawford 544.26) and a quinarius of Augustus in the Zwicker collection at Erlangen University (BMC 647), of a type minted between 29 and 27 BC. In both cases the iron core was surrounded by a layer of silver–copper alloy which originally contained *c.*70% silver.

In view of the fact that the official Republican coinage mainly consisted of essentially fine silver, it is surprising that these silver-on-iron forgeries have a surface layer containing only *c.*70% silver. However, no evidence has been found for the use of a fine silver foil envelope and the metallographic examinations of these seven coins indicate that, in all cases, the silver–copper alloy on the surface was at one stage molten. It is now impossible to know precisely how it was applied; it may have been a conventional soldering process in which hard solder was melted onto the surface using a blowpipe and suitable flux (perhaps borax, see Zwicker 1986), or perhaps the iron blanks were dipped into a molten silver–copper alloy. It is also surprising that the cores were made of iron, which is less dense than copper and would, therefore, be more likely to be detected by weighing. However, the use of an iron core covered with a low melting silver–copper alloy might be connected with the need for the coin to 'ring' when dropped or tapped on a hard surface, as noted in the Fourth Annual Report of the Royal Mint (1870 p. 26) which discussed the manufacture of forgeries of silver coins in 19th-century England. The white colour of the iron core would also be less easily detected by cutting or punching tests than the reddish colour of a copper core.

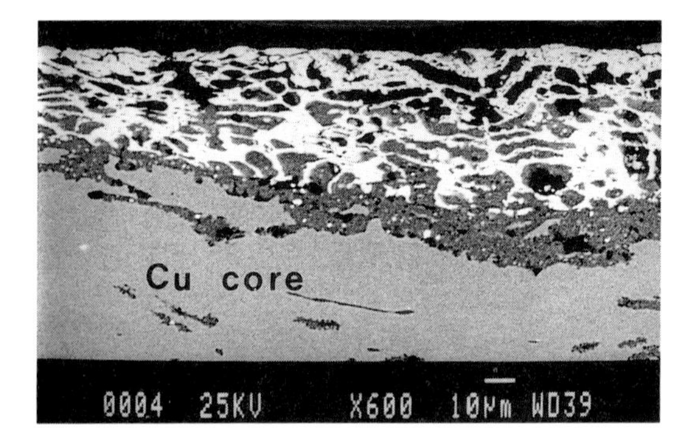

Figure 20.10. *Section through the rim of a serrated denarius copying L. Porcius Licinius (Figure 20.4.6) showing a layer of eutectic on the surface of the copper. Note the continuous nature of the silver-rich phase at the actual surface.*

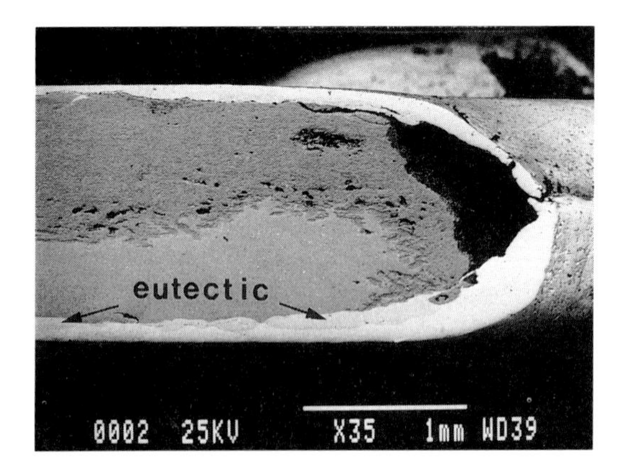

Figure 20.11. *Section through the rim of a forged denarius copying L. Calpurnius Piso (Figure 20.4.8) of 90 BC showing a butt join in the silver foil and a discontinuous layer of eutectic between the silver foil and the copper core.*

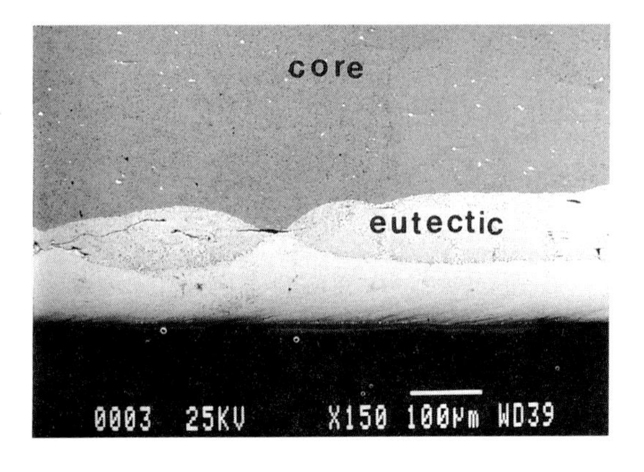

Figure 20.12. *Detail of the section at Figure 20.11.*

Silver-plating-on-copper

The 'earliest' coin to be examined was an anonymous denarius of the period 225–214 BC (Crawford 28.3) (Figure 20.4.1). A layer of silver foil covered the surface and was attached to the core by a layer of silver–copper eutectic. A similar plating was found on another anonymous denarius of the period 206–195 BC (Crawford 114.1) (Figure 20.4.2). A section through the rim of the coin (Figure 20.9) clearly shows the copper core and silver foil, with an intermediate layer of eutectic. At the very edge of the coin the eutectic has penetrated between the join in the two sheets of foil, which butt together rather than overlap, and is visible (under the microscope) on the actual surface.

In most cases, the silver foil does overlap along the edge of the coin. For example, a silver plated denarius copying Sextus Pompeius (Crawford 235.1a) (Figure 20.4.4) showed two layers of foil along the edge, separated from each other and from the copper core by layers of eutectic. Analysis of the silver layer showed it to contain 94.8% silver.

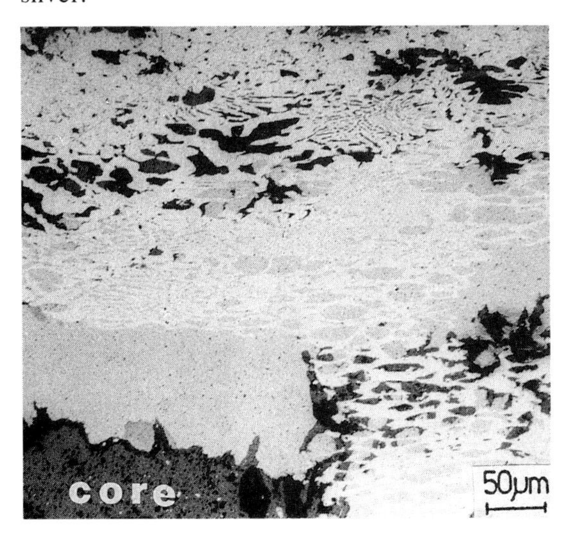

Figure 20.13. *Section through the rim of a forged denarius copying Mark Antony and Octavian (Figure 20.4.12) showing a layer of eutectic on top of the copper core. The surface of the eutectic appears to consist mainly of a silver-rich phase, possibly as a result of deliberate surface enrichment.*

As noted above, some Republican denarii were struck on flans with a serrated edge and these issues were also copied by the forgers. Four silver-plated '*serrati*' were metallographically examined for this study; three copies of types originally struck in 118 BC {L. Porcius Licinius (Crawford 282.5) (Figures 20.4.6 and 20.4.7) and M. Aurelius Scaurus (Crawford 282.1) (Figure 20.4.5)} and the fourth was of a type struck in 68 BC by C. Hosidius (Crawford 407.1) (Figure 20.4.9).

Metallographic examination of the edge of the three 'earlier' *serrati* coins showed only the presence of silver–copper alloy and failed to

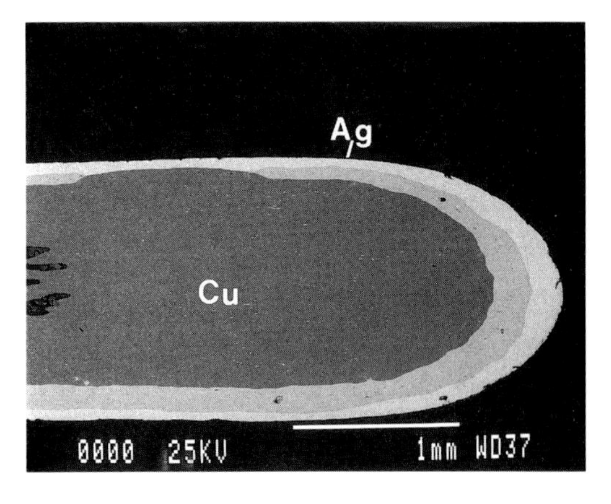

Figure 20.14. *Section through the rim of a forged denarius copying Claudius from the St. Swithin's Lane hoard (Figure 20.5.5) showing layers of silver and eutectic on top of the copper core.*

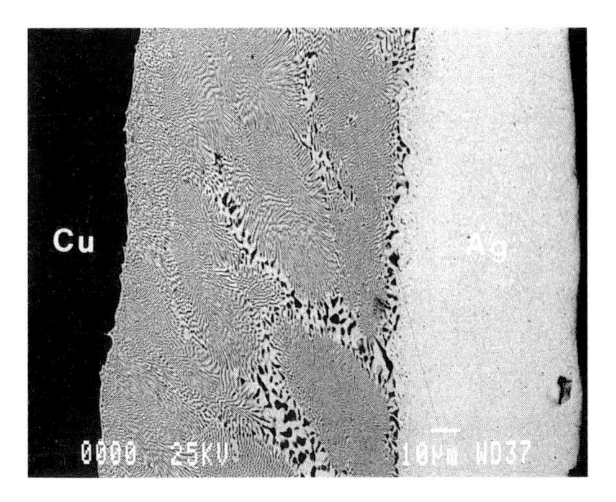

Figure 20.15. *Detail of the section shown at Figure 20.14.*

Table 20.4 Metallographic examination of forged denarii of the Imperial period

Ruler and date of prototype	*Catalogue reference*	*Analyst*	*Metallographic result*	*Comment and figure number*
Tiberius (Pontif Maxim)	RIC.25	C	Foil and Eutectic	Campbell No.37 and 176–8
Claudius AD 41–54	BMC.107	EN	Foil and Eutectic	Figure 20.5.4
Claudius AD 51–52	BMC.69	BM	Foil and Eutectic	St. Swithin's Lane Hoard No. R.10994 Figure 20.5.5
(Civil War) AD 68–69	BMC.34	EN	Eutectic	Figure 20.5.6
Vespasian AD 70	BMC.17	EN	c.85% Ag/Cu alloy	Figure 20.5.7
Domitian AD 81–96		C	Foil and Eutectic	Campbell No. 3 and 22–6
Domitian AD 81–84	BMC.52	BM	Foil and Eutectic	BM 1952.10–11.32 Figure 20.5.8
Domitian AD 82	BMC.32	BM	Foil and Eutectic	BM 1951.5–6.742 Carradice 1984, No.3 Figure 20.5.9
Domitian AD 84	ob:hd left rv:BMC.48	BM	Foil and Eutectic	BM.R.11428 Figure 20.5.10
Hadrian AD 117–138	BMC.1054	EN	c.60% Ag/Cu alloy	NB This is a cistophorus Figure 20.5.11
Septimius Severus AD 196–7	RIC.78a	EN	c.40% Ag/Cu alloy	Figure 20.5.12
Julia Maesa (AD 218–222)	RIC.268	BM	Eutectic	Found at Niska Kamenica BM 1978.9–10.20 Figure 20.5.13

C = Campbell 1933
EN = Analysed at the University of Erlangen-Nürnberg
BM = Analysed at The British Museum

reveal the presence of any silver foil. It would presumably have been easier to coat the serrated edge of the blank with a silver–copper alloy than by using silver foil. Figure 20.10 illustrates a section at the surface of one of the L. Porcius Licinus coins (Figure 20.4.7) where it can be seen that, at the outer surface of the layer of eutectic, the silver-rich phase is more or less continuous, and this would have given the appearance of being struck in fine silver. This effect might have been achieved by pickling the blank in dilute acid to remove some of the copper-rich phase before striking.

In the case of the denarius copying C. Hosidius (Figure 20.4.9), a metallographic section on the rim did show that some silver foil was present, with a composition of >95% silver.

A section through the rim of a denarius copying L. Calpurnius Piso, of a type originally issued in 90 BC (Crawford 340.1) (Figure 20.4.8), again shows that the silver foils butt together at the edge of the coin, without a significant overlap (Figure 20.11, cf. Figure 20.9). Analysis of the foil showed it to contain 97% silver. The growth of the eutectic

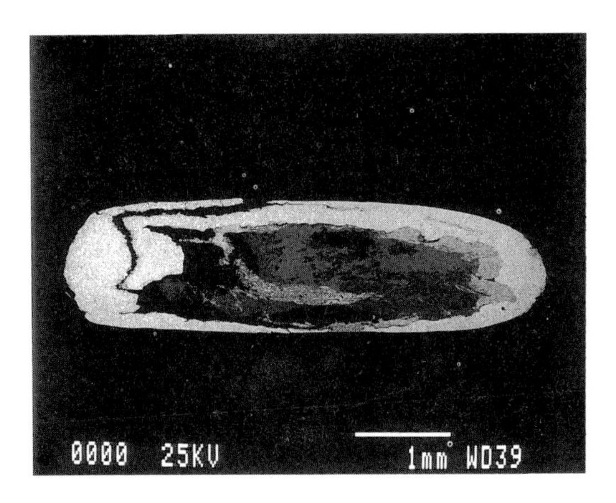

Figure 20.16. *Section through the rim of a forged denarius copying Domitian (Figure 20.5.8) showing how the silver foil has been wrapped round the core.*

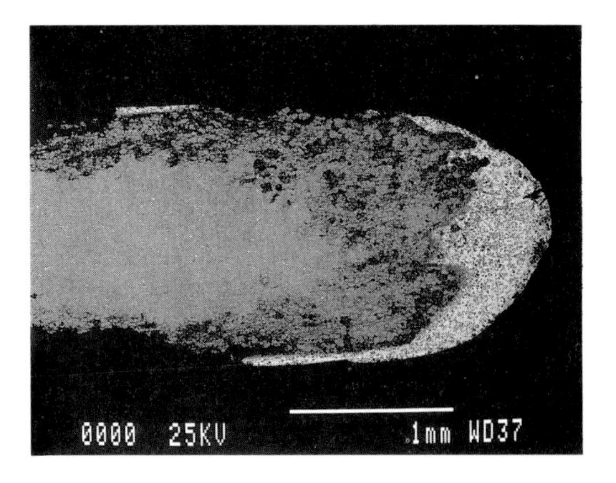

Figure 20.18. *Section through the rim of a forged denarius copying Julia Maesa (Figure 20.5.13) showing that the plating consists of a layer of copper-silver alloy.*

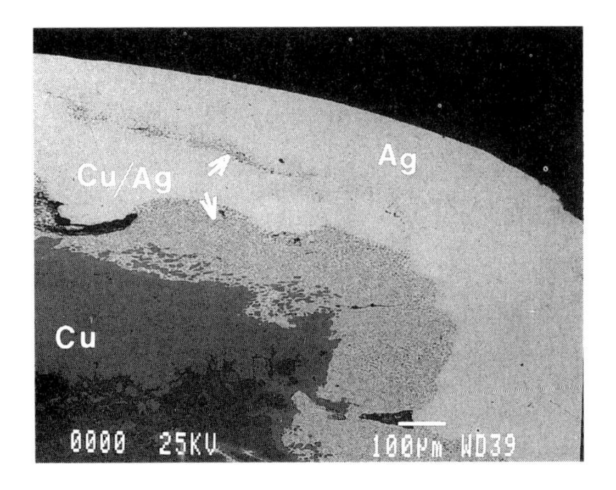

Figure 20.17. *Detail of the section shown at Figure 20.16.*

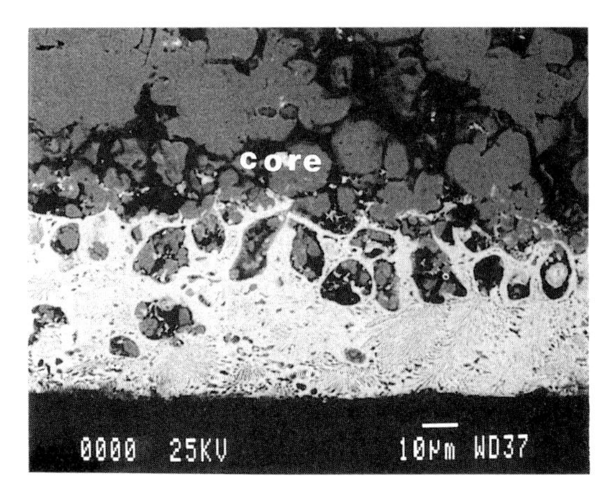

Figure 20.19. *Detail of the section shown at Figure 20.18.*

between the foil and the core was irregular, leading to the formation of a discontinuous layer. This is illustrated in Figure 20.12, where the formation of the eutectic by the dissolution of both copper and silver can clearly be seen.

Metallographic examination of a plated denariuis copying Q. Sicinius, of a type issued in 49 BC (Crawford 440) (Figure 20.4.10), revealed only eutectic on the surface, although a plated denarius copying Q. Metell. Scipio, of a type issued in 47–46 BC (Crawford 461.1) (Figure 20.4.11), retained some of the silver foil which had a composition of more than 95% silver. The reverse of this coin was an incuse version of the obverse.

A plated denarius copying Mark Antony and Octavian in the Zwicker Collection at Erlangen University, of a type issued in 41 BC (Crawford 517.2) (Figure 20.4.12), is another example where the surface has been enriched in silver by dissolution of the copper-rich phase before striking. A micrograph of a section on the edge (Figure 20.13) showed that only silver–copper eutectic is present, but that along the outer surface the silver-rich phase predominates.

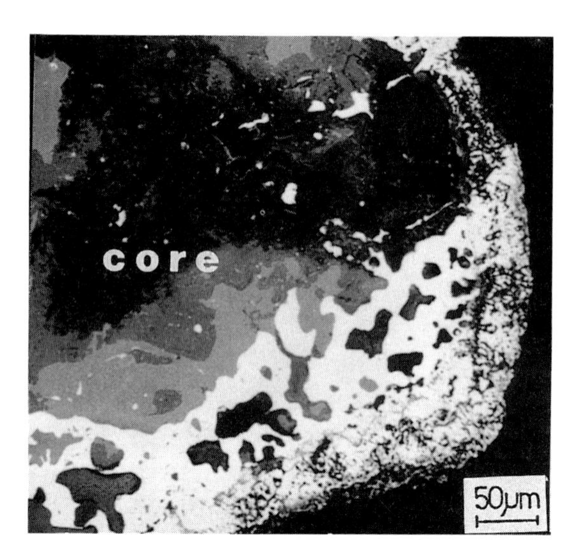

Figure 20.20. *Section through the rim of a silver plated blank from a forgers' workshop excavated in Insula 50 at Augst showing that the plating consists of a layer of silver–copper alloy.*

The final Republican coin which was examined, copying an issue of Octavian (Crawford 538.1) (Figure 20.4.13), retained evidence for the use of silver foil in the silvering process. Analysis showed this to contain 95% silver.

In addition to these Republican forgeries, a number of Imperial period plated forgeries have also been examined metallographically and the results are summarized in Table 20.4. The results indicate that the use of silver foil continued at least until the end of the 1st century AD. Good examples are denarii of Claudius (Figure 20.5.5) and Domitian (Figure 20.5.8). Sections polished on the edge of these coins are illustrated in Figures 20.14, 20.15 and 20.16, 20.17 respectively. In both cases the silver foil envelope completely surrounds the copper core, with an intermediate layer of silver–copper eutectic. Analysis shows that the silver layer on the coin of Claudius contains 98% silver and the eutectic contains 78% silver.

It is interesting to note that in one area of the coin of Claudius (Figure 20.5.5) there is no eutectic layer between the silver foil and the copper core (Figures 20.14 and 20.15). Otherwise the visible eutectic layer is very even. It is, however, impossible to say whether the eutectic layer is the result of a self-soldering (Sheffield plating) process or of the use of a layer of hard solder between the silver and copper. In the case of the coin of Domitian (Figure 20.5.8), the section (Figure 20.16) shows how the silver foil has been wrapped round the copper core. At one end of the section, the overlap is very clear, but it appears to continue across the bottom of the coin where it can be traced as a fine line of silver–copper eutectic. This is more clearly seen in the detail (Figure 20.17), in which the eutectic between the silver surface layer and the copper core is also clearly visible.

As with the Republican forgeries, some of the plated Imperial coins only seem to have silver–copper alloy surrounding the core, with no trace of a 'pure' silver surface layer. It is impossible to know whether these represent forgeries originally made with silver foil, but which have been overheated so that all the foil has dissolved, or whether they were only coated with a silver–copper solder in the first place.

From the time of Vespasian, however, the debasement of the official silver coinage meant that it was no longer necessary to attach a silver foil to the surface of a copper blank in order to make a passable forgery, as coating the surface with a 'solder' made from the debased official coins gives the correct visual appearance. It is also possible to coat the cores with a copper-rich solder, and then pickle the surface before striking in order to artificially enrich the surface, a process apparently observed on a denarius of Mark Antony and Octavian (above p. 239).

A plated denarius of Julia Maesa (Figure 20.5.13) found at Niska Kamenica (BMC 1978. 9–10. 20) is an example of a forgery plated with silver–copper eutectic (Figures 20.18 and 20.19). The surface layer has been partially lost as a result of corrosion, which has penetrated deeply into the core. A detail (Figure 20.19) shows the eutectic structure of the layer and that it has started to dissolve the copper core, which probably explains the uneven profile of the edge of the core seen in Figure 20.18.

The excavation of the forger's workshop in insula 50 at Augst has produced rods of copper for making the blanks, as well as copper

Table 20.5 Key to coins illustrated in Figures 20.4 and 20.5

Figure number	*Moneyer/ruler and date of prototype*	*Collection and inventory number*	*Catalogue reference*
20.4.1	Anonymous 225–212 BC	ER.FOd	C. 28.3
20.4.2	'Rostrum Tridens' 206–195 BC	BM.1951.5–6.7	C.114.1
20.4.3	PINARIVS NATTA 149 BC	BM.1951.5–6.18	C.208.1
20.4.4	SEXTVS POMPEIVS 137 BC	ER.F24b	C.235.la
20.4.5	L. LICINIVS and CN. DOMITIVS with M. AVRELIVS SCAVRVS 118 BC	BM.1951.5–6.91	C.282.1
20.4.6	L. LICINIUS and CN. DOMITIVS with L. PORCIVS LICINIVS 118 BC	BM.1951.5–6.84	C.282.5
20.4.7	L. LICINIUS and CN. DOMITIVS with L. PORCIVS LICINIVS 118 BC	BM.1951.5–6.85	C.282.5
20.4.8	L. PISO FRVGI 90 BC	BM.1951.5–6.205	C.340.1
20.4.9	C. HOSIDIVS C.F GETA 68 BC	ER.F199	C.407.1
20.4.10	Q. SICINIVS 49 BC	BM.1951.5–6.373	C.440.1
20.4.11	Q. METELL. SCIPIO IMP with EPPIVS LEG.F.C 47–46 BC	ER.F260b	C.461.1
20.4.12	C. VIBIVS VARVS 42 BC	ER.F462	C.494.36
20.4.13	IMP. CAESAR DIVI F 37 BC	BM.1951.5–6.499	C.538.1
20.4.14	L. PROCILIVS F 80 BC	RLM.7232	C.379.2
20.4.15	M. SCAVRVS and P. HVPSAEVS 58 BC	BM.1988.9–17.1	C.422.1
20.5.1	Q. SICINIVS and C. COPONIVS 49 BC	RLM.4687	C.444.1a
20.5.2	M. ANTONIVS (LEG. VI) 32–31 BC	AM	C.544.19
20.5.3	M. ANTONIVS (LEG. VI) 32–31 BC	PC	C.544.19–22
20.5.4	Claudius AD 41–54	ER.G251	BMC.107
20.5.5	Claudius AD 51–52	BM.R.10994	BMC.69
20.5.6	(Civil War) AD 68–69	ER.G340	BMC.23
20.5.7	Vespasian AD 70	ER.G451a	BMC.17
20.5.8	Domitian AD 81–84	BM.1952.10–11.32	BMC.52
20.5.9	Domitian AD 82	BM.1951.5–6.742	BMC.32

Table 20.5 ***continued***

Figure number	*Moneyer/ruler and date of prototype*	*Collection and inventory number*	*Catalogue reference*
20.5.10	Domitian AD 84	BM.R.11428	ob: head left rv: BMC.48
20.5.11	Hadrian AD 117–138	ER.895a	BMC.1054
20.5.12	Septimius Severus AD 196–7	ER.H340c	RIC.78a
20.5.13	Julia Maesa AD 218–222	BM.1978.9–10.20	RIC.268
20.5.14	Q. THERM M.F 103 BC	BM.R.10934	C.319.1
20.5.15	MN. ACILIVS 49 BC	BM.R.10923	C.442.1a

BM = British Museum, London
ER = Library of University of Erlangen
AM = Ashmolean Museum, Oxford
RLM = Rheinisches Landesmuseum, Bonn
PC = Private Collection
C = Crawford 1974
BMC = British Museum Catalogue
RIC = Roman Imperial Coinage

blanks, both with and without silver plating (Peter 1990, Peter and Zwicker 1988). In addition, silver-plated forgeries of coins of Commodus were discovered. Figure 20.20 illustrates a section on the rim of a silver-plated blank showing that the silvering consists of a layer of silver–copper solder. Analysis of the core shows that it is copper, containing about 3% tin.

By the reign of Caracalla, early in the 3rd century AD, the silver content of the official coinage had dropped to *c.* 50%, and in the reign of Macrinus it dipped to between 40 and 50% (Walker 1978 Figure 8), so that the silver content of plated layers produced with hard solders might sometimes have been higher than that of the official 'silver' coinage.

As a result of the continuing debasement of the official coinage in the mid-3rd century, the mint itself began to enrich the surfaces of the debased antoniniani by pickling in acid before striking. Considerable work has been done on the composition of these coins (Cope 1972, Hammer 1908, Zwicker 1984), which were so debased by the reign of Postumus that the mint then found it necessary to plate the coins with a very thin 'wash' of silver. This silver wash is a feature of the coinage until after the reform of Diocletian. However, with the reintroduction of a high purity silver coinage (the siliqua) at the end of the 4th century, forgers returned to the use of a silver foil envelope (Zwicker, unpublished analyses).

Conclusion

This survey has confirmed the results of previous investigations in showing that silver-plated forgeries of the Roman period are sometimes covered with a layer of silver foil, attached to the core with a layer of silver–copper eutectic, but sometimes covered only with a layer of silver–copper solder. In most cases the core is copper, but a small number have iron cores. Up to the end of the 1st century AD, when the official silver coinage was struck in high purity silver, about half of the forgeries examined were covered with silver foil of a similar composition to that of the official coinage. In those cases where only a silver–copper alloy has been detected on the surface, it is possible that silver foil was used originally, but that overheating during the manufacture of the forgery caused all the silver to dissolve. However, if such alloy-plated coins were originally made with silver foil to 'match' the composition of the regular coins, the forgers were not deterred from putting them into circulation, even though the surface was more base then originally intended. In fact, the colour and composition of the actual surface could have been improved by pickling the silvered blanks in acid before striking.

After the debasement of the official silver coinage, which began in the later years of the 1st century AD, it is unlikely that silver foil would have been used and the three coins which have been examined were all plated with a silver–copper alloy.

Acknowledgements

We are grateful for technical assistance, for providing coins for analysis, and for discussion, to Mr B.T. Curtis of A.H. Baldwin & Sons Ltd; Dr A. Burnett, Dr R. Bland, Miss C. Enderly and Mr T. Springett, British Museum; Prof. Dr Ch.J. Raub, Institut für Edelmetalle und Metallchemie, Schwäbisch Gmünd; Mr T. Kroha, Münzkabinett, Köln; Dr V. Zedelius, Rheinisches Landesmuseum; Dr K. Wickert and Dr H.O. Keunecke, Bibliothek Universität Erlangen-Nürnberg; Mrs E. Grembler, Mrs B. Röhl, Mr P. Hochholdlinger, Mr H. Lassner, Mr K. Nigge and Mr B. Seitz, Lehrstuhl Werkstoffwissenschaft Metalle, Universität Erlangen-Nürnberg.

References

Akerman, J. Y. (1843–4) On the forgeries of public money. *Numismatic Chronicle*, **6**, 57–82

Bailey, K. C. (1929) *The Elder Pliny's Chapters on Chemical Subjects*, Edward Arnold, London

Bahrfeldt, M. (1884) Die gefütterten Münzen aus der Zeit der Römischen Republik. *Numismatische Zeitschrift*, **16**, 309–366

Bland, R. (1992) *The Chalfont Hoard and other Roman Coin Hoards*, British Museum, London

BMC (1923ff) *Coins of The Roman Empire in The British Museum*, 6 vols., British Museum, London

Campbell, W. (1933) *Greek and Roman Plated Coins*, The American Numismatic Society Numismatic Notes and Monographs No.57, New York

Carradice, I. (1984) Plated Denarii of the Flavian Period: A Supplement. *Numismatic Circular*, **92**(10), 321

Cope, L. H. (1972) Surface-silvered ancient coins. In *Methods of Chemical and Metallurgical Investigation of Ancient Coinage* (eds E. T. Hall and D. M. Metcalf) Royal Numismatic Society Special Publication No.8, 261–278

Crawford, M. H. (1968) Plated coins – false coins. *Numismatic Chronicle* (seventh series), **8**, 55–59

Crawford, M. H. (1974) *Roman Republican Coinage*, Cambridge University Press, Cambridge (reprinted 1989)

Dahl, O. (1931) Die Arbeitsmethoden der antiken Münztechnik, insbesondere der Falschmünzerei. *Z. Metallwirtschaft*, **10**, 653–663

Dahl, O. and Schwartz, N. (1933) Silberüberzug einer antiken Münze. *Berliner Münzblätter*, **53** (363/4), 49–50 and Table 132

Darmstädter, E. (1929) Subaerate Münzen und ihre Herstellung. *Mitteilungen der Bayerischen Numismatischen Gesellschaft*, **47**(7), 27–38

Fried, W. and Hammer, P. (1974) Über einen plattierten römischen Denar. *Numismatische Beiträge*, No. 2, 44–48

Hammer, I. (1908) Der Feingehalt der griechischen und römischen Münzen. *Zeitschrift für Numismatik*, **26**, 99–102

Hammer, P. (1984) Münzen aus plattierten metallen. *Numismatische Beiträge*, **31**(1), 14–28

Kalsch, E. and Zwicker, U. (1968) Untersuchungen über Platierungen antiken Münzen mit Hilfe der Mikrosonde. *Michrochimica Acta* Suppl. III, 210–220

La Niece, S. (1990) Silver plating on copper, bronze and brass. *The Antiquaries Journal*, **70**(1), 102–114

La Niece, S. (1993) Technology of silver-plated coin forgeries. In *Metallurgy in Numismatics III* (eds M. M. Archibald and M. R. Cowell) Royal Numismatic Society Special Publication No.23, 227–236

Lawrence, L. A. (1940) On a hoard of plated Roman Denarii. *Numismatic Chronicle* (5th series), **20**, 185–189

Peter, M. and Zwicker, U. (1988) *Untersuchungen von Fundobjekten von Falschmünzerwerkstätten aus Augst und vergleichende Untersuchungen zur Herstellung von subäraten Denaren*, UB 455/88, Lehrstuhl Werkstoffwissenschaften Metalle, Universität Erlangen Nürnberg

Peter, M. (1990) *Eine Werkstätte zur Herstellung von Subaeraten Denaren in Augusta Raurica*, Studien zu Fundmünzen der Antike Band 7, Berlin

Pinkerton, J. (1789) *An Essay on Medals* (new ed.), J. Edwards and J. Johnson, London

RIC (1923ff) *Roman Imperial Coinage*, vols. 1–9, Spink, London

Royal Mint (1870) *First Annual Report of the Deputy Master of the Mint, 1870*, HMSO, London, 1871

Walker, D. R. (1978) *The Metrology of the Roman Silver Coinage: Part III from Pertinax to Uranius Antoninus*, BAR Supplementary Series No.40, Oxford

Walker, D. R. (1980) The silver content of the Roman Republican coinage. In *Metallurgy in Numismatics I* (eds D. M. Metcalf and W. A. Oddy) the Royal Numismatic Society Special Publication No.13, 55–72

Zedelius, V. (1985) Eisen im Silbergeld: ein Legionsdenar des Marcus Antonius aus Vetera. *Das Rheinische Landesmuseum Bonn, Berichte aus der Arbeit des Museums*, **85**(1), 10–11

Zedelius, V. (1988) Nummi Subferrati. *Rivista Italiana di Numismatica e Science Affini*, **110**, 125–130

Zwicker, U. (1984) Die Münzmetalle, numismatische und metallurgische Probleme am Beispiel antiker und mittelalterlicher Münzen. *Mitteilungen der Universität Clausthal*, **56**, 37–44

Zwicker, U. (1986) Untersuchungen an Silber aus den Grabungen von Ugarit (Ras Shamra, Syrien) und Vergleich mit Silber von Antiken Münzen. *Acta Praehistorica et Archaelogica*, **18**, 157–176

21

Surface characterization of tinned bronze, high-tin bronze, tinned iron and arsenical bronze

Nigel Meeks

Abstract

Tin coated bronze antiquities are found in museum collections on Greek and Etruscan items dating from around the 5th century BC, and later on artefacts throughout the Roman world and Europe and less extensively in various Asian regions. Tinned items of decorative bronze, military ware etc. would resemble silver in colour and be an inexpensive substitute, and tinned bronze mirrors would have been sought for their reflective properties. Copper cooking vessels were tinned to prevent tainting of food. Iron objects protected by tinned surfaces became popular from the medieval period although the earliest tinned iron object is from Spain around 450 BC.

The technological significance of tin-enriched bronze surfaces from the curatorial and conservation viewpoint requires their correct identification, apart from ensuring their distinction from silver. Ambiguity arises because 'tin' on a bronze surface can occur in several ways, either from the tinned surfaces of a low-tin bronze, which can show different compound surfaces corresponding to heat treatment, or the object may actually be a cast high-tin bronze. Additionally tin-sweat may appear on a low tin-bronze casting, and corrosion can enrich surfaces with tin corrosion products and cause colour changes.

Silver coloured surfaces on ancient arsenical copper alloys occur on various artefacts. Their structures have been examined and explanations for their occurrence are discussed.

The characterization of tin-enriched bronze surfaces and arsenic-rich surfaces on antiquities can be achieved by microscopy, microanalysis and X-ray diffraction analysis. Specific intermetallic compound structures are found to occur with different types of surface.

This chapter presents the results of the surface examination of various tinned artefacts, high-tin bronzes, comparative experimental tinned material, and arsenical copper showing the microstructural and analytical characteristics that enable identification to be made of the enriched, silver coloured surfaces on antiquities.

Introduction

Antiquities made of copper–tin alloys are generally classified as either low-tin bronze (up to about 14% tin), or of high-tin bronze (about 19–27% tin) and there are significant differences in physical properties between these alloys. There are a number of reasons why bronze antiquities can possess tin-rich surfaces. These include processes deliberately chosen by the ancient metal craftsman, such as tinning, where tin has been applied to the surface of low-tin bronze to produce a silver coloured surface, and also casting with a high-tin bronze alloy which, when polished, is silver in colour. There are also the natural processes; tin sweat from casting low-tin bronze which causes a silvery surface, and corrosion or patination during burial which can cause various colour changes and selective enrichment of tin on high-tin bronze castings.

The purpose of this chapter is to establish the criteria by which these and other surface phenomena may be distinguished. The technological identification of the type of tin-rich bronze surface is important from both the curatorial and conservation viewpoints, as well assisting in the characterization of manufacturing processes. Furthermore, the ability to distinguish between tin-rich and silvered surfaces is important as these are frequently mis-identified or the terminology is confused.

The paper describes various ways of obtaining tin-rich surfaces on bronze and the effects of patination, and illustrates these processes using examples of antiquities which show the appropriate characteristics. The other related process of tinning iron is included for completeness, but does not have the same difficulties of interpretation. The phenomenon of arsenic enrichment on the surfaces of arsenical bronze is also included because of the similarity in the silvery surfaces, and the metallurgical parallels with inverse segregation of tin-bronze.

History of tinning

Tinning a low-tin bronze creates a silvery coloured, durable, corrosion resistant surface that takes a good polish. It also prevents tainting of food by copper cooking vessels. The earliest Western reference to tinning as a decorative and utilitarian coating on bronze is given by Pliny (*Natural History* Book XXXIV, Chapter XLVIII) – 'When copper vessels are coated with *stagnum* the contents have a more agreeable taste and the formation of destructive verdigris is prevented' and 'A method discovered in the Gallic provinces is to plate bronze articles with white lead (tin) so as to make them almost indistinguishable from silver'. The precise interpretation of '*stagnum*' is unclear but Lang and Hughes (1984) discuss the problems and suggest that it refers to tin.

Figure 21.1. *Roman low-tin bronze mirror in good condition showing the tinned, silver-coloured, reflecting convex surface. British Museum GR. 1904. 2–4. 56. Diameter 106 mm.*

There is extensive archaeological evidence for tinning about 600 years earlier with finds particularly from the Celtic and Classical worlds. In addition, the Early Bronze Age axe from Barton Stacey, Hampshire, dated to about 2000 BC is reported to be tinned (Kinnes *et al.* 1979, Kinnes and Needham 1981, Meeks 1986), and similar axes are found in Scotland (Tylecote 1985, Close-Brooks and Coles 1980). Apart from these axes, the earliest tinned bronze objects so far recognized from museum collections are of about the 5th century BC, for example tinned Celtic bronze spurs found in France (Hedges 1964). The La Tène culture provides examples of decorative tinned bronze discs (Savory 1964). Savory (1966) also reports tinned brooches and bracelets from the Halstatt period at the beginning of the 5th century BC. In the Mediterranean region a cheekpiece from a 5th century BC Greek helmet in the British Museum (GR 1856. 12–26. 616) has recently been found to be tin plated and Craddock (1981) reports a late Etruscan tinned mirror. From late Hellenistic times some low-tin bronze mirrors were certainly tinned (Craddock 1988), as were later Roman low-tin bronze mirrors (Figure 21.1) (Meeks 1988a, b).

The Roman period saw widespread use of tinned bronze for decorative purposes and for the embellishment of military ware, for example horse trappings and helmets were frequently tinned (Lins 1974) and there is a life-size tinned parade mask on display at the British Museum (GR 1919. 12–20. 1). The Romans regularly coated the inside of cooking vessels, and a Romano-British example is on display at the British Museum (PRB 1893. 6–18. 14). Tinned bronze remained popular through the Dark Ages, and Oddy (1980) reports tinning used in conjunction with mercury gilding on the Sutton Hoo shield and helmet. The Merovingians of the 7th century AD have left many examples of highly ornamented grave goods (Bakey-Perjés 1989, Hedges 1964 p. 107); the Ashmolean Museum, Oxford, has a fine collection of Merovingian and Saxon tinned bronzes.

Tinning has not often been extensively reported from cultures outside Europe. Ancient Egypt appears to be devoid of indigenous tinned artefacts although tin metal was being imported into Egypt by the XVIIIth Dynasty, 1580–1350 BC (Lucas 1948, Garland and Bannister 1927). Lucas mentions two (imported) tinned bronze bowls of the Roman period from Nubia. In China mirrors made of high-tin bronze from the Chou dynasty were apparently tinned using a mercury amalgam technique (Needham 1962 pp. 87–97). High-tin bronzes have a long history in India with the use of tin to coat household utensils and cooking vessels reported to have been first used during the Middle Ages (Ray 1956). Similarly in the Islamic world, lidded brass inlay vessels with tinning on their inside surfaces were used in Persia from the 15th century (for example, British Museum OA 78. 12–30. 730) and tinned copper vessels remain common throughout the Islamic world to the present day. This apparent late use of tinned vessels probably reflects museum collections rather than the full history of its use.

Not only copper alloys were tinned; the tinning of iron also has a long history. Harrison (1980) describes a Spanish Iron Age iron dagger with a tinned handle dated to 500–450 BC. (British Museum PRB, 1932, 7–6, 1). A Chinese iron mirror in the British Museum is clearly tinned (OA 1968. 4–25. 1). The earliest British tinned iron object is an Iron Age harness bit from Bredon Hill, Gloucestershire (Henken 1939). The Viking excavations at Coppergate in York have unearthed tinned iron spoons and belt fittings (Corfield 1985), and Jope (1956) cites early examples of tinned iron from other medieval contexts of the 8th–10th centuries AD. By the 11th century dip-tinning of iron was widely used (Jope 1956) and Theophilus in the 11th century gives an account of the technique (*Theophilus, on Divers Arts*). A collection of 10th to 17th century tinned iron spurs from the Oxford district can be seen in the Ashmolean museum and 15 of these have been examined by Jope (1956). Tinning was widely practised in the Middle Ages on a range of iron objects. For example, Hedges notes a tinned 13th century iron key from Margam, South Wales, tinned iron drinking vessels were being imported from Germany in 1493 and tinned wrought iron hinges were found at Great Yarmouth, Norfolk dated to the late 16th–early 17th centuries (Hedges 1964 p. 109–119). The tinning of hammered wrought iron sheet was first practised in Bohemia in about the 14th century and by the end of the 16th century the tinned wrought iron sheet industry was flourishing in Saxony (Hedges 1964 p. 161). With the advent of the rolling mill, the tin plate industry expanded in the mid-18th century with centres such as Pontypool in Wales specializing in decorative ware. The modern tin canning industry for food preservation was founded at the beginning of the 19th century in France and England (ITRI, 1939).

Traditional hot-tinning methods requiring experience and skill are still used throughout the world, particularly for plating and replating baking and kitchen equipment (Dinsdale 1978, Thwaites 1983), both industrially and at a more traditional level. For example, travelling tinners or tin-smiths in the

Middle East re-tin cooking vessels (Wulff 1966) as do nomadic gypsy tinners in America who can trace their origins to Asia (ITRI 1983a, Nemeth 1982). Traditional methods of tinning copper are also practised by Australian coppersmiths making high quality tinned vessels (Sweatman 1981).

Electrolytic deposition of tin on steel was not practised extensively before the late 1920s partly because hot dipping was readily applicable and also the early electrodeposited coatings were very thin and lacked a bright finish (Hedges 1960 pp. 99–114). Electrolytic tin plate is now ubiquitous in the food industry (Hedges 1960 pp. 192–237).

Electrochemical tinning of bronze (Hedges 1960 pp. 141–150) by immersion of bronze objects in a boiling solution of potassium bitartrate, from wine fermentation, containing granules of tin metal was known in the 18th century (Diderot and d'Alembert 1755). Oddy (1980) describes the process.

Tin coatings with their associated intermetallic compound layers are still receiving much attention from research in the canning and electronic component industries some 2500 years after their first appearance on bronze (ITRI 1982, ITRI 1983b).

History of high-tin bronze

High-tin bronze has rather special material properties. It is very hard, but brittle, and when polished it has a bright, highly reflective, silver metallic colour and is corrosion resistant (see also Chapters 5 and 6). These special properties were utilized in antiquity in the manufacture of particular types of cast objects that, although specialized, are widely distributed through the ancient world.

The Chinese from the Chou dynasty had an industry devoted to the production of cast high-tin bronze mirrors of various sizes for cosmetic use and for burial to provide light in the underworld (Needham 1962, Zhu 1986). The Romans also used leaded high-tin bronze of similar composition for one of their two types of mirror, although there are technical differences between the two traditions. The Romans made much thinner mirrors than the Chinese and often drilled or turned decorative patterns on them (Chapter 6, Figure 6.4), while the Chinese were masters of fine cast-in decoration on the backs of their, sometimes massive, mirrors (Chapter 6 Figures, 6.10(a), 6.10(b) and 6.12). The Chinese tradition of bronze mirror production carried on until the 18th century.

Another early use of the alloy was for the 'potin' cast Celtic coins, such as the Cantian coins of 100–50 BC (Van Arsdell 1989). Cowell (personal communication) also reports a 25% tin bronze Celtic coin from a hoard at Donhead St Mary, which is a contemporary forgery simulating debased silver coinage.

During the Dark Ages high-tin bronze was extensively used for the manufacture of buckles and belt fittings where their colour and durability were advantageous. Frankish buckles (Figure 21.2) were cleverly made by casting the brittle high-tin bronze onto a core of wrought iron which provided toughness for practical use. Some buckles have extensive wear showing prolonged use. Ptolemaic rings were cast in high-tin bronze and, although patinated now,

Figure 21.2. *Group of Dark Age high-tin bronze buckles and belt fittings. British Museum. Width of field = 250 mm.*

these presumably looked like silver when first manufactured (Chapter 6, Figure 6.7). Some bronze vessels from Early Islamic Iran were made of 'white bronze', i.e. high-tin bronze, and Melikian-Chirvani (1974) suggests that some of these may have been hot-worked to shape. Similarly, Allen (1979) discusses the use, properties and fabrication of the alloy in this period. Similar vessels have a long history in southern India and south east Asia.

High-tin bronze found another use in the earliest reflecting telescope made by Sir Isaac Newton in 1671, and later Sir William Herschel finally succeeded in making a 48-inch mirror for his 40-foot telescope. The largest high tin-bronze (30%Sn) telescope mirror was made by The Earl of Rosse in 1842: six feet in diameter and weighing four tons (Hedges 1964). Thereafter lighter alloys were used.

Bell metal is a high-tin bronze containing typically 20–24% tin, the balance being copper. Large bronze bells appeared in European churches in the 8th century and by the 9th century they had become common. *Campanile*, specially designed to take heavy bells are recorded from the 9th century in Italian churches. Theophilus in the 11th century describes the manufacture of church bells which replaced earlier 5th-century bells of iron sheet (Hedges 1964 pp. 130–133). The tradition of casting sonorous bells continues to the present day from the smallest hand bells through peals of church bells to monsters such as 'Big Ben' cast at the Whitechapel bell foundry. Early bells such as those of China *c.* 140 BC contain only 12–15% tin (Needham 1962 pp. 194).

The sonorous tone of high-tin bronze was utilized by ancient metalsmiths but in the different and more technically difficult form of hot-worked, quenched high-tin bronze used to make gongs and cymbals from at least the 3rd century BC in Thailand (Seeley and Waranghkan Rajjpitak 1979). This technology is still in use today in the Philippines (Goodway and Conklin 1987).

Examination techniques applied to the study of antiquities

The characterization of tin-enriched bronze surfaces was achieved by microscopy and microanalysis in the analytical scanning electron microscope (SEM) and by some X-ray diffraction analysis (XRD). Many examples of antiquities were examined non-destructively on unprepared surfaces directly in the SEM. Some cross-sections were taken, either as removed samples or as taper sections, particularly on (broken) high-tin bronzes to establish the relationship of surface patination to the body metal.

The physical condition of an object is of importance to the study of the surface structure and composition of antiquities. Stereo optical microscopy is therefore the first technique used to give the overall view of the surface of the object, providing information on metallic sheen, colour, areas of surface plating, underlying and overlying corrosion and surface spalling, cracking, pitting or patination.

Surface X-ray fluorescence analysis (XRF) can only determine qualitatively whether a silvery surface is enriched with tin, arsenic or silver etc. although it is sensitive to low concentrations (*c.* 0.05%). Such analysis may show, for example, high tin levels on the surface of a bronze, but gives no indication of the form of the tin. The form reflects the fabrication history and/or the patination of the bronze, and this can be identified by a combination of optical microscopy, scanning electron microscopy (SEM) plus microanalysis (EDX), and where necessary X-ray diffraction analysis (XRD). Microstructure combined with analysis is the key to interpretation.

Optical microscopy of the surface of relatively flat, polished objects can show eutectoid microstructures of high-tin bronze, although the SEM shows the structure much more easily. The intermetallic compound (δ) associated with this characteristic microstructure is normally corrosion resistant and often clearly appears silver-grey in colour contrasting with the darker corroded interdendritic α bronze. In polished cross-section this, and the other compound layers appear blue-grey in colour, and mineralized surfaces show by colour contrast in dark field illumination and/or polarized light. The same samples can be transferred to the SEM (after carbon coating) for detailed microstructural examination, microphotography and microanalysis.

Fortunately the large size of modern SEM sample chambers means that many objects can now be examined directly, without preparation, apart from wiping away surface dust with lens tissues moistened with industrial methylated spirit. Often a tinned object has high points where the plating has worn away during use, exposing the underlying bronze (or corroded bronze). These form natural taper sections for direct observation and analysis in the SEM of the different layers through the surface to the body metal.

By using compositional contrast imaging in the SEM (otherwise known as atomic number contrast or backscattered electron imaging, e.g. Meeks 1988c) the phase distributions both on the surface of objects and in polished cross-section can be clearly seen and recorded, indeed this type of imaging is essential for non-etched samples in the SEM.

Microanalysis in the SEM can identify the intermetallic compounds associated with tinning because they have specific compositions, as indicated by the copper–tin phase diagram. X-ray diffraction analysis of suitable surface samples can determine the compounds present, although sampling of the hard and thin compounds is not always easy.

Additionally, with sections of corroded or patinated surfaces the technique of digital X-ray mapping can be very effectively used to determine the relative concentrations of the alloy metal elements remaining in the patinated surface layer (e.g. tin, copper and lead) and the distribution of extraneous elements incorporated into the patina from the burial environment (e.g. silicon, aluminium, phosphorus, iron).

With the database of microstructures and analyses derived from previous examinations it is now often possible to categorize clearly and unambiguously the form of the tin-enriched surfaces associated with specific types of object.

Bronze phase diagram: the copper–tin system

The copper–tin phase diagram is very well documented in the metallurgical literature, for example by Hanson and Pell-Walpole (1951). Figure 21.3 shows the copper–tin system after

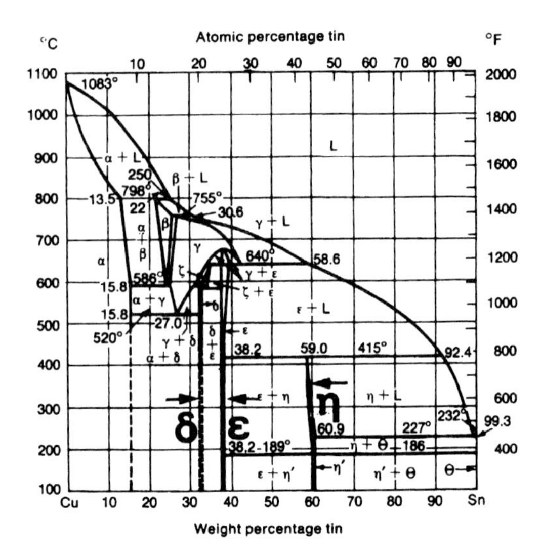

Figure 21.3. *The copper–tin phase diagram showing the three intermetallic compounds, each of specific composition.*

Smith (1973a) with the modification by Hanson and Pell-Walpole (1951 p. 60) to show the α + δ phase field extending down to room temperature; this is considered to be typical for cast bronze, which does not normally reach equilibrium (i.e. α + ε). Various aspects of this system are important, in particular the wide freezing range of low-tin bronze alloys, the presence of intermetallic compounds that occur at higher tin concentrations and the eutectoid transformation that occurs on cooling. These factors influence the structures found on tinned bronze, high-tin bronze and other tin-enriched surfaces.

Low-tin bronze consists mainly of α solid solution of tin in copper. Antiquities made of these alloys can be shaped by hammering. They can be tinned if necessary to provide a protective surface which is also silver in colour. Bronzes in this region normally show either cored, dendritic structures when cast, or equiaxed, twinned grain structures when worked and annealed. Some α + δ eutectoid (of composition 27% tin) can be present in the structure depending on the actual alloy composition (particularly above about 8% tin) and on the casting or working conditions. This low-tin bronze region has a wide solidus/

liquidus separation (i.e. a wide freezing range) which produces the coring within dendrites and can readily lead to tin sweat on the surface of cast bronze (Hanson and Pell-Walpole 1951). Tin sweat actually involves the exudation of tin-enriched alloy and the formation of the $\alpha + \delta$ eutectoid microstructure on the surface, rather than it being tin by itself.

At higher tin concentrations three intermetallic compounds occur, nominally at fixed compositions: δ, $Cu_{31}Sn_8$ containing 32.6% tin; ϵ, Cu_3Sn containing 38.2% tin; and η, Cu_6Sn_5 containing 61.0% tin. These compounds are important because they are hard, brittle, corrosion resistant and can be polished to silvery metallic surfaces.

The δ compound commonly occurs in the body of cast antiquities as the characteristic $\alpha + \delta$ eutectoid microstructure which dominates when the composition is in the region of around 19%-27% tin, and hence has a major influence on the properties of the alloy. This is the high-tin bronze region. The alloys are brittle and cannot be cold worked but the δ compound makes the surface durable and gives optical properties. Tin sweat does not occur in alloys of this composition.

The η and ϵ compounds are only found after deliberate tinning and occur as surface layers formed by interdiffusion of the applied tin with the copper of the underlying low-tin bronze. Heat treatment after tinning allows further diffusion which alters the relationship of compound layers and structures on the surface which are important to interpretation.

Tin metal never survives on the surface of tinned antiquities, either because diffusion during manufacture has produced a fully compound surface, or because corrosion during burial has leached away the excess tin.

Although high-tin bronze cannot be cold worked, hot working is possible in the single-phase β region at around 23%–24% tin and at high temperatures around 650°C–750°C when it becomes soft and malleable (Melikian-Chirvani 1974). Quenching the bronze in water from this high temperature creates a martensitic type structure which is hard and tough rather than brittle, a phenomenon utilized in the production of south east Asian gongs and bowls as mentioned above. The colour of this bronze is golden rather than silvery and has a sonorous tone, and such artefacts often have a smooth, lustrous surface patina from burial.

In summary, the simple addition of tin into copper creates an extraordinary range of alloys with widely differing properties that were fully exploited in the ancient world. Similarly, ancient craftsmen learned that the application of molten tin to the surface of bronze (and iron) creates a protective silvery surface which has corrosion resistance and reflective properties and is also economical in the use of tin.

Tin-enriched surfaces on antiquities

The following sections describe and illustrate the range of 'tin-enriched' surfaces that are found on antiquities and the methods by which they are produced deliberately or, in some cases, how they are derived from natural origins. Five sections below cover methods of deliberately tinning bronze and iron.

(a) Traditional tinning methods on bronze: (i) wipe tinning, (ii) hot-dip tinning, and (iii) electrochemical tinning.
(b) Microstructures associated with deliberate wipe and hot-dip tinning.
(c) Examples of the range of tinned surfaces found on antiquities.
(d) Cassiterite reduction on the surface of bronze.
(e) Tinned iron. (i) hot-tinning, and (ii) electrolytic tinning.

A sixth section describes the natural casting phenomenon.

(f) Tin sweat or inverse segregation on low-tin bronze.

The remaining sections relate to aspects of high-tin bronze and patination.

(g) Surfaces of high-tin bronze antiquities.
(h) Mercury tinning of high-tin bronze Chinese mirrors.
(i) Patination, corrosion and tin-enrichment associated with high-tin bronze and tinned bronze.
(j) Patina of hot-worked, quenched high-tin bronze.

(Characteristics are summarized in Table 21.1.)

Table 21.1 Characteristics of tin-enriched surfaces on antiquities: (a) on bronze, (b) on iron

	η/ε compounds	*Eutectoid*	*Tin metal*	*Patina*
(a)				
1. Traditional wipe or hot-dip tinning	η and/or ε about 2–5 microns thick	Only if overheated to above 520°C (possibly ε overlies eutectoid)	No	Anodic bronze substrate corrodes. Compound layers survive or are physically lost
2. Electrochemical tinning	Probably η very thin *c.* 20 nm?	No	No	No data available
3. (a) Mercury tinning — *cold* as found on antiquities	Probably η (+ Hg) very thin, *c.* 20 nm?	No	No	Loss of evidence during burial (originally very thin)
(b) Mercury tinning — fire-tinned *c.* 375°C, *experimental*	ε *c.* 5 microns thick or more	Only if overheated to above 520°C	No	Assumed to be similar to wipe or hot-dip tinning
4. Cassiterite reduction on bronze *c.* 750°C	No	Yes Variable thickness	No	α corrosion of eutectoid, tin-enriched
5. Tin sweat	No	Yes Variable thickness	No	α corrosion of eutectoid, tin-enriched
6. High-tin bronze, silvery surface	No	Yes Throughout body and at surface	No	α corrosion of eutectoid, Tin enriched
7. High-tin bronze black/green patina	No	Pseudomorphic mineralization of surface eutectoid	No	Mainly SnO_2/SnO + low Cu/Pb oxides + Si, Al, Fe from soil. α and δ corrosion of eutectoid
8. Corroded low-tin bronze (containing some body metal eutectoid)	No	Eutectoid only from original structure remains in situ	No	Mainly copper compounds, but tin can be relatively enriched as oxide in patina
9. Corroded low-tin bronze (containing no body metal eutectoid)	No	No	No	Mainly copper compounds
10. Quenched high-tin bronze	No	Martensitic structure in body metal. Featureless patina	No	Mainly SnO_2/SnO + low Cu/Pb oxides + Si, Al, Fe from soil
(b)				
11. Tinned iron	θ mainly, <1 micron thick; very thin	Not applicable	No	None. Anodic iron substrate corrodes

Traditional tinning methods on bronze

Traditionally only copper or low-tin bronze is tinned (and some brass objects). High-tin bronze is not normally tinned as it is naturally silvery when polished. The case of mercury tinning high-tin bronze Chinese mirrors is considered in Chapters 5 and 6.

Bronze can be tinned in the traditional manner by either: *wiping* molten tin onto bronze; or *hot-dipping* bronze into a molten tin bath. The majority of deliberately tinned antiquities would have been treated by these techniques.

The later technique of *electrochemical* tinning is added for completeness.

Wipe tinning

The bronze surface should be smooth and clean before tinning, for if the bronze is at all oxidized or greasy the tin will not take. The bronze is warmed and fluxed with rosin or other anti-oxidant. Fragments of tin metal or thin tin sheet are melted (232°C) onto the bronze and excess molten tin is physically wiped over the surface with a cloth (Oddy 1980). The tinned surface is silvery in colour but dull and matt with some streaks where excess tin remains. The surface is dull because, when wiped, there is little or no excess tin to form a smooth reflecting surface. The matt surface is caused by angular crystals of the intermetallic compounds which always form in the surface layer as a result of the diffusion of copper from the bronze. Polishing then produces a shiny, reflecting, silvery surface which is durable and scratch resistant.

This is the traditional process used by tinkers to tin cooking vessels, where the tin serves as a barrier to prevent contamination of the food from the copper in the body metal or lead-rich solder.

Hot-dip tinning

Here, clean bronze is warmed, fluxed with rosin and heated to a temperature just above that of molten tin to avoid chill casting excess tin on the surface. The bronze is dipped briefly into a bath of molten tin at about 260°C (Hedges 1960) and excess tin then allowed to drain off. The surface prepared in this manner has a smooth, reflective, silvery appearance because of the presence of an unalloyed tin layer covering the intermetallic compounds below. Leaving the bronze too long in the molten tin allows excessive copper contamination of the tin bath through dissolution of the bronze, and crystals of η compound form in the melt (Hedges 1960, Chapter 6). Tinning by dipping usually results in a thicker layer than that formed by wiping.

On the practical side, Hedges notes that in the hot-tinning process 'tinners rely on their experience and skill to obtain the best and most economic results' and 'the importance of manual skill must not be under-rated'.

Tinning which produces a thin layer of soft tin metal overlying the hard intermetallic compounds will scratch and mark easily. Controlled heating to allow further diffusion consumes the excess tin metal and leaves the hard, silvery compound layer on the surface which can be polished to a bright reflecting surface, as exemplified by the tinned Roman mirrors described later.

Electrochemical tinning

This post-medieval method of tinning is described by Diderot and d'Alembert (1755) for tinning brass pins. The process was carried out by immersion of the copper, bronze or brass objects in a boiling solution of potassium bitartrate, a by-product of wine fermentation, together with pieces of metallic tin or tin foil. This produces an electrochemically deposited silver coloured tin layer which is extremely thin (a few tens or hundreds of nanometres thick), since the deposition ceases when the surface of the object is entirely covered. Dissolution of the article to be coated does occur, but only to the extent equivalent to the tin deposited (Hedges 1960 pp. 142–143). The thinness of the coating is probably its most characteristic feature for identification purposes. Kay and MacKay (1976) discuss the growth of copper–tin compound layers on conventional tinning at ambient temperature over time; the electrochemically deposited tin layer will eventually diffuse to extremely thin compound layers which would be initiated and accelerated by thermal diffusion at the temperature of the boiling solution during application. Oddy and Bimson in a personal communication (1985) were informed that it is possible to distinguish examples of electrochemical tinning microscopically, and it is suggested that the earliest examples thus recognized are from about the 14th century.

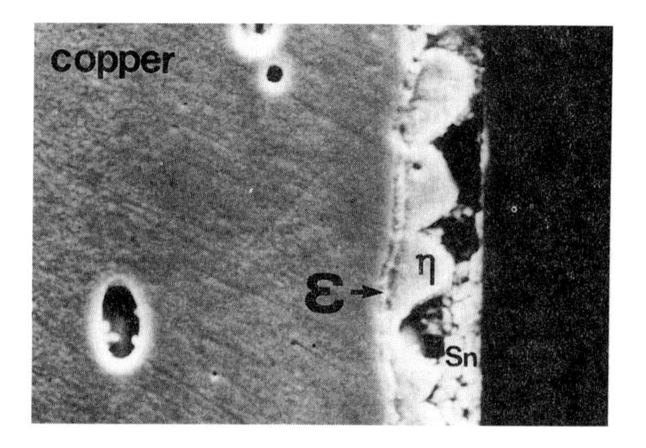

Figure 21.4. *Basic layered surface structures associated with fresh tinning. SEM, cross-section. The tinned layer is 5 microns thick.*

Microstructures associated with deliberate wipe and hot-dip tinning

A newly hot-dip tinned bronze has a silvery metallic surface comprising three well-bonded layers (Tylecote 1985, Meeks 1986, ITRI 1982, Daniels 1936). The outer layer is excess tin metal. Below this, two intermetallic compound layers form sequentially (η, Cu_6Sn_5 on top of ϵ, Cu_3Sn) at the time of dipping by the solid state diffusion of tin into the substrate bronze (Kay and McKay 1976, Hedges 1960).

This basic tinned surface structure is shown in Figure 21.4 which shows the three layers with a total thickness of about 5 microns. The relative thicknesses of these three phases are influenced by the variables of temperature, time of heating and overall thickness of the original tinning layer. Thus, heat treating after tinning can dramatically change the relative thicknesses of the three phases by thermally activated solid state diffusion whereby lower tin phases grow at the expense of the higher tin phases (Mrowec 1980). The growth occurs in steps as illustrated in Figure 21.5, according to the temperatures and phase fields of the phase diagram (Figure 21.3). Thus heating a tinned surface to 250°C quickly allows the excess tin to be consumed by growth of the η compound, and after a short time (a few minutes) this in turn will be consumed by the growth of the ϵ compound. At 350°C only a layer of ϵ compound remains. At 450°C the surface layer becomes solid δ with a thin α diffusion zone in the substrate bronze. After heating to above 520°C the characteristic $\alpha + \delta$ eutectoid microstructure is formed, on cooling. Above 650°C diffusion is so rapid that the tin forms α solid solution in the bronze surface so the original

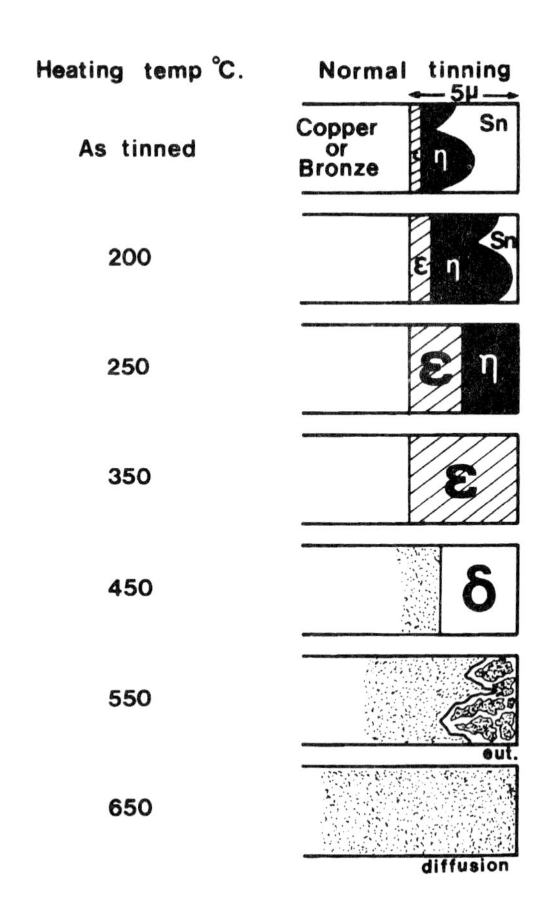

Figure 21.5. *Diagram of the intermetallic compounds associated with tinning. Note the changes that occur by diffusion with increasing temperature. The 'as tinned' structure is comparable to Figure 21.4.*

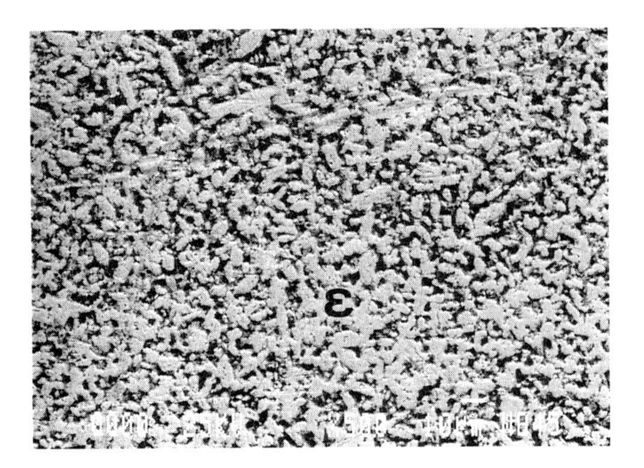

Figure 21.6. *Surface of a tinned Roman mirror showing unpolished crystals of ϵ intermetallic compound. SEM. Width of field = 236 microns.*

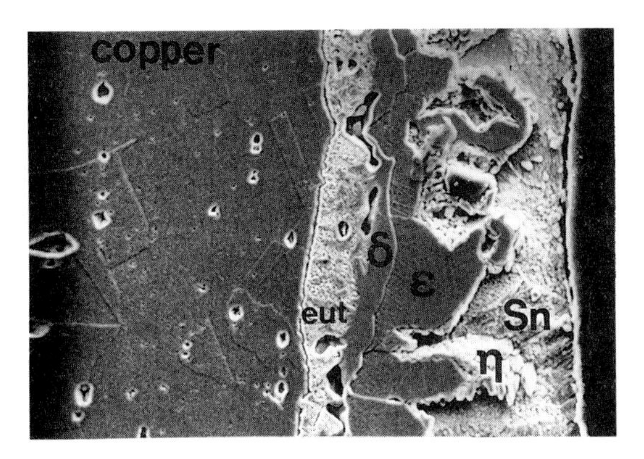

Figure 21.7. *Experimental overheated tinning seen in cross-section showing all possible compound layers for comparison with the phase diagram, Figure 21.3. SEM. The tinned layer is 50 microns thick.*

surface tinning may well be lost, but the surface composition will be slightly higher in tin than the bronze body metal.

As previously mentioned, with the growth of the copper–tin compounds, the surface becomes covered with angular crystals, such as those on the surface of a Roman mirror (Figure 21.6), which require fine abrasion and polishing to produce a reflecting, silver-coloured surface. It is possible to generate the whole range of structures associated with tinning by simply melting a thick layer of tin onto bronze and heating for a few minutes to just above 520°C and then cooling. The effect of this treatment is shown in Figure 21.7. The phases present are, in order, tin metal on the surface, η, ϵ, δ, $\alpha+\delta$ eutectoid and α diffusion zone of tin in the substrate. It is interesting to compare this micrograph with the phase diagram (Figure 21.3) because it exactly replicates the order of the key components of the diagram.

An as-tinned bronze antiquity might therefore be expected to have a surface microstructure similar to that shown in the experimental sample in Figure 21.4. However, wiping the molten tinned surface, heating post-tinning or subsequent polishing removes some if not all of the excess tin metal, leaving only the intermetallic compound layers. Also, corrosion during burial will preferentially remove the anodic tin surface leaving the cathodically protected compound layers (Britton 1975). Similarly the anodic bronze substrate may corrode underneath the compound layers, particularly where the surface is cracked, pitted or damaged. This underlying corrosion can result in further surface cracking and lifting of the brittle compounds and loss of the tinned surface layer.

In practice, the residual evidence for tinning normally found on archaeological material is the presence of compounds η and/or ϵ. The results published by Oddy and Bimson (1985) on Roman and Anglo-Saxon brooches show that about half of the objects analysed by X-ray diffraction show η as the major compound and the other half show ϵ. No unalloyed tin metal has been found on the surface of buried antiquities.

Examples of the range of tinned surfaces found on antiquities

The surface of a tinned Frankish strap end (MLA 1926. 5–11. 9) which clearly shows a very silver coloured surface under the optical microscope, was examined directly in the SEM. Figure 21.8 shows the object at 600× magnification (backscattered electron image) with the the light coloured tinned areas and darker worn areas where the body metal is exposed. This magnification clearly shows the corroded, dendritic, cast body metal of the strap-end and the worn taper section through

Figure 21.8. *Surface of tinned Frankish strap-end seen in the SEM as a natural taper section caused by user wear. The binary layers of η and ϵ compounds overlie the corroding dendritic core metal. SEM. Width of field = 190 microns.*

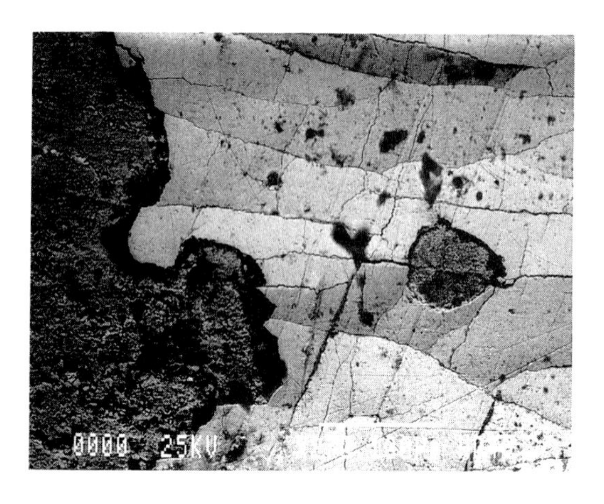

Figure 21.9. *Surface of tinned Roman mirror showing cracking of the ϵ compound layer due to underlying corrosion. SEM. Width of field = 440 microns.*

the tinning. Analysis of the tinning shows the presence of ϵ adjacent to the body metal and η on top of the ϵ. No tin metal remains. This compares with the experimental tinned sample heated to 250°C in the diagram, Figure 21.5.

Similarly, in a recent study of Roman mirrors of the tinned low-tin bronze type, the evidence of tinning on the shiny silver-coloured surfaces was clearly seen in the SEM to consist of the corrosion resistant intermetallic compound layers on top of corroding bronze cores (Meeks 1988a, in press). Often the brittle surface layers were cracked, with regions missing due to lifting by the corrosion below, as illustrated on another Roman mirror GR 1904. 2–4. 55 (Figure 21.9). The whole range of possible tinned surfaces was found on these mirrors including all three intermetallic compounds, η, ϵ and solid δ as well eutectoid type surfaces produced by overheated tinning. These compound surfaces correspond to the temperatures of heat treatment shown in Figure 21.5.

It is concluded that the surfaces found on the Roman tinned mirrors are the result of deliberate heat treatment during manufacture to allow the growth of the intermetallic compound layers. This process consumes the excess surface tin and leaves the hard and durable compound surface. However, it is unlikely that a specific temperature was necessarily chosen in order to form a specific compound layer; there is relatively little difference in colour between the layers and the metalsmith was probably only aware that a harder surface formed on heating. As long as the heating was not excessive, a satisfactory surface was achieved. The surface so formed needed to be finely abraded and polished to give a smooth mirror finish, otherwise the crystalline compound surface layer appears matt. Some Roman mirrors showed regions where the angular crystals were still clearly visible on the surface (Figure 21.6).

One Roman mirror in the British Museum (GR 1926. 4–6. 14) has a complex surface showing three areas of interest with different microstructures. On the reverse side of the mirror large grains of epsilon compound were found, Figure 21.10, corresponding to grain growth during heating in areas of thick tinning. This is obscuring the underlying eutectoid, which was only seen elsewhere where the coating was thinner, for example on the reflecting side where one area is very silvery and a natural taper section reveals epsilon compound overlaying the eutectoid microstructure, with grain boundary δ compound (hypereutectoid) (Figure 21.11). This indicates thick tinning overheated to above 520°C, but for insufficient time to allow

Figure 21.10. *Grain growth of ε compound within the turned decorative lines on the surface of a tinned Roman mirror. SEM. The white scale bar represents 10 microns.*

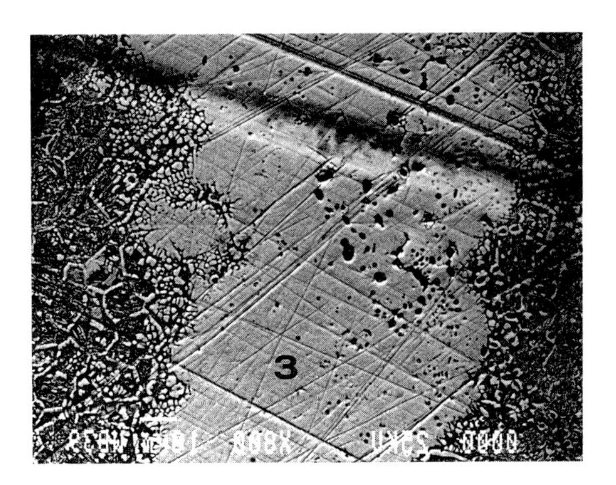

Figure 21.11. *Roman mirror surface showing the ε compound layer overlaying eutectoid, due to overheated tinning, in this natural taper section. SEM. Width of field = 140 microns.*

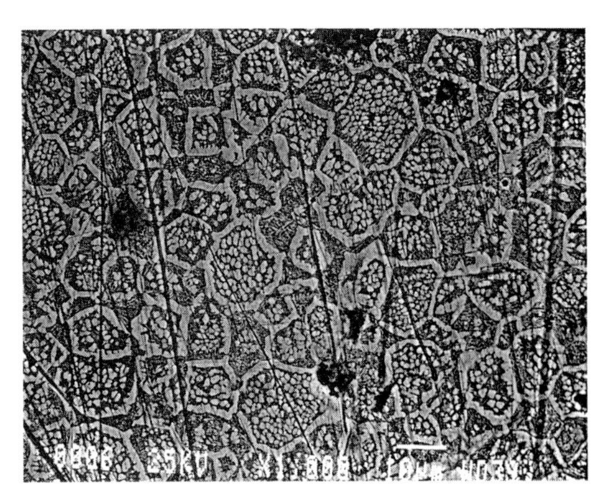

Figure 21.12. *Roman mirror showing extensive eutectoid with grain boundary δ, due to overheated tinning, adjacent to the area in Figure 21.13. Width of field = 115 microns.*

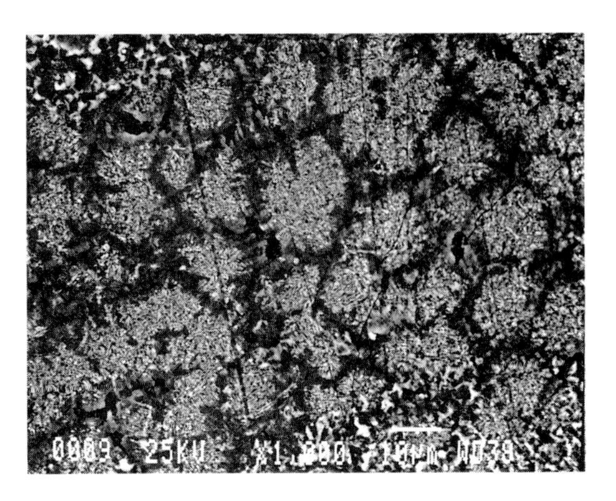

Figure 21.13. *Grain boundary corrosion of δ compound on areas of the mirror surface shown in Figure 21.12, due to local variations in burial environment. SEM. Width of field = 115 microns.*

complete diffusion of the epsilon compound. Adjacent areas show extensive hypereutectoid microstructures where diffusion had progressed further (Figure 21.12). Other areas of the mirror front are covered with layers of azurite and malachite corrosion products. Where these corrosion products are missing the exposed metal surface is matt and dark grey in colour. The SEM clearly showed an unusual corrosion structure where the grain boundary δ phase of the hypereutectoid microstructure has preferentially corroded leaving the eutectoid structures within these grains uncorroded (Figure 21.13). Preferential corrosion of the δ compound phase has been noted elsewhere (Chase and Franklin 1979, Tylecote 1979). Clearly, different corrosion conditions existed locally on this mirror surface.

Although the mirror has a complex surface microstructure it has been shown, using comparisons with the experimental structures,

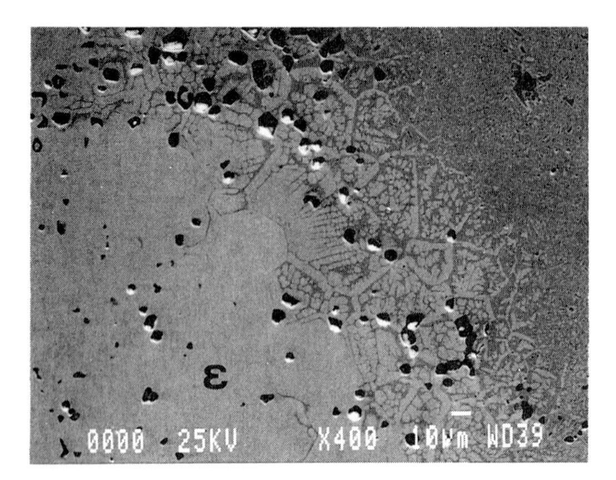

Figure 21.14. *Surface taper section through an experimental overheated tinned copper sample to simulate the effect of eutectoid underlying ϵ compound shown on the mirror in Figure 21.11. SEM. The white scale bar represents 10 microns.*

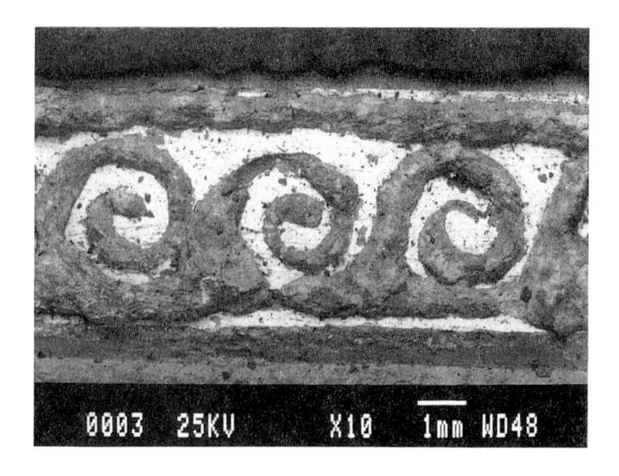

Figure 21.15. *Roman belt fitting from Mansell street showing scrolls of niello (dark) and tinning (light) on the metal surface. Museum of London. SEM. The white scale bar represents 1 mm.*

to be simply a case of thick tinning overheated briefly to above 520°C (Figure 21.15). Several of the Roman mirrors had similar structures, and it is thought that the Romans regularly heat treated mirrors to produce compound layers ready for polishing as part of the manufacturing process.

Tinning was also used as a selective decorative feature, for example on the recently excavated Roman brass belt fittings from Mansell Street, London (Museum of London excavation number MSL87. 593. 325), (Jones and Meeks in press). Scroll designs around the edges (Figure 21.15) were a silver-grey colour which turned out to be η compound crystals (Figure 21.16), indicating low temperature tinning. The dark infill between the scrolls was a silver sulphide niello. Here the use of tinning replaced the more common use of silver which would usually complement a silver niello (La Niece 1983). The low tinning temperature was probably chosen to avoid the risk of decomposing the niello. Similarly, Oddy (1980) notes the association of tinning and mercury amalgam gilding on the Sutton Hoo shield boss, both of which require heat treatment of different durations and temperatures, i.e. the higher temperature gilding must have preceded the tinning.

In summary, on hot-tinned bronze antiquities we generally find evidence of η and/or ϵ compound layers. η and ϵ compounds *only* occur as the result of *tinning*. No unalloyed tin metal is found on antiquities from burial. Occasionally a solid δ layer is found due to overheating the tinning. More often, overheating (above 520°C) creates evidence of eutectoid microstructures which are seen on the surface or at worn taper section areas underlying the ϵ compound surface layer. In the latter case the eutectoid often has excess grain boundary δ compound, which is unlike tin sweat or cassiterite reduction described below.

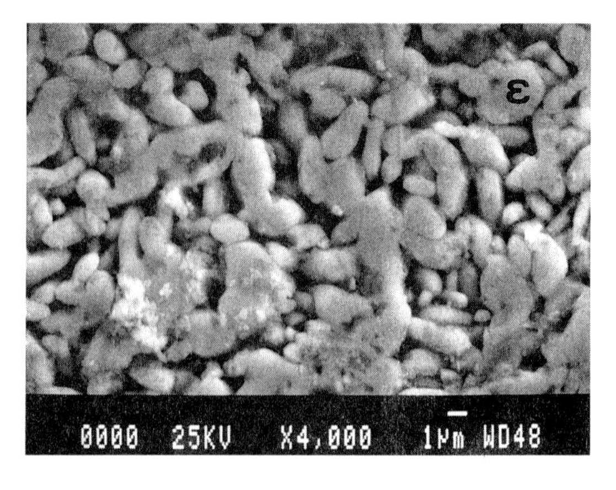

Figure 21.16. *Crystals of ϵ compound on the tinned area of the Mansell street scroll decoration. SEM. The white scale bar represents 1 micron.*

Cassiterite reduction on the surface of bronze

It is possible to reduce cassiterite (SnO_2) on the surface of bronze in a reducing atmosphere at about 750°C. This produces a layer of $\alpha + \delta$ eutectoid (when cool) by diffusion of the released tin metal into the bronze surface. If the reaction is too slow and insufficient tin is released, then only a diffusion zone of tin in α solid solution is produced.

A few Early Bronze Age flat axes have eutectoid surfaces that impart a silvery-grey colour and make them corrosion resistant (Smith 1872, Kinnes *et al.* 1979, Kinnes and Needham 1981, Meeks 1986, Tylecote 1985). If this occurred through extensive tin sweat from casting, it is difficult to explain why *both* sides of the axes, which were cast in open moulds, should be the same. Indeed the top surface of an axe cast in an open mould is so rough through dendritic growth and shrinkage (Meeks 1986 Figure 14), that even if tin sweat could occur on this surface it would never survive the grinding, hammering and polishing needed to finish the axe. The Early Bronze Age flat axe found at Barton Stacey, Hampshire (British Museum, PRB 1979. 6–2. 1) is a heavily worked, homogenized, 12% tin bronze axe with no eutectoid structures within the core. However, it has a polished eutectoid surface over large areas, and extends to regions near the cutting edge where hammering during manufacture would have been greatest (Meeks 1986 Figure 25). On the other hand, plating would be the final operation to be carried out, thus explaining its existence at the cutting edge (Tylecote 1985). It is therefore possible that these axes may have been deliberately tinned by cassiterite reduction using crushed and calcined ore mixed with charcoal.

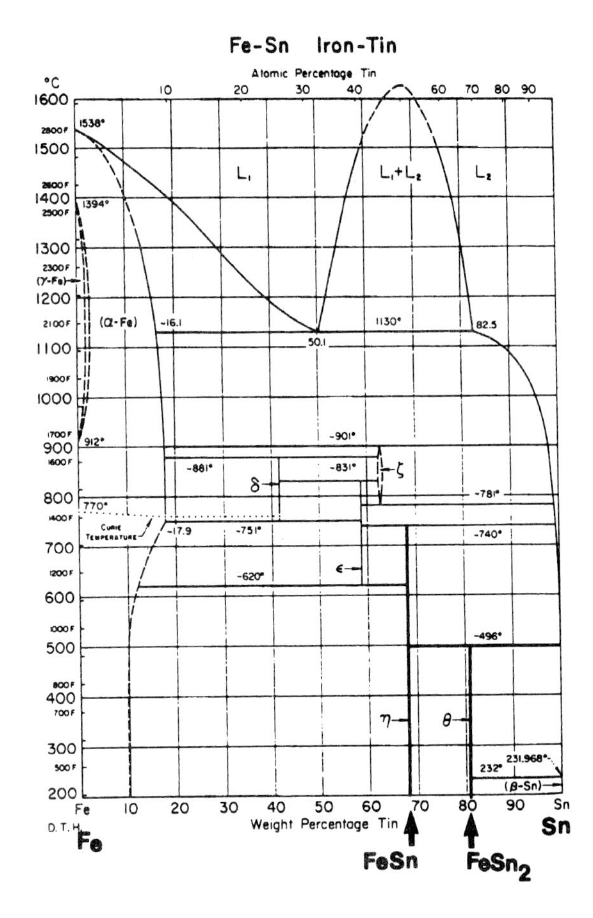

Figure 21.17. *Iron-tin phase diagram showing the intermetallic compounds that form on tinning.*

Tinned iron

Hot-tinning

When clean, fluxed iron is hot-dipped into molten tin, a layer of θ intermetallic compound forms, $FeSn_2$ (81% Sn) by diffusion of tin into the iron, in accordance with the phase diagram (Figure 21.17). Excess tin remains on the surface unless subsequent heat treatment is carried out to allow complete diffusion to the compound. The compound forms a layer of surface crystals in the manner of the bronze compounds (Hedges 1960 Figure 87) and it is this cathodic compound layer which is corrosion resistant. The compound layer is usually very thin, 1 micron or less, because diffusion of tin into iron is slow. The θ compound is stable to 496°C, above which η compound forms, containing 68% Sn. Theoretically, tinning of iron should produce a binary compound layer of θ on η, but θ dominates due to the slow diffusion rate.

The identification of tinned iron is relatively simple, by surface analysis for tin on areas that are still silvery in colour. Iron artefacts are

Figure 21.18. *Tinned iron Chinese mirror. The tinned front surface is well preserved while the unprotected iron of the back is badly corroded. British Museum OA. 1968. 4–25. 1. Diameter 180 mm.*

often heavily corroded and the white metal coating can often be seen against the dark, rusty background. X-ray fluorescence analysis (XRF) can identify the presence of tin, or X-ray diffraction can be used to identify the compound. Unalloyed tin metal would not normally be expected to survive, either because of compound growth from heating during manufacture or because of corrosion during burial.

The Chinese iron mirror shown in Figure 21.18 has a patterned, protective coating of tinning on the back while the other side is not tinned and is now severely rusted. Sometimes the iron corrosion can cover and obscure the plating which will show up by X-radiography (Corfield 1985). Conservation treatment can then reveal the original plated surface.

Jope (1956) notes that some tinned iron spurs have both tin and lead present on the surface and he suspects the use of a soft solder type alloy as the tinning medium.

Electrolytic tinning

This modern process associated with the tin plate industry involves (in its simplest form) the immersion of the iron or steel object in a tin-salt electrolytic bath, and the passage of electric current with the iron object as the cathode. This method produces a very thin (e.g. 0.3 microns) layer of $FeSn_2$ although the thickness can be substantially increased if required (Hedges 1960 pp. 192–237).

Tin sweat or inverse segregation on low-tin bronze

Low-tin bronze is prone to tin sweat or inverse segregation during casting due to the wide freezing range as described earlier. Tin sweat is the exudation from the core of a cast bronze of the last remaining molten alloy, of composition 25.5% tin, which freezes on the surface of the bronze forming a layer of $\alpha + \delta$ eutectoid. It is not the exudation of tin alone. It appears as a silver-grey surface coating which may be localized or extensive depending on the casting conditions and the type of mould used (Hanson and Pell-Walpole 1951). Leaded bronze alloys can also suffer simultaneous lead sweat by the same process; the eutectoid then has larger lead globules dispersed within the same regions. Tin sweat is illustrated in a cross-section in Figure 21.19 taken from an experimental casting showing interdendritic

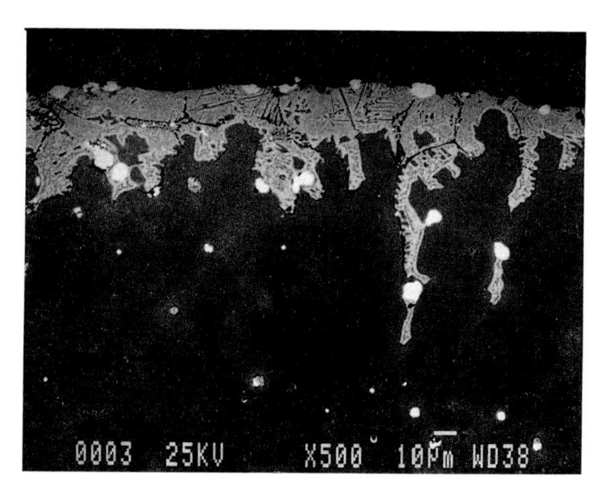

Figure 21.19. *Tin sweat on an experimental casting showing the surface eutectoid and interdendritic feeder, seen in cross-section in the SEM. Note the association of lead globules with the eutectoid. The white scale bar represents 10 microns.*

feeders from the core to the eutectoid surface where it forms a continuous layer, which can be of variable thickness from a few microns upwards. The figure also illustrates the association of the lead globules (white) with the tin sweat. Tin sweat tends to occur with tin contents of about 8–15% (but not in high-tin bronzes above about 20% tin).

Tin sweat originates because of the large α + liquid phase field, indicated by wide separation of the solidus/liquidus lines on the phase diagram. This results in severe coring of the primary α dendrites on solidification and consequent enrichment in tin of the remaining interdendritic molten bronze. The enrichment ultimately approaches eutectoid composition and a solidification temperature of 798°C. Shrinkage of the cooling metal away from the mould causes internal pressure within the bronze. This forces the tin-enriched molten bronze to the surface, via interdendritic feeders, and into the gap between the mould and the casting. Release of dissolved gas during solidification adds to the internal hydrostatic pressure (Hanson and Pell-Walpole 1951, Bailey and Baker 1949, Oya *et al.* 1975). On the surface, the tin-enriched bronze solidifies as a layer of eutectoid that is often substantially thicker (10–50 microns) than a traditionally tinned surface layer. This is sometimes accompanied by the formation of especially thick regions (200 microns) or localized pimples, depending on the feeders to the surface (Meeks 1986 Figure 11). Fast cooling enhances the sweat phenomenon and closed, bivalve ceramic moulds also contribute to the effect, although Hanson notes that gas evolution alone can cause exudation even in slowly cooled, oil-based sand moulds (Hanson and Pell-Walpole 1951 p. 236). In practice it is difficult to eliminate tin sweat completely in bronze of moderately high tin content although it can be reduced by exercising control over the above factors.

It is thought that the open stone moulds used in the Early Bronze Age do not provide the right environment to induce excessive tin-sweat to account for the eutectoid layers found on both sides of some of the finely polished flat axes of the period (Tylecote 1985). Hence the use of cassiterite reduction is suggested (see p. 261 above).

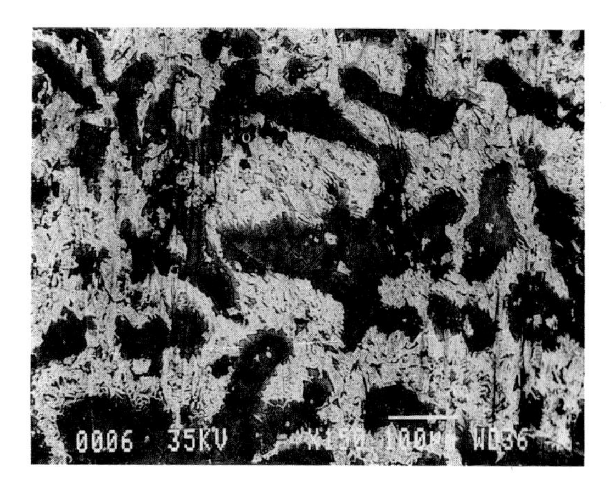

Figure 21.20. *Tin sweat surface eutectoid on the cast bronze stag of the Sutton Hoo sceptre. SEM. Width of field = 750 microns.*

Tin sweat is a problem in bronze founding practice today (Hanson & Pell-Walpole 1951, Hedges 1960), so it is not surprising that it is found on some ancient castings. For example, it occurs on the surface of the bronze stag from the Sutton Hoo sceptre which contains 15% of tin (Bruce-Mitford 1978). Figure 21.20 shows the classic eutectoid surface, with α dendrites, which occurs on the stag in many areas, but does not completely coat it. The eutectoid microstructure of tin sweat can be recognized using a good binocular microscope.

Surfaces of high-tin bronze antiquities

The physical properties of cast bronze change rapidly with increasing tin content above about 10% tin (the limit of solid solubility of tin in copper), although the colour changes less rapidly and does not become particularly silvery until well over 20% tin in a fully homogenized alloy. Roman high-tin bronze mirrors have tin contents generally between 18% and 23% (Craddock 1988), while Chinese mirrors have generally a few per cent more tin, 22%-26% (Barnard 1961). The useful properties of high-tin bronze are due to the increasing presence of the hard, brittle and silver-white δ intermetallic compound which appears in the characteristic α + δ eutectoid microstructural form. This structure dominates

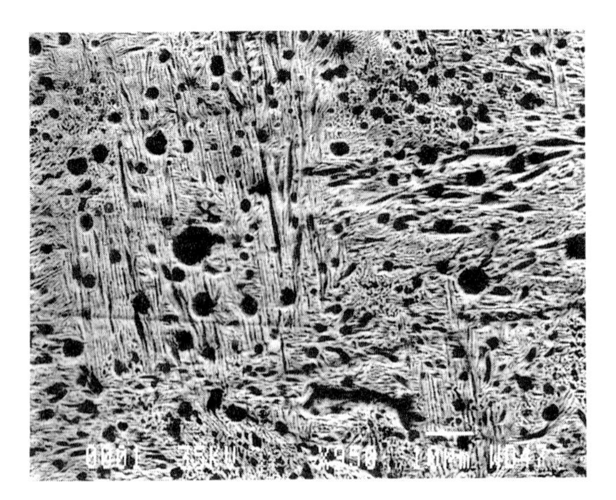

Figure 21.21. *Classic eutectoid microstructure on the surface of a Roman high-tin bronze mirror showing uncorroded δ compound and usual loss of both the α phase and the lead globules, which leave the dark features. SEM. Width of field = 115 microns.*

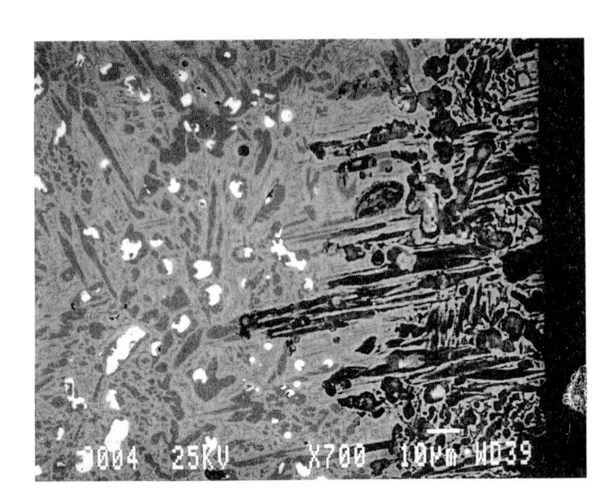

Figure 21.22. *Cross-section of a Dark Age high-tin bronze buckle showing loss of both the α phase and the lead globules at the surface, but no corrosion of the δ compound. The uncorroded core metal is seen to the left. SEM. Width of field = 160 microns.*

the matrix and surface of high-tin bronze and is normally extremely uniform throughout. It is easily observed under magnification, particularly in the SEM.

Most of the Roman, Chinese and Dark Age high-tin bronze castings have a significant lead content of several per cent which improves castability, reduces the melting point and helps fluidity during casting (Staniaszek 1982). The lead resides as tiny droplets and is usually uniformly distributed throughout the eutectoid matrix. Corrosion due to burial will inevitably have removed both the α component of the eutectoid and the lead globules from the surface of high-tin bronze antiquities, leaving the uncorroded δ matrix of the eutectoid at the surface.

A classic example of such a structure is illustrated in Figure 21.21 of the surface of the Roman high-tin bronze mirror (British Museum GR 1975. 9–1. 43). The laths of δ compound show clearly against the corrosion-etched α and round lead porosity. As the tin content approaches the eutectoid composition the microstructure takes on a lath-like appearance, probably due to chill cast conditions (Bailey and Barker 1949), and this is common to most antiquities of similar composition irrespective of provenance or period.

Similarly, a cross-section (Figure 21.22) through another typical high-tin bronze object, a silver-coloured Dark Age buckle, clearly shows both the uncorroded body metal with its classic α + δ eutectoid microstructure, lead globules (white), and corrosion loss of α and lead globules to a depth of about 50 microns. The δ of the eutectoid reaches the very surface with no mineralization. By studying Figures 21.21 and 21.22, which are at right-angles, the uniformity of the structure of cast high-tin bronze in three dimensions can be appreciated.

Mercury tinning of high-tin bronze Chinese mirrors

A detailed account of mercury tinning is given in Chapter 6, but a resume is appropriate here.

Ancient Chinese literature mentions the use of mercury/tin amalgam as part of the polishing process associated with the manufacture of high-tin bronze mirrors (Needham 1962 pp. 87–97, Chase and Franklin 1979, Barnard 1961). However, analysis rarely finds mercury on Chinese mirrors.

The explanation of 'mercury tinning' probably lies in the use of a polishing compound called *Xuan Xi*, containing mercury, tin and other materials (see Chapter 5). Ancient Chinese polishers had a trade in repolishing tarnished mirrors (Needham 1974 pp. 249, Figure 1326).

Experiments to replicate the application of *Xuan Xi* show that rubbing mercury–tin amalgam onto a previously polished high-tin bronze for a few minutes produces an extremely thin layer of tin with some mercury on the bronze surface. The most important feature is that this is a *cold* tinning process resulting in an optically polished, silver-coloured surface. Herein lies the explanation of why mercury is rarely found on Chinese mirrors and why no observable tin–copper compound layer from 'tinning' is ever found. It is simply such a thin layer that it is lost during burial as the mirror surface slowly tarnishes.

Patination, corrosion and tin enrichment associated with tinned objects and high-tin bronze

Some aspects of this topic have been mentioned earlier in this chapter, and the particular problems associated with high-tin bronze, which are of importance to the study of the patination of high-tin bronze mirrors etc., are discussed in Chapter 6.

Corrosion of bronze is often characterized by the apparent enrichment of 'tin' on the surface of an antiquity (Tylecote 1985, Wouters *et al.* 1990). Analysis without microscopic examination has thus often led to the assumption that a bronze is tinned if the surface shows a high concentration of tin, when in fact it could be a high-tin bronze or have a surface layer of corrosion or tin sweat. Corrosion of low-tin bronze can give rise to tin oxide or hydrated tin compounds in banded form in the corroded regions (Werner 1972), or result in tin enrichment on the surfaces of buried antiquities. Alternatively, low-tin bronze corrodes to the familiar insoluble copper oxides, carbonates or chlorides (cuprite, tenorite, malachite, azurite, paratachamite) (Scott 1984, 1985. Cushing 1965). Generally, low-tin bronze is anodic to the tin-copper compounds from tinning: the bronze below this surface layer corrodes when the protective tinning becomes perforated (Britton 1975). The volume expansion associated with these corrosion products often lifts and cracks the overlying tinning. Some of these copper corrosion products have been seen to completely cover tinned surfaces, which may only be apparent after the corrosion products are removed thus exposing the silver-grey coloured tinning below. Often a conservator requires confirmation of the nature of a complex surface before completion of treatment. Various aspects of the corrosion of metal artefacts are discussed with respect to conservators, archaeologists and scientists in a NBS special publication (1977).

When corrosion of bronze leads to the formation of tin oxide there is a tendency towards 'tin enrichment', because tin oxide is very insoluble and stable (Tylecote 1979 p. 351) and remains at the object surface, while copper is often leached away. This enrichment is particularly noticeable on high-tin bronzes as the α phase of the eutectoid is always corroded. When the δ compound is also mineralized, as with patinated high-tin bronze mirrors, etc., the tin enrichment is more noticeable, with severe loss of copper and relative tin concentration of about 60–70%. Gettens (1969) discusses the pseudomorphic corrosion to tin oxide of high-tin bronze Chinese mirrors and other artefacts.

Stannous and stannic oxides have similar stability in a wide range of commonly encountered environments, and either or both may be formed on surfaces exposed to air over time (Soto *et al.* 1983). Once the insoluble, adherent tin oxide layer forms, there is a much lower rate of permeation of corrosive elements and gases through the oxide layer to the metal. Although the bronze surface is actually corroded it is basically a stable oxide system and is in a state of passivity. Hoar (1976) notes that a metal is said to be passive when, although not thermodynamically stable, it remains visibly unchanged for an indefinite period. Hence the often stated 'corrosion resistance' of black patinated Chinese bronze mirrors (see Chapters 5 and 6). The polished, pseudomorphic nature of the patination gives the illusion of enduring corrosion resistance compared to surfaces suffering encrustations and pitting.

In other corrosive conditions destannification of bronze can occur. For example, in conditions where water soluble tin chloride is formed, tin is readily removed from the environment of the object (Weisser 1975), whereas the copper corrosion products malachite and azurite are insoluble and deposit on the corroding bronze (Scott 1985).

Turgoose (1985) discusses the corrosion of tin (and lead) and notes that the assessment of the likely corrosion behaviour of tin is more difficult with the presence of the intermetallic compounds. However, in the presence of certain alkaline treatments during conservation the corrosion of tin (compound) coatings may be rapid. Allen, Britton and Coghlan (1970) report destannification of various Bronze Age weapons and Tylecote (Tylecote 1979 p. 366) notes that the 'evenness of corrosion is striking' and 'the surface does not have the unwelcome deep pitting corrosion'. Chase and Franklin (1979) note that the α bronze/δ compound corrosion couple of high-tin bronze should make the tin-rich phase corrode first. Werner (1972) reviews the situation and concludes that in the case of two-phase bronzes, i.e. with eutectoid present, the situation of δ corrosion attack occurring first appears to predominate in tin-copper alloys containing over 1% of zinc. Werner then notes that when bronze is comparatively pure it is normal for α corrosion attack to occur first, while the δ compound remains unchanged even at the surface of the bronze.

Many illustrations of patinated high-tin bronze, in section, show that corrosion of the α phase does occur first and it advances into the body metal to a depth of up to one or two hundred microns from the surface. This compares to the mineralization of the δ compound which is limited to a depth of only a few microns or less at the surface, as in the case of black-surfaced Chinese mirrors (Chase 1977, Chase and Franklin 1979). An example of a recently excavated black-surfaced mirror from Changsha is illustrated in Chapter 6, Figures 6.18–6.20 [sample courtesy of Professor Zhu, Beijing]. In other cases the δ may not be noticeably affected, even though the surface is optically blackened, because the tin oxide layer is extremely thin, as in the case of some Dark Age buckles and a Roman mirror from St Albans, England (Craddock *et al.* 1989).

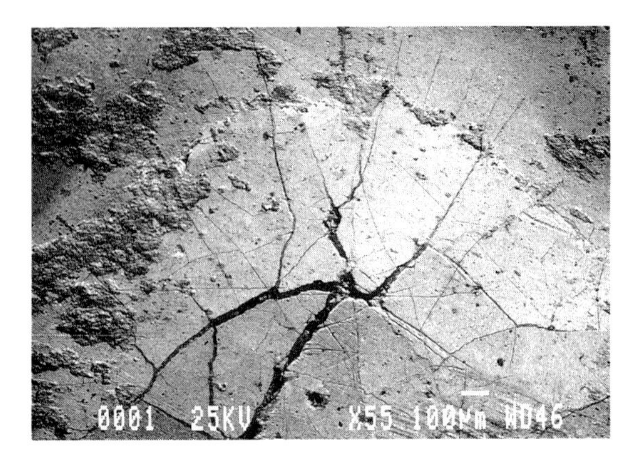

Figure 21.23. *Star shaped eruption on the surface of a silver-coloured high-tin bronze Roman mirror, caused by localized internal corrosion of the copper-rich α phase. The white scale bar represents 100 microns.*

Pitting and encrustations due to corrosion also occur on some high-tin bronzes as do eruptions which raise the surface and produce star shaped cracking, a phenomenon recently examined by McDonnell and Meijers (in press) on the Roman mirror collections catalogued by Lloyd-Morgan (1981). This occurs both on silvery and black objects of high-tin bronze. It appears to be caused by localized internal corrosion of the copper-rich α phase, presumably associated with defects, that produces copper corrosion products of larger volume than the original α within the matrix and eutectoid (Figure 21.23).

Depending on the burial environment, the effects of corrosion may result in a whole range of surfaces on both low and high-tin bronze from smooth patinas to encrustations and deep pitting. These may include a variety of coloured corrosion products, e.g. re-deposited copper, malachite, azurite, cuprite and often black tin oxide deep inside the pits of black surfaced mirrors.

Patina of hot-worked, quenched high-tin bronze

Hot-worked and quenched β phase high-tin bronzes have completely different properties and microstructure to cast high-tin bronze. At

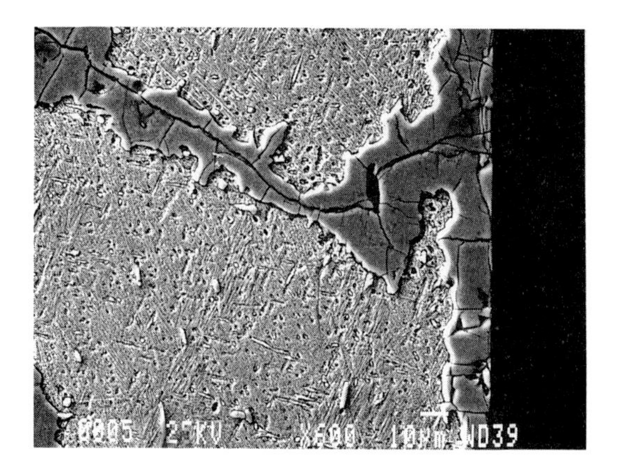

Figure 21.24. *Structure of a quenched high-tin bronze bowl from the Nilgiri Hills, India, seen in cross-section showing the martensitic structure, with a little fine precipitate of intermetallic compound. The mineralized surface patina is seen vertically on the right, and extends to intergranular corrosion in the body metal. SEM. Width of field = 190 microns.*

a temperature between 600 and 750°C, β bronze is a malleable and homogeneous single phase with a composition of 22–24% tin. It can be worked to thin sheet or raised into vessels, but if air cooled it would revert to the usual eutectoid microstructure which would make the object brittle. Hence the objects are quenched in water from above 520°C to produce a martensitic microstructure which is hard and tough, and the metal is a golden colour rather than silver coloured.

Figure 21.24 shows the structure of a quenched high-tin bronze bowl from the Nilgiri Hills, India, which now has a lustrous patina on the surface. The section shows the martensitic matrix, and a little fine precipitate of intermetallic compound. Patination of the surface of this bowl during burial has produced a 10 micron thick layer of mineralization that is also a pseudomorphic replacement of the original as-polished surface, and contains tin oxide and intrusive silicon and other elements from the environment. The colour of the surface is translucent, golden/brown/green, and has a polished 'glassy' appearance like some black Chinese mirrors.

Arsenic sweat on arsenical copper

The metallurgical properties of arsenical copper for industrial use have had early documentation, for example Hanson and Marryat (1927), Gregg (1934). The archaeological interest in arsenical alloys also has a long history for example Charles (1967), McKerrell and Tylecote (1972), and more recently Northover (1989), Zwicker (1990,1991) and Budd (1992). This last section deals with the silvery surface on arsenical copper artefacts (Table 21.2).

Arsenic sweat or inverse segregation occurs readily in low-arsenic copper alloys during solidification after casting, when the remaining low melting point eutectic alloy (21% arsenic) is forced to the surface through interdendritic feeders. It solidifies on the surface forming a

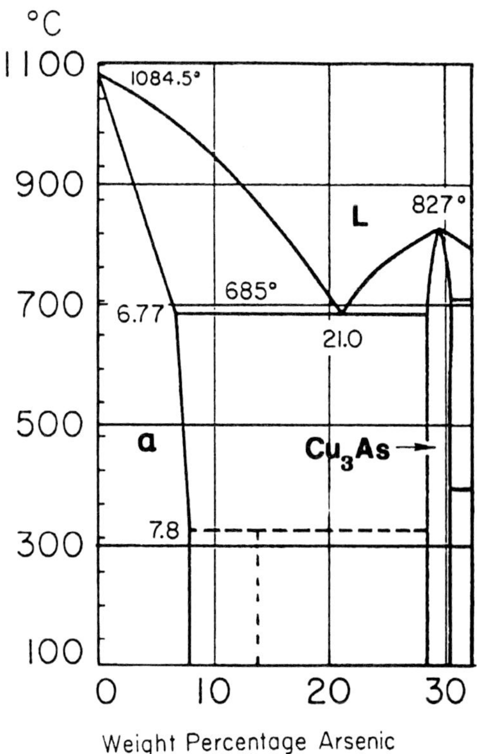

Figure 21.25. *Arsenic–copper phase diagram showing the Cu_3As, γ, intermetallic compound.*

silver coloured layer (McKerrell and Tylecote 1972, La Niece and Carradice 1989). The mechanism is identical to that of tin sweat, resulting from the wide solidus/liquidus region of the phase diagram (Figure 21.25). In the case of arsenic sweat, the inverse segregation occurs at a much lower concentration of arsenic in the body metal than is the case for tin in bronze, probably due to the high vapour pressure of arsenic in the molten alloy. As little as 2% arsenic in copper is sufficient to give some eutectic and hence the potential for sweat (Budd and Ottaway 1991). The higher the arsenic content, the more sweat that can occur and the more extensive and thicker the surface layer can become. It can thus produce a silvery surface over an entire object, as exemplified on some Carthaginian coins containing between 3.1% and 12.2% arsenic, the latter of which is shown in Figure 21.26 (British Museum C.M. 1930. 4–27. 9) (La Niece and Carradice 1989).

There is no doubt that the early metalsmiths would have noticed the effect of arsenic sweat on the colour of such objects, but whether this phenomenon was either deliberately induced, or at least made use of by early metalsmiths is not yet proven.

Arsenic sweat should have a eutectic microstructure comprising α solid solution and γ compound (Cu_3As, containing 29.6% arsenic). However La Niece and Carradice

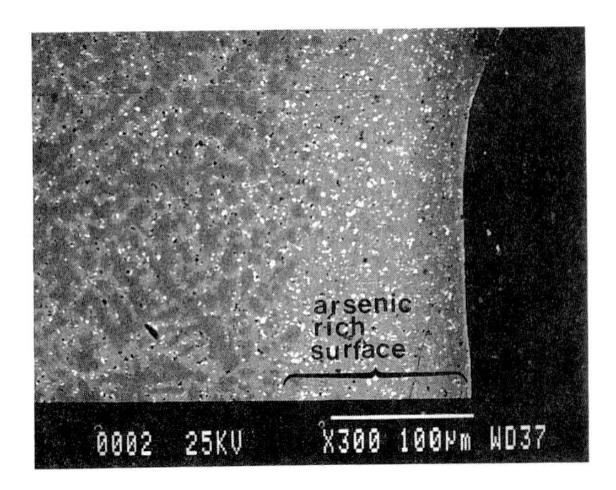

Figure 21.26. *Section of Carthaginian cast coin showing the extensive arsenic sweat forming the surface plating. SEM. The white scale bar represents 100 microns.*

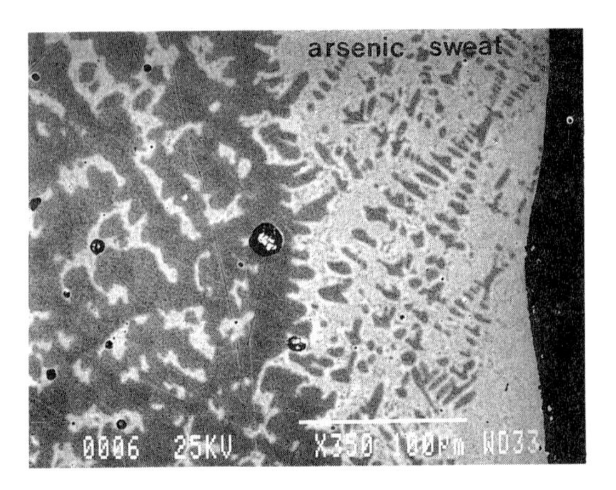

Figure 21.27. *Cast billet of arsenical copper in cross-section showing arsenic sweat on the surface. (Courtesy of Professor R.F. Tylecote.) SEM. The white scale bar represents 10 microns.*

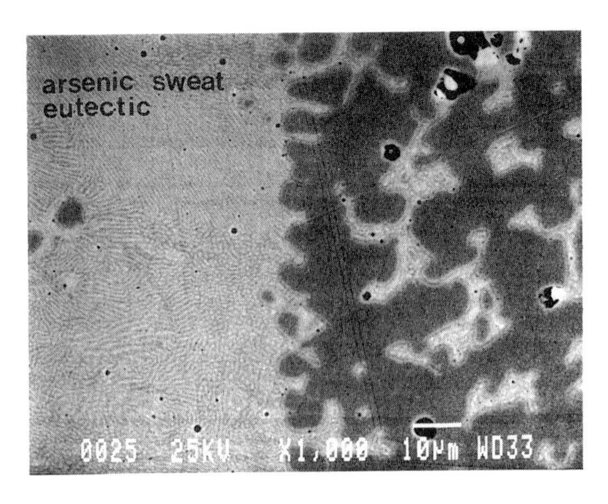

Figure 21.28. *Detail of the arsenic sweat eutectic microstructure on cast billet. SEM. The white scale bar represents 10 microns. (Note that the image is reversed with respect to Figure 21.27.)*

(1989) have shown that the eutectic structure is difficult to resolve under the microscope particularly when impurities such as tin and antimony are present in the alloy. This lack of clear evidence for the eutectic has caused discussion regarding the origin of the surface layer on the Carthaginian coins. However, α dendritic structures within thicker regions on the surface of these coins and interdendritic feeders from the body metal clearly show the cast nature of the surface inverse segregant. McKerrell and Tylecote (1972) have produced cast ingots of

Table 21.2 Characteristics of arsenic-rich surfaces on bronze antiquities

	γ compound	*Eutectic*	*Patina*
(a) Arsenic-rich surface on *cast* Cu–As alloys	γ in eutectic form	Arsenic sweat eutectic, variable thickness on cast body metal	Not corroded
(b) Arsenic-rich surfaces found on some *fully homogenized* alloys	γ compound 1–3 microns + filaments into body metal	No	Not corroded

arsenic–copper of various compositions, but with no tin and antimony, that exhibit inverse segregation (Figure 21.27) that is comparable to that observed on the Carthaginian coin. At high magnification the fine eutectic is resolved on these castings (Figure 21.28).

Two other types of arsenic-rich surface have been identified. The first is that on a bronze bull from Horoztepe, Anatolia 2100 BC, with its decorative and protective areas of arsenic-rich, solid γ compound layer (29.6% arsenic) (Smith 1973b). As the plated areas are in distinctive patterns and the body metal contains no arsenic, arsenic sweat is ruled out. It is thought that the bull has been deliberately plated by a cementation process reducing arsenic-rich minerals onto the surface of the bronze.

The second type consists of an extensive γ compound layer on the surfaces of arsenic–copper alloy blades from Quimperle, France (Briard and Mohen 1974), and the Early Copper Age site of Los Millares, Spain (Hook *et al* 1991). The blade from Los Millares was examined in detail and is shown in cross-section in Figure 21.29. The body metal has 5% arsenic and hence would be very prone to sweating on casting. The whole blade is silver-grey in colour and essentially uncorroded on the surface compared with the many associated finds of lower arsenic content, which are heavily patinated with green copper corrosion products. In cross-section the surface layer of γ compound is seen to be about 8 microns thick and it extends down to the cutting edge

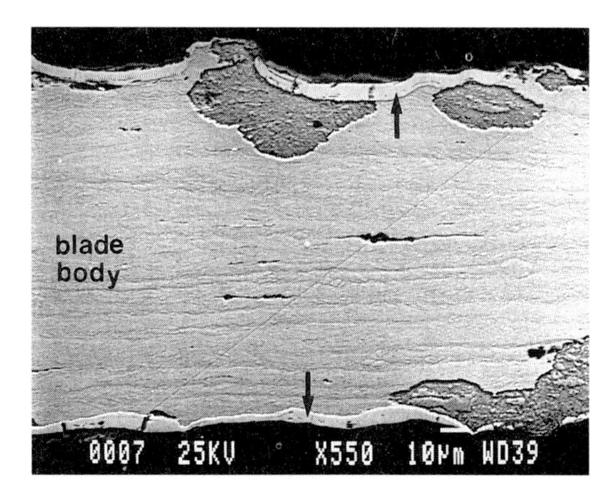

Figure 21.29. *Early Copper Age blade from Los Millares, Spain seen in cross-section with surface coating of Cu3As arsenic–copper compound (arrowed), and homogenized body metal. Corrosion occurs in the body metal where the surface plating is perforated. SEM. Thickness of the blade section = 140 microns.*

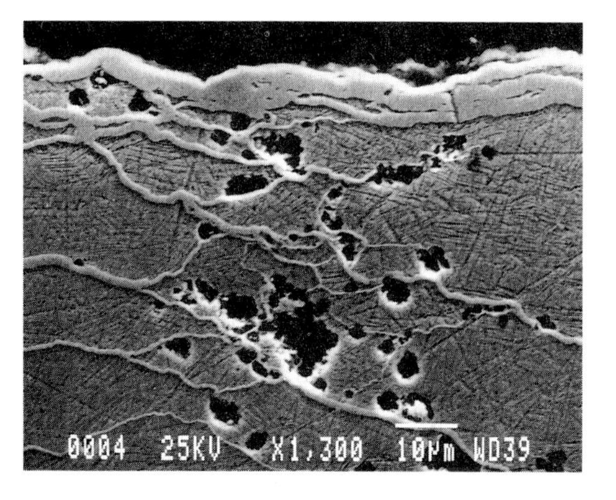

Figure 21.30. *Cross section of Los Millares blade showing detail of the surface layers and body filaments of Cu_3As, arsenic–copper compound. SEM. The white scale bar represents 10 microns.*

of the blade. The body metal grain structure shows that it has been heavily worked, annealed and homogenized. No eutectic is found in the body metal and very little coring remains. It is very difficult to fully homogenize these alloys (Northover 1989) and so this blade must have been heavily worked.

A distinctive feature of this surface layer is the presence of thin bands of γ compound in some regions (Figure 21.30). Accompanying these features in some areas are thin filaments of γ compound running obliquely in the worked direction from the surface into the body metal along the equiaxed grain boundaries. They terminate in grain boundary regions of high arsenic concentration. The filaments do not appear to be directly associated with grain boundary corrosion, which is generally lacking in this blade. Analysis of the filaments clearly identifies the 29.6% arsenic γ compound. Digital X-ray mapping at the ends of these filaments shows the arsenic concentration in solid solution associated with the grain boundaries. These could be described as proto-filament areas.

The blade from Quimperle has similar surface bands of γ compound and filaments running into the body metal, the surface enrichment was thought by the authors (Briard and Mohen 1974) to be a deliberate decorative process, although not understood, rather than one of selective corrosion. Tylecote (1987 p. 238) comments that it is probable that the blades were heated in a powdered arsenic mineral in the manner suggested by C.S. Smith for the Horztepe bull. However, Budd (1991b) records a similar occurrence of filaments of γ compound in blades from Mondsee, Austria, and believes the intergranular γ phase to be precipitated from arsenic-rich solid solution during burial. In a copper–arsenic needle from Peru (Smith 1973b) a different type of precipitation was observed. This was algodomite (Cu_8As, 12% arsenic) at grain boundaries, thought to be electrochemically redeposited, similar to the redeposition of copper in corroding bronze (Gettens 1969, McKerrell and Tylecote 1972). Zwicker (1990) illustrates a piece of native copper from Anarak, Iran, with grain boundary enrichment of 3.5% arsenic appearing as veins. These different phenomena show the natural tendency for arsenic to enrich or precipitate at interfaces.

The origin of the surface plating on the Los Millares blade and similar blades must be considered: is it a natural phenomenon or deliberate and therefore of archaeometallurgical significance with regard to the selection of materials in the Bronze Age? There are two related features that are significant; first, the surface layer and second, the filaments of γ within the body metal. With its high arsenic body composition, the Los Millares blade would almost certainly have originally suffered arsenic sweat during casting. Unlike tin sweat, the γ arsenic compound is relatively ductile (Northover 1989, Smith 1973b) and could well have survived hammering to shape. The surface layer of the Los Millares blade has survived even though it clearly shows evidence of stretching and shearing from hammering (Figure 21.31). Homogenization of the heavily worked metal, and impurities as mentioned earlier, could account for the lack of eutectic structure in this surface. Additionally Budd (1991a, 1992) has shown that the limit of solid solubility of arsenic in copper is much lower at room temperature than the *c.* 6% previously published (Subramanian and Laughlin, 1988), and heavily worked and homogenized high-arsenic alloys would be prone to supersaturation and hence precipitation of γ compound at room temperature.

Thus the grain boundary precipitation of γ compound filaments occurs along the lines of

Figure 21.31. *Surface of Los Millares blade showing the stretched compound layer from working. SEM. Width of field = 750 microns.*

highest concentration of arsenic in solid solution, for example the proto-filament regions in the Los Millares blade. Significantly the filaments reach the surface and may well represent the position of original arsenic sweat feeders which would have provided the excess arsenic before (the incomplete) homogenization. These provide the nucleation points for precipitation of the γ compound over time during burial.

It therefore appears that, in the case of the Los Millares blade, the silvery surface compound layer is probably the result of heavily worked arsenic sweat, which still retains evidence of hammering. It is unlikely to be a deliberate surface plating by a cementation process as applied to the bronze bull from Horoztepe. The filaments of γ compound in the body metal result from post burial precipitation, which probably nucleated at the surface layer.

We may never know whether the ancient metalsmiths wanted the fortuitous plating of arsenic sweat and sought the conditions that induced the phenomenon, although it is perhaps significant that these silvery surfaces appear to be confined to blades in the Bronze Age. Later, arsenical copper alloys were described by the alchemists as false silver, and the Carthaginian coins may be examples of them. The beneficial hardening of copper by arsenic above about 2% would no doubt have been recognized in antiquity and the ancient metalsmiths probably made some selection of materials to give the desired finished product, although the systematic development of the arsenic–copper alloy system is still in question (Budd 1992).

Conclusions

The properties of tin-rich surfaces on bronze have been known for four millennia and have been widely exploited for two and a half millennia. The study of such surfaces on antiquities allows us to characterize the material, follow the use of tinning as a decorative and utilitarian technique applied to low-tin bronze, and to follow the use of high-tin bronze for the manufacture of specific commodities. Recognition of the true nature of these objects and their tin-rich surfaces will ensure correct technological documentation, cataloguing and conservation treatment. The role of arsenic-rich surfaces is becoming clearer.

Acknowledgements

The following colleagues are acknowledged for their collective contributions to the work involved in this study and in the preparation of this chapter. Paul Craddock for overall discussion and comment. Jessica Rawson, Dafydd Kidd and Judith Swaddling for providing the antiquities from the collections of The British Museum that were essential to this study. Professor Zhu from Beijing for his collaboration and provision of newly excavated samples for examination. Susan La Niece for providing figures and data. Peter Northover for providing experimental cast bronzes. Helen Jones for providing the material from the Mansell Street excavation for examination. Paul Budd for discussions on arsenical copper. Tony Milton for processing and printing the micrographs.

References

Allen, I.M., Britton, D. and Coghlan, H.H. (1970) Metallurgical reports on British and Irish Bronze Age implements and weapons in the Pitt Rivers Museum. *Occasional Papers on Technology, Pitt Rivers Museum*, **10**, 73–88, Oxford: Oxford University Press

Allen, J. (1979) *Persian Metal Technology 700–1300AD*, chapter 2, Ithaca Press, London, for the Faculty of Oriental Studies and the Ashmolean Museum, University of Oxford

Bailey, G.L., Baker, W.A. (1949) Melting and casting of non-ferrous metals, in Symposium on metallurgical aspects of non-ferrous metal melting and casting of ingots for working. *Institute of Metals Monograph & Report Series*, **6**, 7–32, Institute of Metals, London

Bakay-Perjés, J. (1989) Technical observation on Merovingian bronze finds. In *Conservation of Metals* (ed. M. Járó) Proceedings of the International Restorer Seminar, Veszprem, Hungary, 1–10th July 1989, pp. 137–141

Barnard, N. (1961) Bronze casting and bronze alloys in ancient China, *Monumenta Serica Monograph XIV*, published jointly by the Australian National University and Monumenta Serica, Tokyo

Briard, J., and Mohen, J-P. (1974) Le tumulus de la forêt de Cannoit a Quimperle (Finistere). *Antiquities Nationales*, **6**, 46–60

Britton, S.C. (1975) *Tin Versus Corrosion*, International Tin Research Institute publication, No. 510, England, pp. 19–25

Bruce-Mitford, R. (1978) *Sutton Hoo Volume 2*, British Museum Publications Ltd., London, pp. 311–393

Budd, P. (1991a) Eneolithic Arsenical Copper: Heat treatment and the metallographic interpretation of manufacturing processes. *Archaeometry 90* (eds E. Pernika, and G.A. Wagner) Proceedings of the Archaeometry Conference, Heidelberg, 1990, Birkhäuser, Verlag, Basel, pp. 35–44

Budd, P. (1991b) A metallographic investigation of Eneolithic arsenical copper artefacts from Mondsee, Austria. *Journal of the Historical Metallurgy Society*, **25**(2), 99–108

Budd, P. (1992) Alloying and metalworking in the Copper Age of Central Europe. *Bulletin of the Metals Museum of Japan*, **17**, 3–14

Budd, P. and Ottaway, B.S. (1991) The properties of arsenical copper alloys: implications for the development of Eneolithic metallurgy. In *Archaeological Sciences 1989* (eds P. Budd, B. Chapman, C. Jackson, R. Janaway and B. Ottaway) Proceedings of a conference on the Application of Scientific Techniques to Archaeology, Bradford, September, Oxbow Monograph 9, pp. 132–142.

Charles, J.A. (1967) Early arsenical bronzes – a metallurgical view in Cycladic metallurgy and the Aegean early Bronze Age, *American Journal of Archaeology*, **71**(1), 21–26

Chase, W.T. (1977) What is the smooth lustrous surface on ancient bronze mirrors? In *Corrosion and Metal Artefacts – A Dialogue between Conservators and Archaeologists and Corrosion Scientists* (eds B.F. Brown, H.C. Burnett, W.T. Chase, M. Goodway, J. Kruger, and M. Pourbaix) National Bureau of Standards special publication 479, Washington, pp. 191–203

Chase, W.T., and Franklin, U.M. (1979) Early Chinese black mirrors and pattern-etched weapons. *Ars Orientalis*, **11**, 215–258

Close-Brookes, J. and Coles, J.M. (1980) Tinned axes. *Antiquity*, **LIV** (212), 228–229

Corfield, M.C. (1985) Tinning of iron. In *Lead and Tin: Studies in Conservation and Technology* (eds G. Miles, and S. Pollard) United Kingdom Institute of Conservation Occasional Papers, No. 3, pp. 40–43

Craddock, P.T. (1981) Report on the composition of 19 Etruscan bronze mirrors from Danish museums. In *Corpus Speculorum Etruscorum, Denmark 1* (ed. H.S. Roberts) The Danish National Museum, Odense University Press, pp. 131–133

Craddock, P.T. (1988) Copper alloys of the Hellenistic and Roman world, new analyses and old authors. In *Aspects of Ancient Mining and Metallurgy: Acta of a British School at Athens centenary conference* (ed. J. Ellis Jones) University College of North Wales, Bangor, 1986, pp. 55–65

Craddock, P.T., Hook, D.R. and Meeks, N.D. (1989) Appendix B, Technical report on Roman mirrors from the Iron Age and Roman cemeteries. In *Verulamium: The King Harry Lane Site, English Heritage Archaeological Report no. 12* (eds I.M. Stead and V. Rigby) pp. 271–272

Cushing, D. (1965) Principles of corrosion applicable to ancient metals and methods of identifying corrosion products. In *Application of Science in Examination of Works of Art,* Proceedings of the Seminar September 7–16th 1965, Museum of Fine Arts, Boston, USA, pp. 53–65

Daniels, E.J. (1936) The hot-tinning of copper: the attack on the basis metal and its effects. *Journal of the Institute of Metals*, **LVIII** (1), 199–210

Diderot and d'Alembert (1755) *Encyclopedie ou dictionnaire raisonne des sciences, des arts et des metiers (Paris) 5*, p. 806

Dinsdale, P.M. (1978) Hot tinning for the food industry. In *Tin and its Uses, No.116*, International Tin Research Institute, pp. 9–11

Garland, H. and Bannister, C.O. (1927) *Ancient Egyptian Metallurgy*, Charles Griffin, London, pp. 28–30

Gettens, R.J. (1969) The Freer Chinese Bronzes, in the Smithsonian Institution, *Freer Gallery of Art, Oriental Studies, No.7, Volume II, Technical Studies*, Washington, pp. 121–139 and 171–195

Goodway, M., and Conklin, H.C. (1987) Quenched high-tin bronzes from the Philippines. *Archaeomaterials*, **2**(1), 1–27

Gregg, J.L. (1934) Arsenical and argentiferous copper. *The American Chemical Society Monograph Series*, The Chemical Catalog Company Inc. New York

Hanson, D. and Pell-Walpole, W.T. (1951) *Chill-cast Tin Bronzes*, Edward Arnold, London

Hanson, D., and Marryat, C. (1927) The effect of arsenic on copper. *The Journal of the Institute of Metals*, **XXXVII**(1), .121–143

Harrison, R.J. (1980) A tin-plated dagger of the Early Iron Age from Spain. *Madrider Mitteilungen*, **21**, 140–146

Hedges, E.S. (1960) *Tin and its Alloys*, Edward Arnold, London

Hedges, E.S. (1964) *Tin in Social and Economic History*, Edward Arnold, London

Henken, T.C. (1939) The excavation of the Iron Age Camp on Bredon Hill, Gloucestershire, 1935–1937. *Archaeological Journal*, **95**, 1–111

Hoar, T.P. (1976) Passivity, passivation, breakdown and pitting. In *Corrosion*, vol 1 (ed. Shreir, L.L.) pp. 114–129

Hook, D.R., Freestone, I.C., Meeks, N.D., Craddock, P.T., and Moreno Onorato, A. (1991) The early production of Copper Alloys in South-East Spain. In *Archaeometry 90* (eds E. Pernika, and G.A. Wagner) Proceedings of the Archaeometry Conference, Heidelberg 1990, Birkhäuser Verlag, Basel, pp. 65–76

ITRI (1939) *Historic Tinned Foods*, International Tin Research and Development Council, No.85 (now International Tin Research Institute)

ITRI (1982) *Metallography of Tin and its Alloys*, International Tin Research Institute, No. 580

ITRI (1983a) Gypsy wipe tinners in America. *Tin and its Uses*, **138**, 15, International Tin Research Institute

ITRI (1983b) *Tin and Alloy Coatings*, International Tin Research Institute, No. 625, pp. 15–19

Jones, H., and Meeks, N.D. (in press) The conservation and technological examination of finds from Mansell Street Roman Cemetery, London. In *Finds from the Imperial West, Proceedings of the Symposium*, September 1991, UK Institute for Conservation

Jope, E.M. (1956) The tinning of iron spurs: a continuous practice from the tenth to the seventeenth century. *Oxoniensia*, **21**, 35–42

Kay, P.J. and MacKay, C.A. (1976) The growth of intermetallic compounds on common basis materials coated with tin and tin-lead alloys. In *Transactions of the Institute of Metal Finishing*, **54**, 68–74 (also ITRI (1976) publication No.517, International Tin Research Institute)

Kinnes, I.A., Craddock, P.T., Needham, S. and Lang, J. (1979) Tin plating in the Early Bronze Age: the Barton Stacey axe. *Antiquity*, **LIII** (208), 141–143

Kinnes, I.A. and Needham, S. (1981) Tinned axes again. *Antiquity*, **LV** (214), 133–134

La Niece, S. (1983) Niello: an historical and technical survey. *The Antiquaries Journal*, **LXIII**, part II, 279–297

La Niece, S., and Carradice, I. (1989) White copper: The arsenical coinage of the Libyan revolt 241–238 BC. *Journal of the Historical Metallurgy Society*, **23**(1), 9–16

Lang, J., and Hughes, M.J. (1984) Soldering of Roman silver plate. *Oxford Journal of Archaeology*, **3**(3), 77–107

Lins, P.A. (1974) A history of metal coatings on metals pre-1800 A.D., p.40, unpublished MSc project, Department of Metallurgy and Materials, Sir John Cass School of Science and Technology, City of London Polytechnic

Lloyd-Morgan, G. (1981) *Description of the Collections in the Rijksmuseum, IX The Mirrors, G.M. Kam, Nijmegen*, The Netherlands, Ministry of Culture, Recreation and Social Welfare, Nijmegen, The Netherlands

Lucas, A. (1948) *Ancient Egyptian Materials and Industries*, Edward Arnold, London

McDonnell, R.D, Meijers, H.J.M. (in press) A study of the composition and microstructure of six fragments of Roman mirrors from Nijmegen, The Netherlands. In *Acta of the 12th International Congress on Ancient Bronzes*, Nijmegen, 1–4 June 1992 (ed. A. Gerhartl-Witteveen) in conjunction with the Provincial Museum G.M. Kam and the Katholieke Universiteit, Nijmegen

McKerrell, H., and Tylecote, R.F. (1972) The working of copper-arsenic alloys in the Early Bronze Age and the effect on the determination of provenance. *Proceedings of the Prehistoric Society*, **38**, 209–218

Meeks, N.D. (1986) Tin-rich surfaces on bronze – some experimental and archaeological considerations. *Archaeometry*, **28**(2), 133–162

Meeks, N.D. (1988a) A technical study of Roman Bronze mirrors. In *Aspects of ancient mining and metallurgy: ACTA of the British School at Athens Centenary Conference, Bangor 1986* (ed. J. Ellis-Jones) University College of North Wales, pp. 66–79

Meeks, N.D. (1988b) Surface studies of Roman bronze mirrors, comparative high tin bronze Dark Age material and black Chinese Mirrors. In *Proceedings of the 26th International Archaeometry Symposium* (eds. R.M. Farquhar, R.G.V. Hancock, L.A. Pavlish) University of Toronto, Canada, May 16–20th 1988, The Archaeometry Laboratory, Toronto, pp. 124–127

Meeks, N.D. (1988c) Backscattered electron imaging of archaeological material. *Scanning Electron Microscopy in Archaeology* (ed. S.L. Olsen) BAR International Series 452, Oxford, pp. 23–44

Meeks, N.D. (in press) A technical study of Roman bronze mirrors. In *Acta of the 12th International Congress on Ancient Bronzes*, Nijmegen, 1–4 June 1992 (ed. A. Gerhartl-Witteveen) in conjunction with the Provincial Museum G.M. Kam and the Katholieke Universiteit, Nijmegen

Melikian-Chirvani, A.S. (1974) The White Bronzes of Early Islamic Iran. *Metropolitan Museums Journal*, **9**, 123–151

Mrowec, S. (1980) Defects and dffusion in solids, an introduction, *Materials Science Monograph*, vol.

5, Elsevier Scientific Publishing Company, Amsterdam, The Netherlands, pp. 283–28

NBS (1977) *Corrosion and Metal Artifacts – A Dialogue between Conservators and Archaeologists and Corrosion Scientists* (eds B.F. Brown, H.C. Burnett, W.T. Chase, M. Goodway, J. Kruger, and M. Pourbaix) NBS special publication 479, Washington

Needham, J. (1962) *Science and Civilisation in China*, vol. 4, part 1, Cambridge University Press, pp. 87–97

Needham, J. (1974) *Science and Civilisation in China*, vol. 3, part 2, Cambridge University Press

Nemeth, D. (1982) A gypsy wipe tinner and his work. *Journal of the Gypsy Lore Society*, **2**, 30–53

Northover, J.P. (1989) Properties and use of arsenic-copper alloys. In *Old World Archaeometallurgy*, Proceedings of the International Symposium, Heidelburg 1987 (eds A. Hauptman, E. Pernicka, and G.A. Wagner) Selbstverlag des Deuschen Bergbau-Museums, Bochum, pp. 111–118

Oddy, W.A. (1980) Gilding and tinning in Anglo-Saxon England. In *Aspects of Early Metallurgy*, British Museum Occasional Paper no 17, London, pp. 129–134

Oddy, W.A. and Bimson, M. (1985) Tinned bronze in antiquity. In *Lead and Tin: Studies in Conservation and Technology*, United Kingdom Institute of Conservation, Occasional Paper No.3, pp. 33–39

Oya, S., Nakano, K., Takada, T., and Suzuki, K. (1975) Role of absorbed gases in inverse segregation in the Cu-Sn alloys. In *Report of the Castings Research Laboratory*, **26**, 1–9, Waseda University

Pliny, *Historia Naturalis*, book XXXIV, ch.XLVIII (translation by H. Rackham (1952), Vol.IX, William Heinemann, London, pp. 243–245

Ray, P. (1956) *History of Chemistry in Ancient and Medieval India*, Indian Chemical Society, Calcutta, p. 216

Savory, H.N. (1964) A new hoard of La Tène metalwork from Merionethshire. *Bulletin of Celtic Studies*, **20**(4), 449–475

Savory, H.N. (1966) Further notes on the Tal-y-Llyn (Mer.) hoard of La Tène metalwork. *Bulletin of Celtic Studies*, **22**(1), 88–103

Scott, D. (1984) *The Conservation of Metallic Artefacts*, course notes: C45, Department of Archaeological Conservation and Materials Science, Institute of Archaeology, London

Scott, D (1985) Periodic corrosion phenomena in bronze antiquities. *Studies in Conservtion*, **30**(2), 49–57

Seeley, N.J., and Waranghkan Rajjpitak (1979) The bowls from Ban Don Ta Phet, Thailand: an enigma of prehistoric metallurgy. *World Archaeology*, **11**(1), 26–32

Smith, C.S. (1973a) The interpretation of microstructures of metallic artefacts. In *Application of Science in Examination of Works of Art*, Proceedings of Seminar in 1970 (ed. W.J. Young) Museum of Fine Arts, Boston, p. 27, Fig. 10

Smith, C.S. (1973b) An examination of the arsenic-rich coating on a Bronze Bull from Horoztepe. In *Application of Science in Examination of Works of Art*, Proceedings of Seminar in 1970 (ed. W.J. Young) Museum of Fine Arts, Boston, pp. 96–102

Smith, J.A. (1872) Notice of bronze celts or axe heads, which have apparently been tinned. In *Proceedings of the Sussex Archaeological Society*, **IX**(2), 433

Soto, L., Franey, J.P., Graedel, T.E., and Kammlott, G.W. (1983) On the corrosion resistance of certain ancient Chinese bronze artefacts. *Corrosion Science*, **23**(3), 241–250

Staniaszek, B.E.P. (1982) An investigation of the use of leaded bronzes in Bronze Age castings. Thesis submitted for the examination of Metallurgy and Science of Materials, part II, Department of Metallurgy and Science of Materials, University of Oxford

Subramanian, P.R., and Laughlin, D.E. (1988) The As-Cu (arsenic–copper) system. *Bulletin of Alloy Phase Diagrams*, **9**(5), 605–617

Sweatman, K. (1981) Tinned copperware from Australia. *Tin and its Uses*, **128**, 15–16, International Tin Research Institute

Theophilus, *On Divers Arts* (translated by J.G. Hawthorne and C.S. Smith 1979) Dover Publications, New York, p. 187

Thwaites, C.J. (1983) *Practical Hot Tinning*, International Tin Research Institute, publication no. 575, London

Turgoose, S. (1985) The corrosion of lead and tin before and after excavation. In *Lead and Tin: Studies in Conservation and Technology*, United Kingdom Institute of Conservation, Occasional Paper No. 3, pp. 15–26

Tylecote, R.F. (1979) The effect of soil conditions on the long term corrosion of tin-bronze and copper. *Journal of Archaeological Science*, **6**, 345–368

Tylecote, R.F. (1985) The apparent tinning of bronze axes and other artefacts. *Journal of the Historical Metallurgy Society*, **19**(2). 169–175

Tylecote R.F. (1987) *The Early History of Metallurgy in Europe*, Longman, London, p. 238

Van Arsdell, R.D. (1989) *Celtic Coinage of Britain*, Spink, London

Weisser, T. (1975) The de-alloying of copper alloys.

In *Conservation in Archaeology and the Applied Arts*, IIC Congress, Stockholm, pp. 207–214

Werner, O. (1972) *Spektralanalytische und Metallurgische Untersuchungen an Indischen Bronzen*, E.J. Brill, Leiden, The Netherlands

Wouters, H.J., Butaye, L.A., Adams, F.C., and Van Espen, P.E. (1990) Spetroscopic investigation of tin-rich surfaces on Bronze Age copper alloys. In *Archaeometry 90*, Proceedings of the International Symposium on Archaeometry, Heidelberg, 1990 (eds E. Pernika, and G.A. Wagner) Birkhäuser Verlag, Basel, pp. 145–153

Wulff, H.E. (1966) *The Traditional Crafts of Persia*, M.I.T. Press, London, p. 131

Zhu, Shoukang (1986. Ancient metallurgy of non-ferrous metals in China. In *Proceedings of Symposium on the Early Metallurgy in Japan and the Surrounding Area*, October 1986, Special issue Vol. 11, Bulletin of the Metals Museum, pp. 1–13

Zwicker, U. (1990) Archaeometallurgical investigation on the copper and copper alloy production in the area of the Mediterranean Sea (700–1000 BC). *Bulletin of the Metals Museum*, **15**, 3–32

Zwicker, U. (1991) Natural copper-arsenic alloys and smelted arsenic bronzes in early metal production. In *Decouverte du Metal*, Amis du Musee des Antiquites Nationales, Millenaires, dossier 2 (eds C. Palet, and E. Vagende) Picard, Paris, pp. 331–340

22

Copper plating on iron

Michael Corfield

Abstract

The use of modern cleaning techniques has revealed a much wider use of copper alloy plating on iron than had previously been recognized. This paper examines the evidence for plating, the type of artefact on which it is found and the techniques used for plating.

Introduction

Iron is a valuable commodity which has been in continuous use since the second millennium BC. Despite the development of new materials, as we approach the end of the second millennium AD the position of iron as a major strategic metal remains unchallenged. The value of iron is its versatility; it can be used as a structural material of great strength and it can equally be used to fabricate the smallest of components and even fine jewellery. It was this range of use and comparative ease of working which doubtless attracted early technologists searching for an improved and more readily available material to replace the various copper alloys which hitherto predominated.

For all its positive attributes, iron suffers from one major flaw; the ease with which it corrodes demands that it be carefully protected from the elements. If objects fabricated from iron are to be used in situations where they will be exposed to the weather, then some form of protection is needed to prevent their rapid corrosion. No doubt for some objects coatings of wax or grease would have sufficed; sheepskin liners in sword scabbards seem to have been used for their natural oils throughout the Iron Age, Roman and medieval periods (Watson and Edwards 1990). For objects where a coating of grease or wax would be inappropriate, other forms of protection were required. Tinning seems to have been a widely used technique, described in the 1st century AD (though not on iron) by Pliny (Rackham 1952) and in Theophilus's great medieval text on technology (Smith and Hawthorne 1963). Tinning is a well-recognized feature of archaeological iron and is readily observed on X-radiographs (Corfield 1982) and on cleaned specimens (Corfield 1985, see also this volume Chapter 21).

A second and less well-documented method of enhancing the visual appearance of iron and improving its protection against corrosion is the use of copper alloys as the plating medium (Plate 13.2). Copper alloy plating has long been recognized on iron objects, particularly from the Iron Age and especially on bridle bits, examples of which are widespread. The type site of La Tène yielded one example (Vouga 1925), and similar examples were found at Glastonbury Lake Village (Bulleid and St George Grey 1917), Mere Lake Village

(Bulleid and St George Grey 1948), Arras and Drifield in Yorkshire, Walthamstow in London and Bigbury in Kent (Ward Perkins 1939). The Llyn Cerrig Bach, Anglesey hoard included eleven bridle bits or parts of bits including at least two which were plated with copper alloy (Fox 1946). More recently a copper alloy plated side link from a bit found at Gussage All Saints has been the subject of intensive examination (Wainwright 1979) (see below).

Copper alloy plating has been recognized on other major groups of objects. Iron bells were commonly plated, though here it has been suggested by Arwidsson and Berg (1983) that the primary reason that plating was applied was to improve the resonance of the bell rather than for protection. Padlocks have long been recognized as examples of copper alloy plating; the mechanical function of the lock demanded a high degree of protection from the weather that could only be provided by a substantial coating. An additional reason for the use of copper as the plating medium was that the components of the lock were joined by brazing and so presumably it would be a relatively simple operation to extend the brazing to coat the entire object. There is a fine example of a barrel padlock and its bolt, both of which are plated, in Perth Museum. A third group is represented by the elaborately decorated Viking stirrups which additionally were inlaid with a contrasting metal (Seaby and Woodfield 1980).

Recently, improved conservation methods have yielded a great diversity of copper alloy plated objects for which analyses are now becoming available. From Roman Caerleon copper plating has recently been identified on a fragment of mail originally excavated in 1929, the piece was required for the new exhibitions in the Roman Legionary Museum, and on preparing it, traces of copper corrosion were noted. Further investigation revealed the presence of copper plating, which was present as a uniform layer on all of the links, though it was not possible to reveal it in its entirety due to the fragility of the iron substrate. It was possible to remove a small fragment which was mounted and polished (Figure 22.1), this shows that the plating was continuous and not an accident of burial. Analysis shows that the plating is an alloy of copper and zinc with tin and the metallographic structure of the copper alloy shows it to be as-cast. The manufacture of this mail must have required a number of stages: first the iron wire would have been made; next, copper alloy must have been applied by dipping; as the structure is as-cast there was presumably no further working and the plated wire was used directly to form the rings. Such mail would have considerable advantages, combining the strength of iron mail with the brilliant effect of copper alloy.

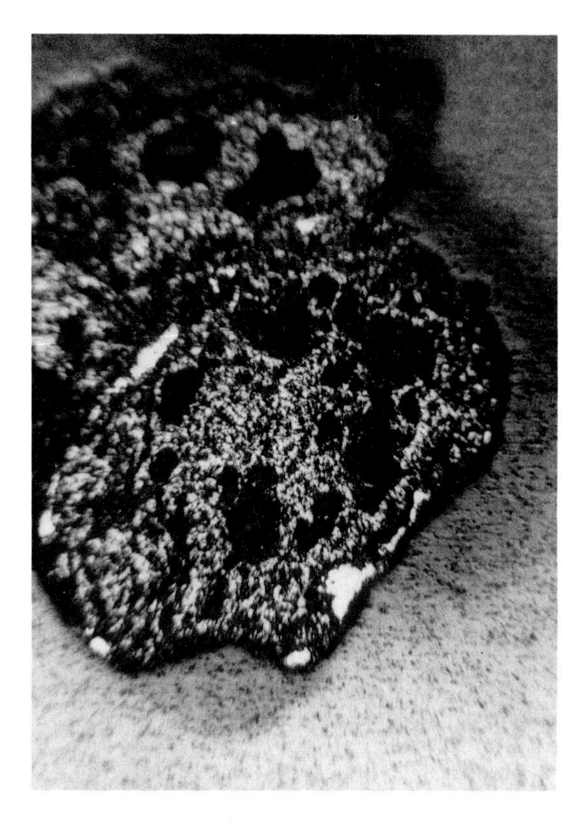

Figure 22.1. *Section across a link of iron mail from Caerleon plated with copper alloy. The plating shows as bright white. (Photo courtesy of the National Museum of Wales.)*

Also from Caerleon a copper alloy plated key bit has recently been identified during conservation, here the plating appears to have been hammered on, distinct overlaps can be seen on the plating suggesting that it is not continuous as would be the case with dipping methods (Figure 22.2).

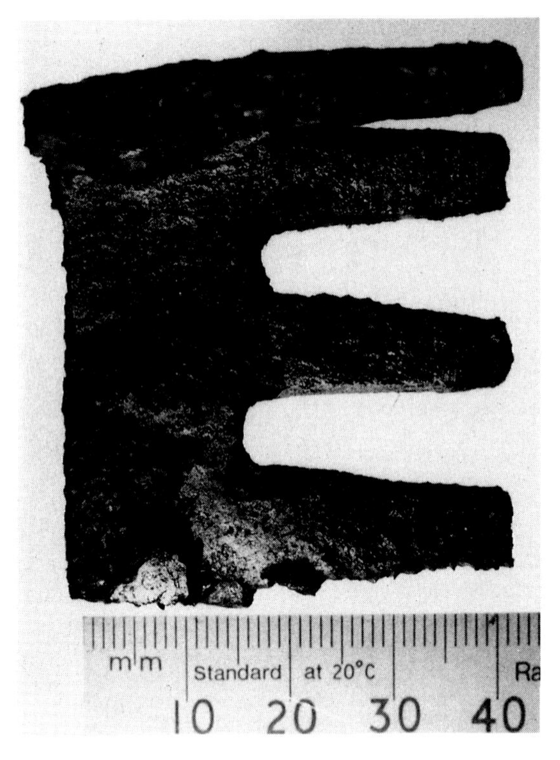

Figure 22.2. *Plated key bit from Caerwent showing at the left side the overlap of layers of copper alloy. (Photo courtesy of National Museum of Wales.)*

Current work shows the medieval period to have been the most prolific production period for copper alloy plated objects. The range of objects found and their geographical spread suggest that by this time it had become a common technique. The largest identified group of objects is from the Anglo-Scandinavian levels at Coppergate, York (Table 22.1). The twelve items include three keys or parts of keys, a padlock, two bells, a spur and several other objects (Ottaway 1992 p. 490). During the conservation of objects from Victoria Street and Jewry Street in Winchester, copper alloy plating was found on a similar range of objects including a key, knife handle and barrel padlock. A copper alloy plated chape and ferrule was found in Newbury and from Wiltshire several arrow-heads have been found with thin copper alloy platings (personal communication, Sarah Watkins). The London area has yielded a plated bell from a Saxon context as well as numerous barrel padlocks (personal communication, Katherine Starling).

A distinction must be drawn here between plating (that is a uniform thin layer applied overall to the object for protection), casting on, and the application of bronze repoussé

Table 22.1 Copper alloy plated iron finds from Coppergate: Analysis of platings (undertaken by Paul Wilthew, Ancient Monuments Laboratory)

Ref. No.	*Item*	*Cu*	*Sn*	*Zn*	*Pb*	*Description*
7788	Bell	***	*	*	*	Copper with a little lead, tin and zinc
6599	Bell	***		**	tr.	Brass with a trace of lead
11758	Padlock	***	*		*	Copper and tin with a little lead
11384	Padlock	***	*	*	*	Copper with a little zinc, tin and lead
6474	Key	***				Copper
7864	Key	***	tr.	**	*	Brass with a little lead and a trace of tin
12795	Key	***	tr.		tr.	Copper with a trace of tin and lead
8805	Spur	***	**		tr.	Bronze with a trace of lead
12988	Knife	***				Copper
15329	Knife	***			tr.	Copper with a trace of lead
10828	Scale pan	***	?*	*	*	Copper with zinc, lead and possibly a little tin
10015	Tubular handle	***	**			Bronze
7018	Pin head	***	**	**	*	Leaded gunmetal
5545	Strip	***	*		*	Copper with a little tin and lead
746	Rod	***		*	*	Copper and zinc with a little lead
8471	Tube	***		**		Brass

plaques for decorative effect. A substantial coating, or even additional parts could be cast on to iron components. For example, some early Iron Age smiths appear to have found it difficult to create parts such as sockets; spearheads are found with a copper alloy socket cast onto an iron blade. An unpublished example of this technique from Melksham, Wiltshire is in the Devizes Museum. Similarly, substantial copper alloy handles were cast onto swords, for example an Iron Age anthropoid sword from Salon in France or another fine example from Worton Lancs, both illustrated in the British Museum Catalogue of 1925. More commonly, copper alloy was cast onto the iron shafts of terret rings, lynch pins and so on. Jacobsthal (1944) cites numerous examples of repoussé decorated bronze work applied to iron substrates, for example on helmets, scabbards, brooches, lynch pins and so on. These are not examples of true plating but rather, as Jacobsthal describes, a thin embossed veneer on an iron core.

Figure 22.3. *Bridle bit from Llyn Cerrig Bach, Anglesey, showing copper plating on the side link and the crimped effect on the loop caused by the drawing together of the plating metal. (Photo courtesy of the National Musem of Wales.)*

Techniques of copper alloy plating

There are several methods that could be used for copper alloy plating.

1. The first method is not true plating but rather casing. The sheet metal is wrapped around the iron substrate and, where the edges meet, they are drawn up to a slight flange. A punch is then used to draw the flanges tightly together by striking from either side. Bridle bit No. 49 from Llyn Cerrig Bach is an example of this method, the drawing in of the flanges being seen as a slightly wavy seam (Figure 22.3).
2. The plating could be applied by hammering the copper alloy onto the prepared surface of the iron. Keys would be raised on the substrate to form a physical bond with the plating medium. It is possible that the substantial plating on the base terminals of the Viking stirrups were applied by this method.
3. Spratling *et al.* (1980) reports a pin from Gussage All Saints encased in bronze; the bronze was said to have the original cast structure, and the hammering was said not to have created a homogenized plating. The original report is not clear and in any case it is unlikely that this object would have represented a widely used technique, and like the bridle bit described below probably was no more than the experimental tinkering of an apprentice.
4. Fusion plating was commonly practised; Fox (1946) says that the method was used to plate early Christian bells from Ireland. If the bells were plated by immersion in molten alloy their size would require the melting of large volumes of metal; by employing fusion plating only the metal that was to be used was melted. The method is described by Theophilus as part of the manufacturing process of barrel padlocks:

> ....Now make an alloy of two parts of copper and a third of tin and crush it to a powder with a hammer in an iron pot; then burn some argol [potassium carbonate, as a flux] add a little salt to it, mix it with water, smear it all around, and sprinkle the powder

Figure 22.4 *Reconstruction of the cow bell forge, from Kramer in Grund, near Hilchenbach, Siegerland. Built in 1861, demolished in 1968 and rebuilt at the Westfalisches Freilichtmuseum, Hagen. (Photo by Roy Day.)*

> around it. When it is dry, smear the mixture around it again more thickly and put it on live coals. Then carefully cover it all around [with charcoal] as in the case of the silver above and braze it in the same way; let it cool by itself and wash it. In this way you can braze anything of iron that you want, but it cannot be gilded in any way (Smith and Hawthorne 1963).

Arwidsson and Berg (1983) also describe the method drawing on Scandinavian sources. Bells were forged from iron and were then plated by a process known as *brassning*. To *brassa* the bell it was lined inside and out with strips of copper alloy then covered with clay to which had been added horse dung to make it more porous. This mass was then heated with the aid of bellows, turning it periodically to achieve an even surface. The turning was continued as the clay and metal cooled, after which the mass could be broken open and the bell finished. The copper and iron fuse without the need to add fluxes. This method continues to this day for the manufacture of cow bells in Switzerland and in the remote Siegerland hills of Germany. A cow bell forge of the 1860s has been carefully reconstructed at the Westfalisches Freilichtmuseum, Hagen, in the Ruhr, and described in the museum's handbook (Dauskardt 1990 p. 85). After fabrication from sheet iron the bells are wrapped in brass sheet, packed in clay and heated in a furnace (Figure 22.4).

5. Copper alloys could be applied simply by dipping. Herbert Maryon and Harold Plenderlieth demonstrated the method to Fox during the latter's research of the Llyn Cerrig Bach hoard. In the demonstration, a file was cleaned and dipped into molten brass, as soon as the fusion temperature was reached the brass flushed over the iron. When cool, the plating could be filed and polished (Fox 1946). Spon (1873) recommends that when iron is plated by dipping into molten copper, the iron should be

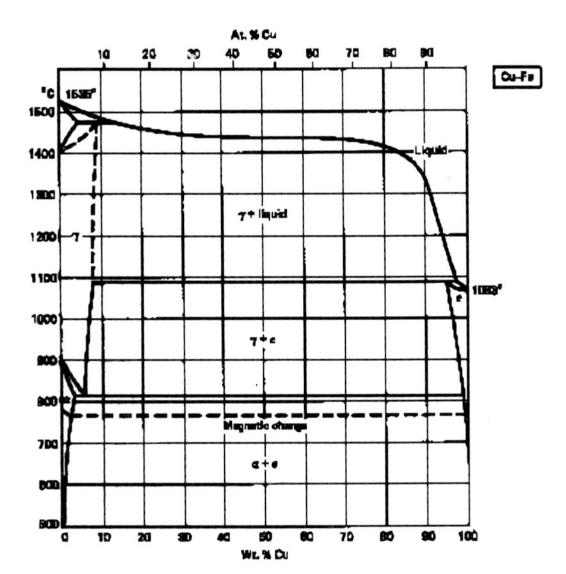

Figure 22.5. *Copper/iron equilibrium diagram (after Brandes 1983).*

at the same temperature as the bath and that the surface of the melted copper should be protected with a layer of melted cryolite and phosphoric acid to prevent oxidation of the metal. Numerous other recipes are given for producing copper, bronze and brass platings.

A variant of this method has been proposed by Spratling *et al.* (1980). The copper alloy plated side link of a bridle bit from Gussage All Saints was sectioned and analysed. An assumed sequence of manufacture was revealed: a thin layer of tin was identified immediately adjacent to the steel core, over this was a substantial layer of bronze and finally two small rings of iron were brazed onto either side of the main loop of the link. Spratling suggests that the purpose of the tin layer was to prevent the dissolution of the iron core into the molten bronze during the plating process. There is no reason to suppose that the tin layer would in fact serve any beneficial purpose, the supporting evidence that Spratling offers, that an iron bar dissolved when held in molten copper alloy for 30–45 minutes has no relevance. It has already been noted above that only a very brief immersion would be required to effect plating, and this is acknowledged by Spratling himself. Equally, the tin would not provide any barrier to the dissolution of the iron as it is soluble in both copper and iron at 1000°C.

True copper plating by dipping or by fusion plating produces a coherent coating to the iron in which the copper alloy is firmly bonded to the iron by intermetallic compounds. Craddock and Meeks (1987) have described the ease with which iron dissolves in molten copper: 2.8% at 1100°C, 6% at 1200°C rising to 20% at 1400°C (see Figure 22.5). It will be seen from the phase diagram that a range of iron–copper compounds can be expected; this gives further credence to the supposition that the technological processes proposed by Spratling are incorrect.

Owners of Iron Age bridle bits are, not unnaturally, reluctant to allow researchers to cut them up, however, a recent find of three copper alloy plated bridle bits at Maiden Castle has been subjected to analysis; no comparable tin layer has been identified (C.J. Salter personal communication). Tinned bridle bits have been found (Henken 1938) and Spratling notes this as significant; however, there is no reason to suppose that this example was left unfinished or that it was intended to be copper alloy plated. It would appear altogether more feasible that the Gussage piece was either a replating of a piece that was previously tinned, accidentally coppered, or it was another bit of experimental work to see what would happen.

6. Iron could be readily plated simply by immersion in solutions of copper salts. Spon (1873) describes numerous methods of applying copper plating to iron objects by reaction with chemical solutions. A solution of copper sulphate and sulphuric acid was said to impart a coating in a short space of time although the plated surface did not have a good adhesion to the iron. The cohesion of the copper particles could be improved by compression. Small articles could be plated by agitating them about in sand, bran or sawdust impregnated with the copper sulphate solution.

7. Finally, no dicussion of the methods of plating copper onto iron would be complete without considering electroplating. Electroplating of copper alloys onto iron began soon after the discovery of the principle of electrodeposition by Michael Faraday. Cast

ironwork could be readily plated. Phillips (1911) describes the process for brass plating: a cyanide based electrolyte of copper and zinc salts is used, the current is passed and the copper and zinc deposited proportionately. The colour of the plating could be varied by increasing or decreasing the surface area of the brass anode immersed in the electrolyte.

Phillips mentions a patent of M.M. Pearson and Sire for obtaining a thick coating of brass on iron by alternately depositing layers of copper and zinc and then heating the plated object to alloy the metals. Jones (n.d.) describes a similar process for brass plating, but also describes gunmetal plating using a tertiary electrolyte in which half the zinc salt is replaced by tin chloride. Copper plating is also described, but is said to be used only as a base for silver or nickel plating.

The implications for conservators are clear, dipped or fusion plated specimens may be expected to have well-bonded coatings, those seen on objects where the plating is mechanically applied, or on electroplated objects will be less coherent and will need special care. Where the platings are defective accelerated corrosion may be expected as the exposed iron will have a lower potential than the copper and will thus become the anode in a galvanic cell, therefore, care will be needed to ensure that the exposed iron is passivated by means of a substantial coating of lacquer, or that the object is kept below 20% R.H.

Analyses

I am grateful to Justine Bayley of English Heritage and Patrick Ottaway of the York Archaeological Trust for permission to use Paul Wilthew's analyses of copper-plated finds from Coppergate (Table 22.1). With the exception of two specimens plated with pure copper, each plating was of a unique composition with bronze, brass and ternary alloys represented suggesting that any odd scraps of copper alloy were used. There may be significance in the frequent occurrence of lead, which could have been added to improve the flow of the melted metal over the surface of the object.

Analysis of a Roman mail fragment carried out by Peter Northover has shown it to be copper with zinc, tin and lead, and X-ray fluorescence analysis of the plating on the Seagry stirrup (Plate 22.1) identified a ternary alloy of copper, zinc and tin.

Conclusions

The evidence shows that copper alloy plating was undertaken from the earliest use of iron, continuing through Roman and medieval times and beyond. Long recognized on particular objects; bridle bits from the Iron Age, early medieval bells, Viking stirrups and padlocks, it is now becoming apparent that copper alloy plating was used on a much wider range of material. Generally the evidence seems to point to its use as a protective coating applied to objects which would be exposed to the elements and where a more durable layer was required than could be provided by tin or tin alloys. The copper plating on bells, although probably intended initially as a protective coat, was found to improve their tone and this became a more important reason for its use.

Fox (1946) is somewhat disparaging of the people who commissioned the Llyn Cerrig Bach hoard. He implies that they suffered a poverty of bronze and therefore had to make the most of what they had by coating their iron harness fittings to simulate those cast in solid bronze. A truer picture would be a people who recognized the superior qualities of iron but who also understood its limitations, thus they produced a combination of the best qualities of both iron and copper alloys.

Copper alloy plating was a simple process. As described by Arwidsson and Berg it would not be beyond the capabilities of any competent metalworker, and even if the more complex method described by Spratling were to have been used, it would have presented few problems.

Like so many processes the identification of copper alloy plating depends on the conservator maintaining a vigilance during the removal of corrosion and other accretions from iron objects. Sometimes the radiographic evidence will indicate plating, more often the surviving plating will be fragmentary and only found by careful observation during the cleaning process.

Acknowledgements

This work was carried out while the author was employed at the National Museum of Wales, and the examples from the museum include work done by Jane Henderson. I am grateful to the Keeper of Archaeology, Dr Stephen Aldhouse-Green for allowing me to publish the results of my investigations, particularly of the Roman mail from Caerleon. I am grateful to Paul Craddock for providing additional information on the *brassning* of cow bells and to Roy Day for providing the photograph (Figure 22.4). I must also thank my many colleagues who generously assisted in the compilation of the examples of copper plating; individual names have been incorporated into the text.

References

Arwidsson, G. and Berg, G.G. (1983) *The Mastermyr Find: A Viking Age Tool Chest from Gøtland.* Almquist and Witsell International (for the Kungl Vitterhets Historie och Antikvitets Akadamien), Stockholm

British Museum (1925) *Guide to the Antiquities of the Prehistoric and Medieval Department*, London

Bulleid, A. and George, H. St G. (1917) *Glastonbury Lake Village*, vol. II, Glastonbury Antiquarian Society

Bulleid, A. and George, H. St G. (1948) *Mere Lake Village*, vol I, Somerset Archaeological and Natural History Society

Brandes, E.A. (1983) *Smithells Metals Reference Book*, sixth edition, Butterworths, London

Corfield, M.C. (1982) Radiography of archaeological ironwork. In *Conservation of Iron* (eds R.W. Clarke and S.M. Blackshaw) Maritime Monographs and Reports no 53

Corfield, M. (1985) Tinning of iron. In *Lead and Tin, Studies in Conservation and Technology* (eds G. Miles and S. Pollard) UKIC Occasional Paper no 3, UKIC London

Craddock, P.T., and Meeks, N.D. (1987) Iron in ancient copper. *Archaeometry*, **29**, 187–204

Dauskardt, M. (1990) *Museum Guide*, Westfalisches Freilichtmuseum, Hagen

Fox, Sir Cyril (1946) *A Find of the Early Iron Age from Llyn Cerrig Bach, Anglesey*, National Museum of Wales, Cardiff

Henken, T.C. (1938) The excavation of the Iron Age Camp on Bredon Hill, Gloucestershire, 1935–1937. *Archaeological Journal*, **XCV**, 1–111

Jacobsthal, P. (1944) *Early Celtic Art*, OUP, Oxford

Jones, B.E. (n.d.) *The Amateur Mechanic*, The Waverley Book Co, London

Ottaway, P. (1992) Anglo-Scandinavian Ironwork from Coppergate. *Archaeology of York*, **17**(6), CBA, London

Phillips, A. (1911) *The Electro-plating and Electro-refining of Metals*, Crosby Lockwood and Son, London

Rackham, H. (trans.) (1952) *Pliny, Natural History*, Book XXXIV. Heinemann, London

Seaby, W.A. and Woodfield, P. (1980) Viking stirrups from England and their Background. *Medieval Archaeology*, **24**, 87–122

Smith, C.S. and Hawthorne, J.G. (1963) *Theophilus, On Divers Arts*, Translation from the Medieval Latin, University of Chicago Press, Chicago

Spon, E. (1873) *Workshop Receipts for Manufacturers, Mechanics and Scientific Amateurs*, E and F.N. Spon, London

Spratling, M.G., Tylecote, R.F., Kay, P.J., Jones, L., Wilson, C.M., Pettifer, K., Osbourne, G., Craddock, P.T. and Biek, L. (1980) An Iron Age Bronze Foundry at Gussage All Saints, Dorset; Preliminary Assessment of Technology. In *Proceedings of the 16th International Symposium on Archaeometry and Archaeological Prospection 1976* (eds E.A. Slater and J.O. Tate) National Museum of Antiquities, Edinburgh, pp. 268–292

Vouga, P. (1925) *La Tène, (Monograph of the Station)*, Commission for the Excavation of La Tène, Leipzig

Wainwright, J.G. (1979) *Gussage All Saints, An Iron Age Settlement in Dorset*, DoE Archaeological Report No. 10, HMSO, London

Ward Perkins, J.B. (1939) Iron Age metal horses' bits of the British Isles. *Proceedings of the Prehistoric Society*, New Series, **V**, 173–198

Watson, J. and Edwards, G. (1990) Conservation of material from Anglo-Saxon Cemeteries. In *Anglo-Saxon Cemeteries, A Reappraisal* (ed. E. Southworth) Alan Sutton, Stroud.

23

The history of electroplating

Christoph Raub

Abstract

The exact date of the first electroplating experiment is debatable, but most agree that in 1772 Beccaria was the first to successfully deposit metal, by discharging a Leyden bottle and using the spark to decompose metal salts. Development accelerated after 1791 when Galvani discovered the physiological effects of electricity on frog legs and called the 'matter' generating this effect 'an own fluidum identical with electricity'. In 1796 his pupil, Volta, demonstrated the pile, which consisted of two dissimilar metals producing electric energy via a separating electrolyte. The voltaic pile, made up from hundreds of individual elements, generated electricity which was identical to Galvani's 'animalic electricity'. It served for many years as a source of electricity for electroplating on a commercial scale, until it was replaced by the electromagnetic generator.

Brugnatelli, a friend of Volta, was the first to cover a significant area by successfully gold-plating a silver coin. In 1807 Davy isolated sodium by electrolytic decomposition of molten salts and later Faraday established the scientific basis of electrochemistry. At the same time, the notorious battles over the industrial application of the process started between the Elkingtons in England, von Siemens in Germany, Ruolz and Christofle in France, de la Rive in Switzerland and Jacobi in Russia. Commercial companies began producing a range of items by electroforming. Few new developments occurred until the second half of the 20th century, when technological changes led to more emphasis on the functional, rather than the artistic, applications of electroplating.

Introduction

From antiquity, mercury amalgam gilding (Struve 1886) was arguably the most important method of gilding available because it gives a good and durable finish (see Chapter 15), but the dangers from poisonous fumes produced during the gilding process became known. Consequently, electroplating was eagerly seized on as a safer alternative. Electroplating is the deposition of a plating layer of metal from an aqueous chemical solution by the action of electricity. It was first applied to gilding and silvering on base metals, though non-conductive surfaces can also be electroplated if metallized first. Electroforming is an electroplating method by which a relatively thick-walled positive or negative replica of any metallic or metallized surface of an object is generated. Both processes began in the 1840s and are still used today on an industrial scale in the manufacture of a wide range of goods, from electronics to cars, and from jewellery to

Figure 23.1. *Electroform of a Roman silver bowl from the Hildesheim treasure. Städtisches Museum Schwäbisch Gmünd. (Photo from Johannes Schüle.)*

replicas of archaeological objects (Figure 23.1).

The beginnings of electroplating are closely linked to the discovery of electricity. As with many of the inventions which our technological society takes for granted, it is not certain when the first electroplating experiment took place. This is partly due to the mystery which has surrounded electricity right from its beginnings through to atomic power and solar energy today. Some secrecy is due to economic considerations: the possibility that electroplating could be exploited for monetary advantage was quickly realized.

Early stages in the search for electroplating

The development of electroplating has been described in detail by Hunt (1973), Krämer, Weiner and Fett (1959) and Pavlova (1963). An aqueous solution of metal salts and a source of electricity are the chief requirements for electroplating, so the early efforts were concentrated on the search for a good electrical source.

The commonly accepted opinion is that the first man to deposit metal from its chemical compounds was Professor G.B. Beccaria in 1772, back in the days of the phlogiston theory (phlogiston being something which was thought to be needed to generate metals). The energy, or electricity, was produced by a Leyden jar, which acted as a capacitor, and provided a means of storing electricity, but could only supply weak, short pulses of current.

In 1791, Luigi Galvani of the old and famous university of Bologna discovered that muscle tissue of frogs reacts to pulses of electric current. He observed that a frog's muscle suspended on a copper ring convulses on contact with iron. The matter generating these convulsions was named *fluidum* and mistakenly thought to be a property of the animal tissue itself, but identical with electricity. The effect was named *galvanismus* after Galvani.

A better source of strong and continuous electricity was needed. This was discovered in 1796 by Alessandro Volta, a pupil of Galvani. Volta's famous 'pile' consisted of dissimilar metals, e.g. silver and zinc or copper and iron. Each pair was separated by materials like wood, paper, cloth, etc., impregnated with a solution of electrolyte. These voltaic piles generated electricity, identical to the 'animalic electricity' first found by Volta. After this, rapid advances were made, not least because the scientifically minded community were often rich nobility in the courts of Europe who found it entertaining to produce unexpected miracles. It was soon discovered that the energy of these piles not only depended on the kind of metals and the separating electrolyte but also on the number of elements. The number and variety of elements in the piles multiplied into the hundreds during the following years, and they made enough energy for production plants to be based on them. However, the voltaic pile did not prove to be suitable as an energy source for commercial electroplating because of its complicated construction and relatively short life when required to produce large currents. This drawback provided the impetus for the discovery of the electromagnetic principle by Werner von Siemens, and for the development of its theory by James Clark Maxwell, leading to the manufacturing of the first electricity generators and motors by Siemens, Gramme and Westinghouse and others.

The discovery of the voltaic pile had an effect on science to be compared only with the effect of high-temperature superconductivity today. Everybody studied electricity, either out of general interest, or scientific curiosity. Volta reported his experiments in a long letter to the president of the Royal Society, Sir Joseph Banks, on 20 March 1800. He sent his letter in two parts, the second part following 3 months after the first, as the mail had to pass through the war zone of the Napoleonic Wars and might easily have got lost. This was why the letter as a whole was not read to the Society until June 1800.

Before the official announcement, rumours had already spread and experiments and results poured out from the European centres of learning: in Britain, William Nicholson (patent lawyer and mathematician), Sir Anthony Carlisle (anatomist) and William Cruickshank (chemist) discovered the decomposition of water into hydrogen and oxygen, which was so important later for electroplating. Cruickshank published, in German, the deposition of the metals lead, silver and copper from the aqueous solutions of some of their salts (Cruickshank 1800 and Cruickshank and Woolrich 1801). In 1801, J.L.W. Gruner, the court pharmacist from Hannover, deposited silver from aqueous electrolytes, with a voltaic pile consisting of 80 silver–zinc elements. K.W. Boeckmann used 100 silver–zinc elements, separated by cardboard impregnated with electrolyte to precipitate copper, silver and tin. He discovered that metals separate only at one electrode (the cathode). He also stated that the current efficiency of metal deposition is slightly less than 100% (Boeckmann 1801). Boeckmann was a physics teacher at a Karlsruhe high-school: his paper was 25 pages in length as compared to Cruickshank's single page!

The Italians, like the British, concentrated on pure science rather than its practical application, but the first successful gilding of a silver coin using a voltaic pile was published in 1801 by Luigi Brugnatelli. This was the first decorative use of electroplating.

The next decades were mostly taken up with the discovery of new effects of electrodeposition. England lead the way, with Sir Humphry Davy and his even more famous pupil Michael Faraday. Sir Humphry described, in his 2nd Baktrian lecture in 1807, the electrolytic preparation of sodium and potassium from a fused salt melt; 'when metallic solutions were employed, metallic crystals and depositions were formed, as is common in galvanic experiments, on the negative wire'. Sir Humphry Davy's cousin, Edward Davy, a professor to the Royal Dublin Society, proposed that electrochemical methods might be used for analytical purposes. Michael Faraday built his first galvanic pile before he was 21, in his private laboratory at the back of the bookbinder's shop where he was apprenticed. He only recommenced his experiments some 20 years later, in 1832, finally putting electrochemistry on a sound scientific basis. He coined the terms 'electrolysis' and 'electrolyte' and in an exchange of letters between Faraday and William Whewell of Trinity College Cambridge, Whewell suggested the expressions 'anode' and 'cathode' instead of 'Voltode' and 'Galvanode' (Hunt 1973).

The great breakthrough occurred in 1836, based primarily on the work of two men: Professor J.F. Daniell invented the self-polarizing cell, which was more reliable as a source of electricity, and George Richards Elkington from Birmingham filed a patent for 'An improved Method of Gilding Copper, Brass and Other Metals or Alloys of Metals', an immersion process working without an outside current source (Hunt 1973).

The following years illustrate the colourful development of electroplating. In Hunt's words:

> it must also come close to world record standards for the number and diversity of scientists and dilettante whose contributions eventually led to success; not only were there numerous professors and medical men involved but a curious collection that included an unsuccessful French opera composer (Henri-Catherine-Camille Ruolz); a leading English astronomer (de la Rue), a St. Petersburg dentist (Briant); the son-in-law of the Tsar of Russia (Herzog von Leuchtenberg), one [really two] of the famous Siemens family (Werner von Siemens and his brother Sir William Siemens, born Wilhelm von Siemens) and the founder of [what was later to become] the University of Birmingham.

The Elkington cousins

The story of the Elkington cousins, George Richards and Henry Elkington, is very complicated and even L.B. Hunt had problems unravelling it. G.R. Elkington referred to himself as a 'Gilt Toy Maker'; he and Henry made small articles like military badges, buttons and snuff boxes. He was interested in replacing the dangerous amalgamation process for gilding with something less poisonous and easier to handle. The earlier patents of the Elkingtons mostly covered immersion gilding processes but on March 25th, 1840 the cousins filed a patent (B.P. 8447) 'Improvements in Colouring, Covering, or Plating certain Metals', detailing silver and gold solutions in connection with the application of current. The source of current was very important: '... a solution of chloride of sodium ... into this a cylinder of zinc is immersed, with a wire of copper ...', clearly it is an electrochemical battery. This patent claim gave Werner von Siemens an opening. He sent his younger brother Wilhelm to England to negotiate his own patent claims with the Elkingtons. This ultimately led to Wilhelm becoming famous as Sir William Siemens, industrialist and scientist, and honoured with a memorial window in Westminster Cathedral.

Another famous assistant of the Elkingtons was Alexander Parkes, the later inventor of the first plastic celluloid, of phosphor-bronze and of the Parkes Process of desilvering lead with zinc. He was hired by the Elkingtons to develop thicker and more coherent electrodeposits of gold and silver.

Later in 1840 the Elkingtons met Dr John Wright, a former veterinary surgeon who had entered into a partnership (later Henry Wiggin & Co. of INCO, Ltd) to manufacture German silver. John Wright was fascinated by chemistry in connection with electricity. From reading the work of the German–Swedish chemist, Carl Wilhelm Scheele, on cyanide complexes of gold and silver in 1783, Wright apparently got the idea of using such solutions as electrolytes. The Elkingtons, having seen Wright's deposits, entered into agreement with him and 'hastily had the cyanides written in' (to their patent) (Hunt 1972). It is interesting to note, however, that J.J. Berzelius (1836), in his well-known textbook on inorganic chemistry, as translated by his equally famous pupil, F. Wöhler, described in detail the preparation and properties of the cyanide-complexes of gold and silver. The history of cyanides and electrolytes has been described in detail by Williams (1978). Because of their electrochemical properties, cyanide based electrolytes form the basis of many commercially applied electrolytes today.

Christofle and De Ruolz

The Elkington patent aroused great excitement in France, Germany and in Russia, and heated discussions followed. In France, for example, Comte De Ruolz Montchal (a pupil of the composer, Rossini), and a Professor A.A. De La Rive of Geneva, filed patents claiming 'the employment, for the first time, of the compounds of cyanogen with gold and silver' as well as of the battery for plating. The great potential of Ruolz's patent was realized by Charles Christofle, the founder of a small but enterprising jewellery manufacturing concern in Paris, and up to the present the leading name in silver tableware (Figure 23.2). In 1842 Christofle bought the rights to Ruolz's and Elkington's patents. A famous battle with Ruolz followed, provoking Christofle to publish his version of the story in 400 pages in 1851, with a foreword by Elkington.

Like Elkington in Britain, Christofle developed electroplating and electroforming of various metals into a big business, and became well established as the leading gold and silver plater in France (Hunt 1973).

Werner von Siemens

In England the story continued with ups and downs, involving many names later to become famous in British industrial chemistry and metallurgy. The controversies did not even end with the Elkington cousins' deaths (Henry in 1852 and George in 1865).

The history in Germany is even more fascinating since it centered on a famous name in Germany industry, Werner von Siemens (1816–1882). As a young artillery officer he

Figure 23.2. *Memorial plaque to Christofle. (Photo Kollar, courtesy of Musée Bouilhet-Christofle.)*

was imprisoned for assisting in a forbidden duel between an artillery and an infantry officer. While in the Magdeburg military prison he was allowed to continue his chemical experiments in *meiner vergitterten aber geräumigen Zelle* (in my fenced but spacious cell). He and his brother-in-law Himly, a professor of chemistry at the university of Göttingen, experimented with Daguerre's invention of dissolving gold and silver salts by *unterschwefelgsaurem Natron* (sodium sulphite).The solution proved to be a good electrolyte '*Iche glaube es war eine der größten Freuden, als ein neusilberner Teelöffel, den ich mit dem Zinkpol eines Daniellschen Elementes verbunden, in einen Becher tauchte ... sich schon nach einigen Minuten in einen goldenen Löffel von schönstem, reinem Goldglanze verwandelte*'. (I believe it was one of the biggest pleasures in my life when a teaspoon made from German-silver, which I connected to the zinc-pole of a daniell cell and dipped into a beaker... transformed itself after a few minutes into a golden spoon of the prettiest, purest gold sparkle). He was given a pardon but wanted to stay in his gaol laboratory to finish his studies (Siemens 1922). He sold the patent to Neusilberfabrik J. Henninger, which as a result became the first electroplating shop in Germany, and used the money for his studies in electricity, eventually founding the modern Siemens AG. As mentioned above, he sent his brother Wilhelm to England to negotiate with the Elkingtons. The Elkingtons told him at first that he could not use his patent in England since their patent covered the use of electric currents produced by 'galvanic batteries or induction'. However, Werner von Siemens had been clever enough to include in his application 'thermo-electric currents', which they had not covered. After giving experimental proof by depositing gold and silver from sulphite electrolytes with a current from a multiple iron-german silver thermocouple, Wilhelm obtained the patent and sold it to the Elkingtons for 1500 louis d'or. As Werner von Siemens wrote: '*Dies war für unsere damaligen Verhältnisse eine kolossale*

Summe, die unserer Finanznot für einige Zeit ein Ende machte.' (In those days this was a tremendous sum which for some time put an end to our financial worries).

Wilhelm returned to Germany, *'fand aber an den dortigen kleinen Verhältnissen keinen rechten Geschmack mehr, nachdem er die Großartigkeit der englischen Industrie kennen gelernt und das Leben in England ihm gefallen hatte'* (he did not find the narrow minded conditions there to his taste after having realized the grandeur of British industry and enjoyed the British way of life).

Jacobi and electroforming

In Russia there are two names which dominated the beginning of Russian electrochemistry in science and industry: Moritz Hermann von Jacobi (1801–1874) and Maximilian Herzog von Leuchtenberg (1817–1852) (Pavlova 1963). Moritz Hermann von Jacobi (1801–1874) or Boris Semeniovich Jacobi, physicist and electrochemist, was born in Potsdam (Germany) and educated at the university of Göttingen. From 1831 to 1833 he worked as an architect in Potsdam, then he settled in Königsberg. From 1835 to 1837 he was a professor at Dorpat university, then he worked in St Petersburg and in 1839 he became a member of the St Petersburg Academy of Sciences. Until 1855 he worked on electric machines, the electric telegraph, electricity in mines, electrochemistry and electrical measurements. In 1859 he studied platinum and his last years he devoted to the study of meteorology. He was a scientific advisor to the Duke Maximilian von Leuchtenberg.

Duke Maximilian Herzog von Leuchtenberg (1817–1852) married the daughter of Czar Nicholas I and settled in Russia. From 1839 onwards he was an honorary member of the Academy of Sciences and, from 1843, President of the Academy of Arts. He was also what we might today call an industrial manager, founding and heading The St. Petersburg Electroforming, Casting and Mechanical Plant. The company first produced electroforms for printing state papers, and later, using copper electroforms, reproduced art objects such as statues and bas-reliefs for the *Isaakievskii sobor* (St Isaac's Cathedral), copper horses for the Bolshoi theatre, and statues for the Hermitage. In addition to copper plating and electroforming, he worked with silver and gold on a large scale (Pavlova 1963).

In 1853 his plant consisted of three shops: the bronze department with 210 workers, the mechanical and iron forging department with 540 workers, and the 'new silver shop' with 86 workers. The first order was for the electrolytic copper plating of 4000 eagles for cuirassier helmets with a total surface of 9 square metres. The shops used 110 438 kg of copper for statues and bas-reliefs for St Isaac's Cathedral, the Hermitage, the Bolshoi Theatre in Moscow, the Winter Palace, the Peter and Paul Cathedral and several others, and 750 kg of gold to cover the domes of the Church of the Saviour in Moscow, St Isaac's Cathedral (Plate 23.1), and several other domes and objects (Pavlova 1963).

Von Leuchtenberg plated the dome of the church of the Redeemer in Moscow, according to the first industrial specification, with 28.44 grams of gold per square metre. The specifications were set up and controlled by Struve, another illustrious Baltic German Russian, of the family of the founder of the famous astronomic observatory at Poltava and editor of a star atlas. In the specification of 1854, only a 20% deviation was permitted. It took 3 years to fulfil this order, using electrolytes of nearly 6000 litres volume; the total amount of gold deposited being 493.1 kg. The typical electrolyte composition was: 1 part by weight of gold as gold chloride, 1 part by weight of KOH (no concentration given) and 2.5 parts by weight of potassium cyanide. Potassium cyanide was produced directly in the plant. In the course of 3 years, 280 kg of gold were plated and the daily consumption was between 8 and 12 kg. Insoluble platinum anodes were used (Pavlova 1963). The art of the Russian electroformers became so renowned that many of their sculptures were exported to other parts of Europe, e.g. France, Italy and Germany.

Conclusion

The history of electroplating is a fascinating and colourful story of the interweaving of arts, craft and industry and also of innovative creativity and industrial venture on a European scale, from England to Russia. May it be an example for the Europe of today and tomorrow.

Acknowledgements

This paper would have been impossible without the co-operation of my friends and colleagues in the countries involved: I especially want to express my thanks and gratitude to Professor Dr V.N. Kudryavtsev of the Institute of Physical Chemistry of the Academy of Sciences, Moscow, Professor Dr P. Cavlotti of the Politecnico di Milano, Institute of Electrochemistry and Electrometallurgy, Milano, Italy, Dr Ch. Eluère, Musée des Antiquites Nationales, St. Germain-en-Laye and the company of Ch. Christofle, Paris and to Mrs M. Hunt, London, as well as to Dr J.A. Chaldecott, Eastbourne, UK. I am very much obliged to Mrs S. La Niece for her careful and patient work on my manuscript drafts.

References

Berzelius, J.J. (1836) *Lehrbuch der Chemie, übersetzt v. F. Wöhler, 4,* verbesserte Original Auflage Bd. 4. Dresden und Leipzig, pp. 623–699

Boeckmann, K.W. (1801) Versuche und Beobachtungen über die Wirkungen der galvanischen Electrizität durch Voltas Säule. *Annalen der Physik*, **8**, 137–162

Brugnatelli, L. (1801) Chemische Bemerkungen über die elektrische Säule. *Annalen der Physik*, **8**, 284–299

Cruickshank W. (1800) Beschreibung des neuen elektrischen oder galvanischen Apparates Alexander Voltas und einiger damit angestellter Versuche. *Annalen der Physik*, **6**, 340–341

Cruickshank, W. and Woolrich, W. (1801) Versuche und Beobachtungen über chemische Wirkungen der galvanischen Elektrizität. *Annalen der Physik*, **6**, 360–366 and 7105–7106.

Hunt L.B. (1973) The early history of gold plating. A tangled tale of disputed priorities. *Gold Bulletin*, **6**, 16–27

Krämer, O.P., Weiner, R. and Fett, M. (1959) *Die Geschichte der Galvanotechnik*, Eugen G. Leuze Verlag Saulgau

Pavlova, O.I. (1963) *Electrodeposition of metals, a historical survey* (ed. S.A. Pogodin) Moscow. English translation, US Department of commerce, Israel Programme of Scientific Translations, Jerusalem 1968

Siemens W. von (1922) *Lebenserinnerungen*, **12**, Auflage, Berlin Julius Springer

Struve, H. (1886) Die Vergoldung im Feuer mit Hilfe des Quecksilbers und die galvanische Vergoldung. *Bulletin de l'Academy de Science, St. Petersburg*, col. 49–59

Williams G. (1978) The cyanides of gold, the history of their key role in electroplating. *Gold Bulletin*, **11**, 56–59

24

Modern electroplating and electrofinishing techniques

Robert Child

Abstract

Electroplating has changed little since the time of Faraday, when developments in the generation of electricity paralleled those of electrometallurgy. This paper illustrates the current commercial practice of electroplating with nickel, chromium, zinc, and the nobler metals. It then considers the appearance and properties of the plated finish as aids to their identification in museum objects.

Electroplating

The principle of electroplating is essentially simple (Figure 24.1). Metal salts dissolved in an aqueous solution are reduced by a circuit current of electricity that is passed between two electrodes immersed in the solution. The negative electrode (cathode), has the metal plated onto it while the positive electrode (anode), dissolves into the solution (the electrolyte), replenishing the metal ions in solution. For instance, in nickel plating the anodes are made of nickel, the conducting medium is an aqueous solution of nickel salts and the object to be plated is the cathode (Napier 1857) (Figure 24.2). The relationship between the amount of metal deposited and the amount of electric current used is given by Faraday's Laws:

> The weight of deposited metal is proportional to the amount of electricity passed through the solution, and to the chemical equivalents of the plated metal or metals (Watt 1937).

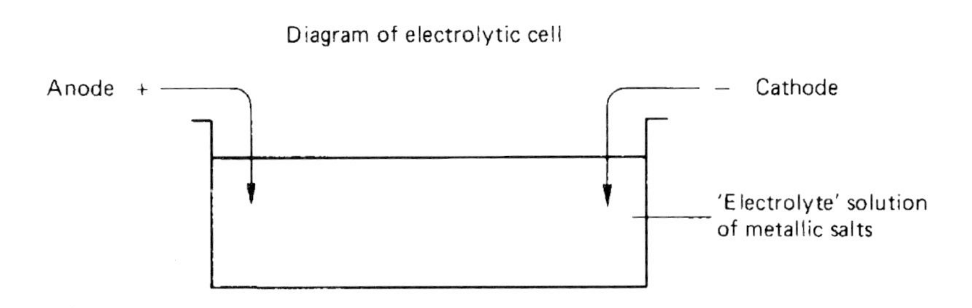

Figure 24.1. *Diagram of Electrolytic cell.*

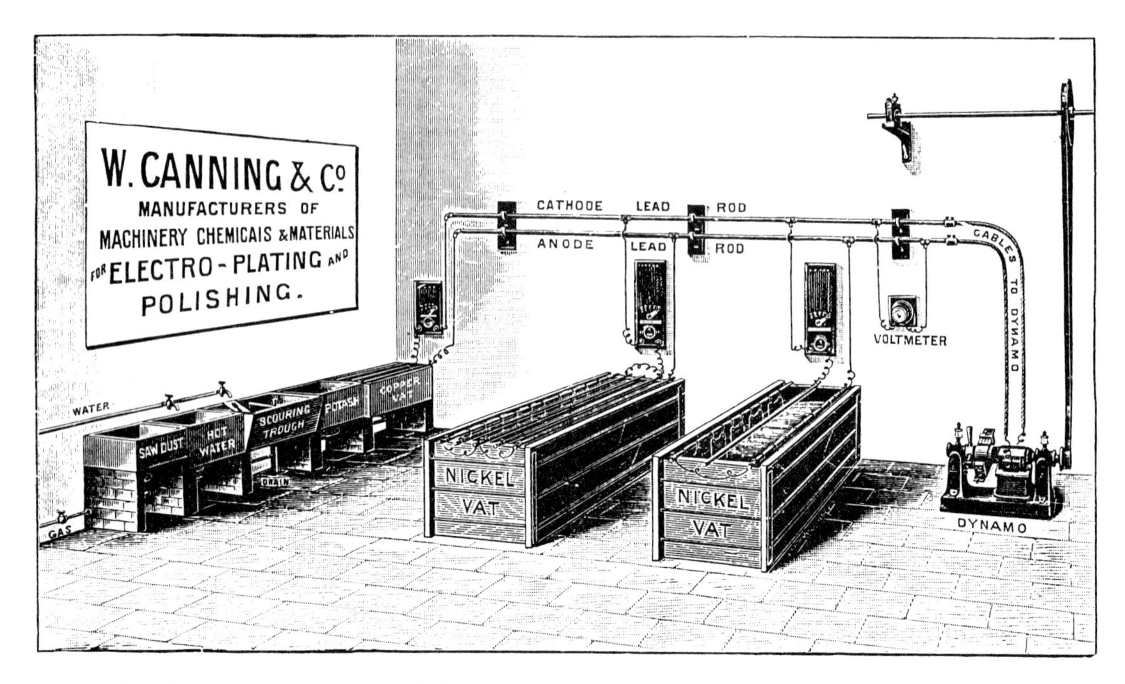

Figure 24.2. *Schematic representation of plating shop (from Canning's 1922 Handbook).*

Electroless deposition

An alternative method of electrodeposition is electroless deposition, where no external electrical source is used and the metal plating is deposited by displacement (for example, a steel penknife blade is coated with metallic copper when dipped in a solution of copper sulphate). Coatings can also be formed by chemical reaction, which cause a metal plating to be deposited on the substrate metal.

Electroless deposition is of great importance in the electroplating industry as it gives even thicknesses of coating on awkward or convoluted shaped objects. Conventional electroplating would give uneven thicknesses as the rate of deposition depends on the distance between the anode and the cathodic object (Burns and Bradley 1967).

Uses of electroplating

Originally electroplating was seen as a scientific novelty, whose principle application was in electrotyping, i.e. for making metal facsimiles of small original objects such as coins and medals. A Mr C.J. Jordan wrote in the London Mechanics Magazine on the 8th of June 1839:

> It appears, therefore, that this discovery may be turned to some practical account. It may be taken advantage of in procuring casts from various metals as previously alluded to; for instance, a copper die may be formed from a cast of a coin or a medal, in silver, type metal, or lead, &c, which may be employed in striking impressions in soft metals. (Jordan 1839).

It was not long, however, before the decided advantages of being able to cover cheap metal objects with a thin layer of a more precious metal became obvious. Former mechanical and chemical plating methods, such as Sheffield plate and gold amalgamation, rapidly became redundant, as they were superseded by the more economic and efficient electroplating techniques. The principle advantages were the ability to plate the finished article (thus avoiding the complicated manufacturing and fabricating techniques of Sheffield Plate), and the controllable thickness of the electroplated surface. Platings can be made microscopically thin, allowing the fast and economic production of decorative objects.

In the 20th century the predominate use of electroplating is for corrosion protection of base metals as in tin plating, galvanizing and cadmium plating. Former hot dip methods whereby items to be protected were dipped into molten tin or zinc, have been replaced by the more efficient and economical electroplating methods. Modern techniques can now give protective platings an aesthetically pleasing appearance in contrast to their former utilitarian look. The use of new metals such as chromium and rhodium combine good looks with corrosion resistance (Barratt and Massalski 1980).

Specialist coatings have been developed to impart specific properties to plated metal films, such as hardness, electrical conductivity, wear resistance, low friction, etc. The substrate onto which the coating is plated does not have to be metal, and is now commonly plastic (Dennis and Such 1986).

A brief history of electroplating

The history of electroplating closely parallels the history of the battery. Electroplating requires a low continuous, steady direct current, and the early batteries provided just that.

The first breakthrough in the art and science of electrochemistry, came in 1800 with Volta's invention of the voltaic pile. This was a column of discs (about 3 cms in diameter) of zinc and silver, separated by discs of pasteboard soaked in common salt; with up to a hundred such discs in a column. A wire was attached to the bottom zinc disc, and another to the top silver one, to allow this first battery to perform experiments in electrochemistry (Figure 24.3).

Improvements on this first battery, and investigations using it by serious scientists as well as dilettantes, demonstrated that many solutions of metallic salts can be broken down by an electric current. In some cases it was discovered that the metal was deposited on to the negative pole of the battery. However, the economic importance of these discoveries was not

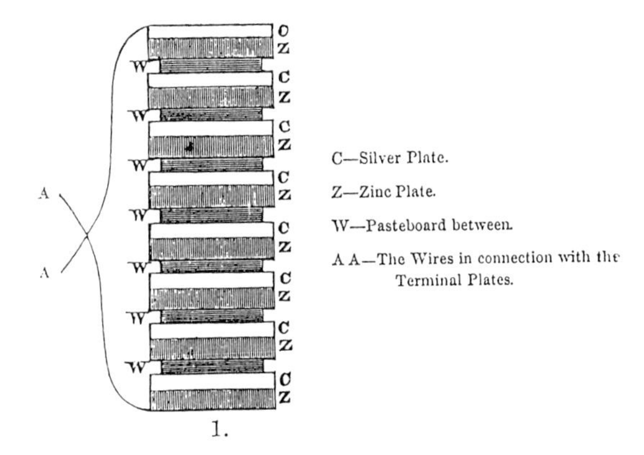

Figure 24.3. *Voltaic pile (from Napier 1857).*

immediately appreciated, and remained a scientific curiosity for many years (see Chapter 23).

'Electro-metallurgy may be said to have its origin in the discovery in the constant battery by Professor Daniell'; so said Alfred Smee in his 'Elements of Electro-metallurgy' (1841). It was the improved batteries of Daniell, Bunsen and Smee himself that laid the foundation of electroplating, and allowed much of the theory to be established. The dawning of the importance of the discovery caused a flood of patents to be taken out on the application of electrometallurgy to a variety of uses. Many of the patents were commercial failures, but in 1840 the patents taken out by the Elkington cousins on silver and gold plating with an electric current, virtually monopolized the emerging industry for decades.

The work of the scientist Michael Faraday in the 1830s, in generating electricity by electromagnetic dynamo was also initially unappreciated commercially. At a meeting in 1850 of the British Association, the Elkington cousins still expressed a preference for batteries over the dynamo, even though they had the patents for dynamos since 1846. However, their statements were possibly mere polemic as large Magneto machines were installed at that time in their Newhall Street, Birmingham works, with a capacity to deposit up to 50 ounces of silver in an hour (Figure 24.4).

The early commercial years of electroplating were concerned with the silver and gold plating of decorative and better quality utilitarian objects such as tea-pots and vases. Base metal

Figure 24.4. *Plating shop at Elkington's Newhall St works showing the new dynamo. (From Cassell's* Illustrated Exhibitor, **I**, *1852, guide to the Great Exhibition).*

artefacts, usually of brass, copper, nickel-brass (also known as nickel-silver) and Britannia metal (an alloy of antimony and tin) were all easily fabricated by casting and hand-forming and were electroplated with a thin coating of silver. By the 1860s silver electroplating was a commonplace finishing process to many metal artefacts (Bury nd). By the end of the 19th century, iron and copper-based alloys were being plated with corrosion resistant coatings of nickel in the manufacture of medical equipment, industrial items and many domestic objects such as bicycle lamps (Brannt 1891).

Since steel is difficult to electroplate directly with nickel, an intermediate plating of copper is used to smooth out any irregularities in the steel surface and to bond with both the steel substrate and the final nickel coating.

Nickel platings were a common surface finish on many steel and other base metal objects, as they were corrosion resistant and initially shiny although they soon tarnished out of doors. Through the late 19th century there was an intensive search for a plating that would remain bright in outdoor conditions. The corrosion resistant properties of chromium were well known but not until the 1920s were the problems of making a durable plating finally resolved (Dupernell 1977). Within a very short time chromium plated fittings became standard on motor vehicles (Figure 24.5). Indeed the introduction of chromium plating may be said to have radically changed the whole appearance of the motor car (Miller 1961) as well as having a lasting influence on modern design generally, with the Bauhaus and Art Deco styles rapidly adopting the new material.

In recent years hard zinc electroplating has given finishes attractive enough for use in domestic items such as refrigerator shelves, bicycle mudguards and steel fixtures and fittings.

The commercial practice of electroplating

Since the basic concepts of electroplating are simple, the necessary apparatus easy to obtain, and the practice simple to carry out, it is perhaps not surprising that the commercial

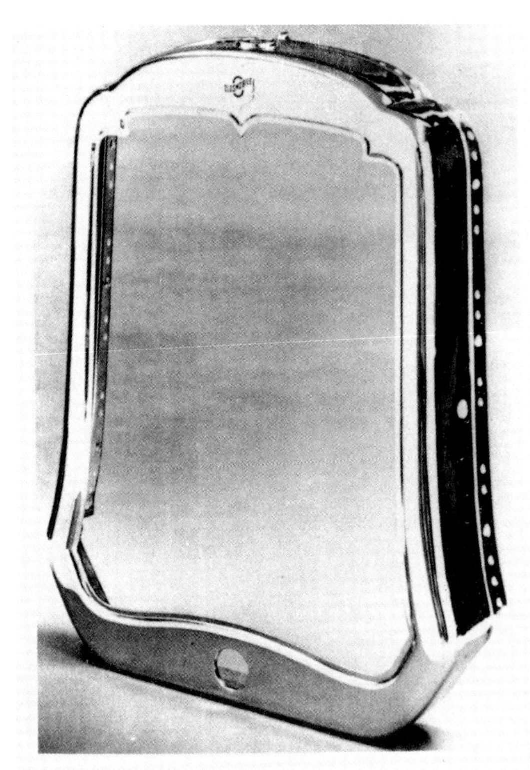

Figure 24.5. *Radiator shell for 1926 Oldsmobile. Chromium plating was to rapidly change the appearance of the motor car (from Dubpernell 1977).*

Figure 24.6. *Elkington's electroplating shop at Newhall St, Birmingham. 1855. Major works such as this were established within a few years as electroplating swept all before it (from Cornish's Guide to Birmingham.) The 1858 edition of the Guide shows the works expanded by about one third.*

Figure 24.7. *Typical late 19th-century plating shop (from Brannt 1891).*

Figure 24.8. *Typical late 20th-century plating shop (Photo by permission of Reeline Electro Plating, Cardiff).*

enterprise has remained essentially unchanged for 150 years (Figures 24.6, 24.7 and 24.8). The Indian craftsman silvering his copper jewellery using a car battery, scrap silver and simple chemicals would understand the problems of the sophisticated electrochemist working in a modern electroplating works. The day-to-day problems of surface preparation, cleanliness, thickness of deposit, avoidance of contamination etc, would be familiar to both.

The following description of standard current commercial electroplating techniques, is thus not unique to UK practice but has parallels from Albania to Alberta.

The chrome plating process has been used to illustrate the process as it covers most of the commercial practice (Tegart 1959).

Figure 24.9. *Scratch brushing (from Canning 1922).*

Chrome plating process

Objects to be chrome plated are usually fabricated in mild steel, with little regard for a fine finish. In order to produce a smooth brightly plated surface using the minimum of expensive chromium a first electroplating with nickel is carried out. The succession of operations is as follows (Canning 1937, Dubpernell 1977):

Surface preparation

The object to be plated is first buffed on rotating cloth mops impregnated with an abrasive polish to remove any superficial rusting and to smooth the surface. It is then washed in soap and hot water to remove any grease residues.

A further intensive cleaning process is carried out by 'anodic cleaning' where the object is made the anode and immersed in an electrolyte, typically of sodium hydroxide/ sodium metaphosphate solution with added surfactants. Slight etching of the material occurs as the surface is electrolytically dissolved, leaving a chemically clean face which is finally acid pickled in hydrochloric acid and rinsed in clean water (Figure 24.9).

Nickel plating

The clean objects are wired to a conducting frame and immersed in the electrolyte bath of nickel salts. Typically nickel sulphate, nickel chloride, boric acid and organic compounds to brighten the plated finish, are used. The anodes are pure nickel chips held in titanium baskets (for ease of replacement and handling), with a canvas cover to prevent possible debris contaminating the plated finish. Using a current density of 1.1 Amps/dm^2 a deposit 0.025 mm thick will be plated in about 2 hours.

The initial plating of nickel is used to level the metal surface and provide a corrosion resistant substrate for the decorative chrome. Owing to the expense of both nickel and chrome, if much levelling is necessary, a first deposit of copper is often used.

Chrome plating

After the nickel plating process the plated objects are washed in water 'swill" baths and transferred to the chrome plating tanks. Typically chromic acid electrolytes are used with chromium anodes at a current density of 35–40 Amps/dm^2, to produce plating thicknesses in the region of 0.0025 mm (Figure 24.10).

The plated objects are inspected and given a final buffing to remove any surface defects.

Zinc plating process

Traditionally zinc coatings have been applied to ferrous surfaces to improve their corrosion resistance, as it is anodic to the iron/steel substrate and will corrode preferentially to it. Formerly the zinc was applied solely by hot-dip galvanizing where the steel was dipped in molten zinc. Electrodepositing of zinc allows better control and thinner coatings to be used to provide good corrosion resistance and an aesthetically pleasing appearance. Better anti-

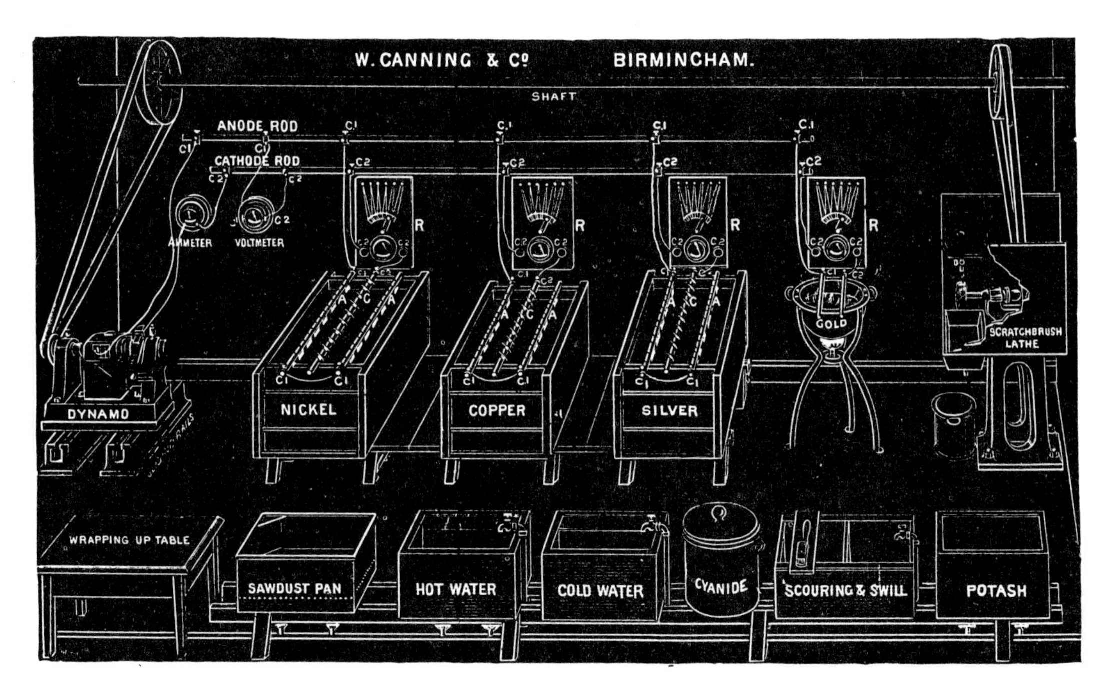

Figure 24.10. *Plating shop (from Canning 1922).*

corrosion properties are provided by 'passivating' or 'chromating' the surface after electroplating (HMSO 1951).

The process is similar to nickel plating, starting with surface preparation by washing, cleaning and acid pickling to remove any residual iron oxides.

The electrolytic bath was formerly based on cyanides but is now normally an acid solution, giving deposit thicknesses of 0.0075 mm to 0.038 mm. The finish is hard and shiny needing no further treatment and is commonly used on commercial and domestic items such as refrigerator shelves, bicycle parts and fixtures and fittings where a non-rusting reasonably attractive appearance is desired. Where a highly corrosion resistant surface is required as for instance with screws, nuts and bolts, the zinc-plated objects are dipped into a chromate passivating bath giving them a distinctive yellow tinge (De Garmo *et al.* 1984).

'Sherardizing' is an alternative form of zinc coating of steel items, formed by rotating the items in a heated drum with zinc powder, until a surface protective zinc–steel alloy is formed. Mechanical plating of small steel objects by cold welding is also routinely carried out by rotating them in zinc powder. In neither of these processes is electroplating involved (Metallurgist 1954, Ross 1988).

Other electroplating processes

Many other metals can and are used in the electroplating industry. The noble metals gold, silver, rhodium, etc. are used on various base metals as decorative finishes or when certain properties such as reflectivity are required. Where corrosion resistance of steel is necessary as in many industrial applications, cadmium plating is preferred especially in marine environments (Rudzki 1983).

Alloy plating where two (or more) metals are simultaneously deposited, is increasingly used for decorative finishes. For instance steel can be plated with brass by using an electrolyte of copper and zinc salts in the ratio 60:40 dissolved in a solution of sodium cyanide. Similarly other alloys such as bronze, speculum, lead-tin, etc. can be produced as platings (Silman *et al.* 1978).

Silver plating

All forms of silver plating are designed either to produce cheaper copies of solid silver originals, or to decorate utilitarian but unattractive materials such as iron. Early methods such as Sheffield Plate heat welded a thin silver sheet onto a block of copper and then rolled the block to the desired thickness. Objects were then fabricated from the rolled sheet; parts were joined by soldering, and edges which showed the copper core were hidden by soldering silver wire over them (Turner, this volume, Chapter 19, Bradbury 1912). Other methods included 'close plating' (see pp. 211–212 and Chapter 18, this volume).

The invention of silver electroplating allowed finished articles to be covered with a calculable and thus costable layer of silver. The processes perfected in the 1840s are essentially unchanged today (Bury nd). Absolute cleanliness of chemicals, equipment and objects to be plated are essential precursors to successful plating.

Initially platers would make up their own solutions to ensure quality, by precipitating pure silver chloride from an impure solution, dissolving the precipitate in nitric acid and crystallizing out silver nitrate. The pure silver nitrate was then dissolved in a solution of potassium cyanide. A typical solution for silver plating would be:

20 g silver nitrate
36 g potassium cyanide
22 g potassium carbonate
1 litre of water

Base metals such as Britannia metal and pewter were plated in solutions with an excess of cyanide ions, while iron and steel were first plated with copper or brass before being silver plated. The silver plated finish achieved by such materials was a white matt one, that was burnished and polished to produce the shiny highly reflective surface required. By the end of the 19th century it was found that adding certain organic 'brighteners' such as carbon disulphide to the plating bath caused the plated finish to be shiny and need little final polishing.

Gold plating

Scrap gold was dissolved in aqua regia (a mixture of nitric and hydrochloric acids), and the chloride precipitated with sodium chloride. The washed precipitate was dissolved in a solution of potassium cyanide and was then ready to use as the plating bath, usually with an inert anode, which necessitated replenishing the gold content of the bath periodically.

Gold alloys can be electroplated by adding other metal cyanides to the plating solution. Twenty-two carat gold (alloyed with silver) gives a hard yellow/green plating which has an attractive colour and is commonly found on hard wearing surfaces such as watchcases. Addition of small amounts of copper give a 'rose' or 'pink' gold alloy.

Much cheap modern 'gold' trinkets are base metal castings of zinc or brass, bright nickel plated, and then flash-plated with a microscopically thin layer of 24 carat gold (Selwyn 1945).

Anodizing

The creation of a thick oxide film electrolytically on aluminium finds widespread application mainly to increase corrosion resistance, but also to produce a variety of pleasing surface colours and textures (Anon 1952, Wernick and Pinner 1959).

Basically the cleaned aluminium objects form the anodes of electrolytic cells, the electrolyte being dilute solutions of sulphuric, chromic or oxalic acids, either alone or in combination. On application of a current the 'anodized' layer of Al_2O_3 forms. The phenomena was first noted by H. Buff in 1857, but commercial application as a finish only began in the 1920s.

With super purity aluminium alone or alloyed with magnesium the oxide film is transparent and thus a highly polished finish can be preserved. This treatment has been widely used since the 1950s and has replaced chromium plating in many areas (Figure 24.11).

A range of finishes can be obtained by etching, shot blasting, scratch brushing, etc., seen to best effect under transparent films. A

Figure 24.11. *Bright anodised aluminium replaced chromium plated fittings for many uses from the 1950s including bumpers, radiator grills, and hub caps on the bus illustrated here (from Anon 1952).*

variety of colours can be obtained by use of different aluminium alloys, producing tints ranging through yellow brown, as well as shades from milky white to black. As the oxide film is porous it can be easily and permanently dyed with a range of inorganic and organic pigments.

Electrofinishing

Electrodeposition of metals tends to accentuate any flaws and imperfections in the underlying metal as the current concentrates on ridges and peaks and is reduced in holes and troughs. In order to produce a smooth, shiny finish the metal substrate normally needs to be highly polished before plating, and repolished after.

Levelling compounds are chemicals added to the electrolyte which cause a polarizing effect on ridges and peaks reducing the electroplating there and thus evening out the deposit to give a smoother finish.

Brightening agents are organic compounds added to the electrolyte bath which cause the deposited metal to plate out as a shiny surface.

Electropolishing is a method of producing smooth shiny platings by making the plated object anodic. Ridges and peaks on the surface will be preferentially dissolved, thus levelling it out. This principle is much used in the brightening of aluminium for reflectors and decorative purposes (Wernick 1948).

Identifying features of electroplates

Electroplated metals can be deposited as pure metals, as successions of metals, or as alloys. They can be deposited as a matt surface or a shiny surface and with some metals such as nickel, a matt black finish can be obtained electrochemically. Any conducting surface can be electroplated including base metals (e.g. mild steel, nickel, brass, copper) and specially prepared plastics.

On new, unworn and undamaged material, it is often difficult to determine the substrate material using non-intrusive means, however, an indication can often be obtained from the following:

- Response to a magnet indicating an iron-based core;
- Weight and density can distinguish between core metals where the surface area to volume ratio is small;
- Trade markings such as E P N S (Electro Plated Nickel Silver).

Where material is old and worn, the substrate can sometimes be presumptively identified by its colour, or the colour of any corrosion. However, care is needed to distinguish between those of the base metals and of the platings.

Identification of the electroplates is normally done from their colour compared with reference standards. As a rough guide the following colours apply:

Cadmium	Silvery-white. Industrial use only
Chrome	Silvery with a bluish tinge. Highly reflective
Nickel	Silvery with a yellowish tinge
Silver	Silver highly reflective. Tarnishes
Tin	Silvery with a white tinge. Goes grey slowly
Zinc	Silvery blue. Usually not highly polished

Corrosion colours can be misleading as the direct source of the corrosion can often be difficult to identify. Chemical spot tests and modern instrumentation methods are valuable aids to identification as the choice of alternatives is usually small (Child and Townsend 1988).

Acknowledgement

I am grateful to Paul Craddock for additional information on chrome plating and anodizing and for selecting and supplying the illustrations.

References

Anon (1952) *Anodising Aluminium*, Aluminium Development Association, London

Barratt, C. and Massalski, T.B. (1980) *Structure of Metals*, Pergamon Press, London

Bradbury, F. (1912) *History of Old Sheffield Plate*, Macmillan, London

Brannt, W-T. (1891) English Edition of *Langbeins Treatise* on Electro-deposition of Metals, Henry Carey Board, London

Burns, R.M. and Bradley, W.W. (1967) *Protective Coatings for Metals*, 3rd edition, Reinhold Publishing Corporation, USA

Bury, S. (nd) *Victorian Electroplate*, Country Life, London

Canning, W. (1922) *Handbook on Electroplating*, 8th edition, Birmingham

Canning, W. (1937) *The Canning Practical Handbook on Electroplating*, W. Canning and Co., Birmingham

Child, R.E., and Townsend, J.M. (1988) *Modern Metals in Museums*, I.A.P., London

Cornish (1958) *Guide to Birmingham*

De Garmo, E.P., Temple Black, J., and Kohser, R.A. (1984) *Materials and Processes in Manufacturing*, 6th edition, Macmillan, London

Dennis, J.K. and Such, T.E. (1986) *Nickel and Chromium Plating*, 2nd edition, Butterworths, London, pp. 256–314

Dubpernell, G. (1977) *Electrodeposition of Chromium from Chromic Acid Solutions*, Pergamon, London

HMSO (1951) *Protection and Electro-deposition of Metals*, Vol. 3, Selected Government Research Reports

Jordan, C.J. (1839) *London Mechanics Magazine*, June, London

Metallurgist (1954) *Sherardizing*, Wolverhampton

Miller, W.G.T. (1961) Acceptance requirements for nickel–chromium plating. In *Nickel–Chromium Plating*, Robert Draper, Teddington

Napier, J. (1857) *A Manual of Electro-metallurgy*, 3rd edition, Griffin Scientific Manuals, London and Glasgow

Ross, R.B. (1988) *Handbook of Metal Treatments and Testing*, Chapman and Hall, London

Rudzki, G.J. (1983) *Surface Finishing Systems*, American Society for Metals, USA

Selwyn, A. (1945) *The Retail Jeweller's Handbook*, Heywood and Co., London

Silman, H., Isserlis, G., and Averill, A.F. (1978) *Protective and Decorative Coatings for Metals*, Finishing Publications Ltd., London

Smee, A. (1841) *Elements of Electro Metallurgy*, London

Tegart, W.J. (1959) *The Electrolytic and Chemical Polishing of Metals*, Pergamon, London

Watt, A. (1937) *Electroplating*, The Technical Press Ltd., London

Wernick, S. (1948) *Electrolytic Polishing and Bright Plating of Metals*, Alvin Redman Ltd., London

Wernick, S., and Pinner, R. (1959) *The Surface Treatment and Finishing of Aluminium and its Alloys*, Robert Draper, London

Index